Biosafety in Microbiological and Biomedical Laboratories

5th Edition

U.S. Department of Health and Human Services
Public Health Service
Centers for Disease Control and Prevention
National Institutes of Health

HHS Publication No. (CDC) 21-1112
Revised December 2009

Published by Books Express Publishing
Copyright © Books Express, 2010
ISBN 978-1-83931-000-3
To purchase copies at discounted prices please contact
info@books-express.com

Foreword

Biosafety in Microbiological and Biomedical Laboratories (BMBL) quickly became the cornerstone of biosafety practice and policy in the United States upon first publication in 1984. Historically, the information in this publication has been advisory is nature even though legislation and regulation, in some circumstances, have overtaken it and made compliance with the guidance provided mandatory. We wish to emphasize that the 5th edition of the BMBL remains an advisory document recommending best practices for the safe conduct of work in biomedical and clinical laboratories from a biosafety perspective, and is not intended as a regulatory document though we recognize that it will be used that way by some.

This edition of the BMBL includes additional sections, expanded sections on the principles and practices of biosafety and risk assessment; and revised agent summary statements and appendices. We worked to harmonize the recommendations included in this edition with guidance issued and regulations promulgated by other federal agencies. Wherever possible, we clarified both the language and intent of the information provided. The events of September 11, 2001, and the anthrax attacks in October of that year re-shaped and changed, forever, the way we manage and conduct work in biological and clinical laboratories and drew into focus the need for inclusion of additional information in the BMBL. To better serve the needs of our community in this new era, this edition includes information on the following topics:

- Occupational medicine and immunization
- Decontamination and sterilization
- Laboratory biosecurity and risk assessment
- Biosafety level 3 (Ag) laboratories
- Agent summary statements for some agricultural pathogens
- Biological toxins

At last count, over two hundred of our scientific and professional colleagues have assisted in the preparation of the 5th edition through participation in technical working groups, serving as reviewers and guest editors, and as subject matter experts. We wish to thank them all for their dedication and hard work for without them the 5th edition of the BMBL would not be possible. We also recognize the hard work and contributions made by all who participated in preparation of the previous editions of the BMBL; we have built on their solid work and commitment. It is impossible to publish this revision without recognizing the visionary leadership of the previous BMBL editors, Drs. John Richardson and W. Emmett Barkley, and Drs. Jonathan Richmond and Robert W. McKinney, without whom the BMBL would not be the widely and well-regarded resource it is today. The Executive Steering Committee did a stellar job in shepherding this massive revision effort and not without many bumps and bruises along the way. It is through their absolute commitment to quality, technical accuracy, and dedication to the professional practice of biosafety that the 5th edition is born. We are truly grateful to Ms. Kerstin Traum, Council Rock Consulting for her expertise, keen eye for detail and seemingly tireless efforts in performing the duties of technical writer-editor. We also gratefully acknowledge Ms. Cheryl Warfield of Proven Practices, LLC for her copy-editing and formatting skills that significantly enhanced this edition's readability and ease of use.

Finally, without the superb project management abilities and leadership of Dr. Joseph McDade and the technical/scientific editing expertise of Dr. Karl Johnson, especially in virology, the 5th edition of the BMBL would not be possible.

We hope you find this 5th edition of *Biosafety in Microbiological and Biomedical Laboratories* complete, timely and most of all, easy to use. Thank you for your patience and understanding during the long and comprehensive revision process. We believe you will find it was well worth the wait.

Deborah E. Wilson, DrPH, CBSP	L. Casey Chosewood, M.D.
Director	Director
Division of Occupational	Office of Health and Safety
Health and Safety	Centers for Disease Control
National Institutes of Health	and Prevention
Bethesda, Maryland	Atlanta, Georgia

September 2009

Editors

L. Casey Chosewood, MD
Director, Office of Health and Safety
Centers for Disease Control and Prevention
Atlanta, GA 30333

Deborah E. Wilson, DrPH, CBSP
Director, Division of Occupational Health and Safety
Office of the Director
National Institutes of Health
Bethesda, MD 20892

Steering Committee

Shanna Nesby-O'Dell, DVM, MPH
Chief, External Activities Program and
     WHO Collaborating Center for Biosafety and Training
Office of Health and Safety
Centers for Disease Control and Prevention

Robbin S. Weyant, PhD
Chief, Laboratory Safety Branch
Office of Health and Safety
Centers for Disease Control and Prevention

Martin L. Sanders, PhD, CBSP, RBP
Deputy Director
Office of Health and Safety
Centers for Disease Control and Prevention

Deborah E. Wilson, DrPH, CBSP
Director, Division of Occupational Health and Safety
Office of the Director
National Institutes of Health

Guest Editors

Matthew J. Arduino, MS, DrPH
Chief, Environmental and Applied Microbiology Section
Division of Healthcare Quality Promotion
Centers for Disease Control and Prevention

W. Emmett Barkley, PhD
President
Proven Practices, LLC

Mark Q. Benedict
Division of Parasitic Diseases
Centers for Disease Control and Prevention

Louisa Chapman, MD, MSPH
Assistant to the Director for Immunization Policy
National Immunization Program
Centers for Disease Control and Prevention

Dennis M. Dixon, PhD
Chief, Bacteriology and Mycology Branch
Division of Microbiology and Infectious Diseases
National Institutes of Health

Mark L. Eberhard, PhD
Director, Division of Parasitic Diseases
Centers for Disease Control and Prevention

Martin S. Favero, PhD
Director, Scientific and Clinical Affairs
Advanced Sterilization Products
Johnson and Johnson, Inc.

Susan Gorsky
Regulations Officer, Office of Hazardous Materials Standards
Pipeline and Hazardous Materials Safety Administration
United States Department of Transportation

Mary E. Groesch, PhD
Senior Advisor for Science Policy
Office of Science Policy, Office of the Director
National Institutes of Health

Ted Hackstadt, PhD
Laboratory of Intracellular Parasites
National Institute of Allergy and Infectious Diseases
Rocky Mountain Laboratory

Robert A. Heckert, DVM, PhD
National Program Leader, Animal Health
USDA, Agriculture Research Service
Beltsville, MD

Mark L. Hemphill, MS
Chief of Policy
Select Agent Program
Centers for Disease Control and Prevention

Barbara L. Herwaldt, MD, MPH
Medical Officer
Parasitic Diseases Branch
Division of Parasitic Diseases
Centers for Disease Control and Prevention

Julia K. Hilliard, PhD
Viral Immunology Center,
Department of Biology
Georgia State University

William C. Howard, MS
Industrial Hygiene/Safety Manager
Office of Health and Safety
Centers for Disease Control and Prevention

Herbert Jacobi
Deputy Director
Division of Occupational Health and Safety
National Institutes of Health

Rachel E. Levinson, MA
Director
Government and Industry Liaison Office
The Biodesign Institute
Arizona State University

Brian W. J. Mahy, MA, PhD, ScD, DSc
Senior Scientific Advisor
National Center for Infectious Diseases
Centers for Disease Control and Prevention

Charles B. Millard, PhD
Lieutenant Colonel, U.S. Army
Director, Division of Biochemistry
Walter Reed Army Institute of Research

Shanna Nesby-O'Dell, DVM, MPH
Chief, External Activities Program and
        WHO Collaborating Center for Biosafety and Training
Office of Health and Safety
Centers for Disease Control and Prevention

Amy P. Patterson, MD
Director
Office of Biotechnology Activities/Office of Science Policy
Office of the Director
National Institutes of Health

Jonathan Y. Richmond, PhD
Biosafety Consultant
Jonathan Richmond and Associates
Southport, North Carolina

Martin Sanders, PhD, CBSP, RBP
Deputy Director
Office of Health and Safety
Centers for Disease Control and Prevention

James Schmitt, MD, MS
Medical Director
Occupational Medical Service
Division of Occupational Health and Safety
National Institutes of Health

Frank Simione, MS
American Type Culture Collection
Manassas, VA

David G. Stuart, PhD
Microbiologist
The Baker Company

Lee Ann Thomas, DVM
Director
Select Agent, Organisms and Vectors, and Animals
Animal and Plant Health Inspection Service
U.S. Department of Agriculture

Larry G. Thompson
Veterinary Diagnostic Laboratory
Tifton, GA

Robbin S. Weyant, PhD
Chief, Laboratory Safety Branch
Office of Health and Safety
Centers for Disease Control and Prevention

Contributors

BMBL is by its nature a continuously revised manual, and each revision refines and extends the contributions to previous editions. Since 1984, when the first edition of BMBL was published, many scientists and biosafety specialists have contributed to this important reference work. The 5th edition is no exception, as specialists in multiple disciplines generously provided their considerable expertise to this revision. The Editors and Steering Committee gratefully acknowledge the contributions of all of these many contributors over the life of the BMBL, especially contributors to the current edition, who are listed below.

Contributors to BMBL 5th Edition

David Abraham
L. Garry Adams
Michael Adler
Lee Alderman
Christopher E. Ansell
Amy Barringer
Ronald A. Barry
Raymond L. Beets
Ermias Belay
Kathryn Bernard
Carolyn Black
Walter Bond
Cheryl Bopp
Sandra Bragg
David Bressler
Charles Brokopp
Elizabeth J. Browder
Bobby Gene Brown
Corrie Brown
Douglas R. Brown
Michael Buchmeier
Robert Bull
Karen B. Byers
Jane Caputo
Arturo Casadevall
Christina Cassetti
Byron Caughey
Vishnu Chaturvedi
Louisa Chapman
Bruce Chesebro
May Chu

Jeffrey I. Cohen
Murray Cohen
L. Eugene Cole II
Chester Cooper
David Cox
Nancy Cox
Rebecca Cox
Jon Crane
Jack Crawford
Inger Damon
Charles L. Divan
Walter Dowdle
Dennis Eagleson
Eileen Edmondson
Carol L. Eisenhauer
Ana Espinel-Ingroff
Joseph Esposito
Michael T. Fallon
Heinz Feldmann
Barry Fields
Howard Fields
Michael J. Firko
Collette Fitzgerald
Diane O. Fleming
Thomas Folks
Ken Gage
John Galgiani
G. Gale Galland
Leslie Garry Adams
Mahmoud Ghannoum
Mark Gibson

Chester Gipson
Irene Glowinski
Richard Green
William Grizzle
Mary E. Groesch
Ted Hackstadt
Ted Hadfield
Susan B. Harper
Lynn Harding
Kathryn L. Harris
Robert J. Hawley
Mark L. Hemphill
David K. Henderson
Richard Henkel
Sherry Henry
Robert H. Hill
Julia Hilliard
Alex Hoffmaster
James D. Holt
William C. Howard
Melony Ihrig
Eddie L. Jackson
Peter Jahrling
Robert C. Jambou
J. Michael Janda
Jackie Katz
Carole Kauffman
Molly Kellum
Fred Khoshbin
Maxim Kiefer
Michael P. Kiley
Susan K. Kingston
Bruce Klein
Phillip H. Klesius
Joseph P. Kozlovac
Randy Kray
Katrina Kretsinger
Steve Kridel
Thomas G. Ksiazek
Robert Lamb
Linda Lambert
Ross D. LeClaire
Geoffrey J. Letchworth
Paul N. Levett

Randall Levings
Stuart Levitz
Douglas Luster
Keither Mansfield
Dale Martin
Al Mateczun
Henry Mathews
Michael McGinnis
John J. McGlone
Farhad Memarzadeh
Claudia A. Mickelson
Barry Miller
John G. Miller
Thomas L. Miller
Douglas M. Moore
Mario Morales
Ronald L. Morgan
Rand Mortimer
Bernard Moss
Waleid I. Muhmed
Brian Murphy
Irving Nachamkin
Janet K. A. Nicholson
Alison D. O'Brien
Marina O'Reilly
Peter Palese
Ross Pastel
Bill Peters
CJ Peters
Jeannine Petersen
Janet S. Peterson
Louise M. Pitt
Mark A. Poli
Tanja Popovic
Nathaniel Powell
Ann M. Powers
Suzette Priola
Robert Purcell
Greg Raymond
Yvonne M. Reid
Eric Resultan
Judith Rhodes
Robert L. Rice
Jonathan Richmond

Betty Robertson
Luis Rodriguez
Pierre Rollin
Nancy Rosenstein
Eugene Rosenthal
Michael D. Ruff
Charles Rupprecht
Scott Rusk
Janice M. Rusnak
Elliot Ryser
Reynolds M. Salerno
Jim Samuel
Gary Sanden
Thomas Sawicki
Michelle Saylor
Linda R. K. Schlater
Connie Schmaljohn
James J. Schmidt
Beverly Schmitt
James M. Schmitt
Gregg N. Schuiling
Lynne Sehulster
Dennis Senne
Tom Shih
Allan Shipp
Lance Simpson
Theresa J. Smith
Christine Spiropolou
Gregory J. Stewart
Yvonne J. Stifel
Rudy Stotz
Janet E. Stout
Nancy Strockbine
Kanta Subbarao
Bala Swaminathan
David Swayne
James R. Swearengen
Bill Switzer
Mallory K. Tate
James F. Taylor
Louise D. Teel
Robert B. Tesh
Larry Thompson
Alfonso Torres

Kerstin E. Traum
Theodore J. Traum
David Trees
Charles Trimarchi
Robert G. Ulrich
Cristina Vargas
Don Vesley
Paul E. Vinson
David Waag
Robert W. Wannemacher
Cheryl Warfield
David Warnock
William Watson
Mike Weathers
Robert Webster
Elizabeth Weirich
Louis S. Welker
Diane L. Whipple
Denise Whitby
Richard Whitley
Catherine L. Wilhelmsen
Axel Wolff
David Woods
Scott D. Wright
Jack Wunder
Robert Yarchoan
Uri Yokel
Lisa Young
Gary L. Zackowitz

Contents

Tables

Figures

Section I—Introduction

Over the past two decades, *Biosafety in Microbiological and Biomedical Laboratories* (BMBL) has become the code of practice for biosafety—the discipline addressing the safe handling and containment of infectious microorganisms and hazardous biological materials. The principles of biosafety introduced in 1984 in the first edition of BMBL[1] and carried through in this fifth edition remain steadfast. These principles are containment and risk assessment. The fundamentals of containment include the microbiological practices, safety equipment, and facility safeguards that protect laboratory workers, the environment, and the public from exposure to infectious microorganisms that are handled and stored in the laboratory. Risk assessment is the process that enables the appropriate selection of microbiological practices, safety equipment, and facility safeguards that can prevent laboratory-associated infections (LAI). The purpose of periodic updates of BMBL is to refine guidance based on new knowledge and experiences and to address contemporary issues that present new risks that confront laboratory workers and the public health. In this way the code of practice will continue to serve the microbiological and biomedical community as a relevant and valuable authoritative reference.

We are living in an era of uncertainty and change. New infectious agents and diseases have emerged. Work with infectious agents in public and private research, public health, clinical and diagnostic laboratories, and in animal care facilities has expanded. Recent world events have demonstrated new threats of bioterrorism. For these reasons, organizations and laboratory directors are compelled to evaluate and ensure the effectiveness of their biosafety programs, the proficiency of their workers, as well as the capability of equipment, facilities, and management practices to provide containment and security of microbiological agents. Similarly, individual workers who handle pathogenic microorganisms must understand the containment conditions under which infectious agents can be safely manipulated and secured. Application of this knowledge and the use of appropriate techniques and equipment will enable the microbiological and biomedical community to prevent personal, laboratory and environmental exposure to potentially infectious agents or biohazards.

The Occurrence of Laboratory-Associated Infections

Published reports of LAIs first appeared around the start of the twentieth century. By 1978, four studies by Pike and Sulkin collectively identified 4,079 LAIs resulting in 168 deaths occurring between 1930 and 1978.[2-5] These studies found that the ten most common causative agents of overt infections among workers were *Brucella* spp., *Coxiella burnetii,* hepatitis B virus (HBV), *Salmonella typhi, Francisella tularensis, Mycobacterium tuberculosis, Blastomyces dermatitidis,* Venezuelan equine encephalitis virus, *Chlamydia psittaci,* and *Coccidioides immitis.* The authors acknowledged that the 4,079 cases did not represent all LAIs that occurred during this period since many laboratories chose not to report overt cases or conduct surveillance programs to identify sub-clinical or asymptomatic infections.

In addition, reports of LAIs seldom provided data sufficient to determine incidence rates, complicating quantitative assessments of risk. Similarly, there were no distinguishable accidents or exposure events identified in more than 80% of the LAIs reported before 1978. Studies did show that in many cases the infected person worked with a microbiological agent or was in the vicinity of another person handling an agent.[2-6]

During the 20 years following the Pike and Sulkin publications, a worldwide literature search by Harding and Byers revealed 1,267 overt infections with 22 deaths.[7] Five deaths were of fetuses aborted as the consequence of a maternal LAI. *Mycobacterium tuberculosis, Coxiella burnetii,* hantavirus, arboviruses, HBV, *Brucella* spp., *Salmonella* spp., *Shigella* spp., hepatitis C virus, and *Cryptosporidium* spp. accounted for 1,074 of the 1,267 infections. The authors also identified an additional 663 cases that presented as sub-clinical infections. Like Pike and Sulkin, Harding and Byers reported that only a small number of the LAI involved a specific incident. The non-specific associations reported most often by these authors were working with a microbiological agent, being in or around the laboratory, or being around infected animals.

The findings of Harding and Byers indicated that clinical (diagnostic) and research laboratories accounted for 45% and 51%, respectively, of the total LAIs reported. This is a marked difference from the LAIs reported by Pike and Sulkin prior to 1979, which indicated that clinical and research laboratories accounted for 17% and 59%, respectively. The relative increase of LAIs in clinical laboratories may be due in part to improved employee health surveillance programs that are able to detect sub-clinical infections, or to the use of inadequate containment procedures during the early stages of culture identification.

Comparison of the more recent LAIs reported by Harding and Byers with those reported by Pike and Sulkin suggests that the number is decreasing. Harding and Byers note that improvements in containment equipment, engineering controls, and greater emphasis on safety training may be contributing factors to the apparent reduction in LAIs over two decades. However, due to the lack of information on the actual numbers of infections and the population at risk, it is difficult to determine the true incidence of LAIs with any degree of certainty.

Publication of the occurrence of LAIs provides an invaluable resource for the microbiological and biomedical community. For example, one report of occupational exposures associated with *Brucella melitensis,* an organism capable of transmission by the aerosol route, described how a staff member in a clinical microbiology laboratory accidentally sub-cultured *B. melitensis* on the open bench.[8] This error and a breech in containment practices resulted in eight LAIs with *B. melitensis* among 26 laboratory members, an attack rate of 31%.

Reports of LAIs can serve as lessons in the importance of maintaining safe conditions in biological research.

Evolution of National Biosafety Guidelines

National biosafety guidelines evolved from the efforts of the microbiological and biomedical community to promote the use of safe microbiological practices, safety equipment and facility safeguards that will reduce LAIs and protect the public health and environment. The historical accounts of LAIs raised awareness about the hazards of infectious microorganisms and the health risks to laboratory workers who handle them. Many published accounts suggested practices and methods that might prevent LAIs.[9] Arnold G. Wedum was the Director of Industrial Health and Safety at the United States Army Biological Research Laboratories, Fort Detrick from 1944 to 1969. His pioneering work in biosafety provided the foundation for evaluating the risks of handling infectious microorganisms and for recognizing biological hazards and developing practices, equipment, and facility safeguards for their control. Fort Detrick also advanced the field by aiding the development of biosafety programs at the United States Department of Agriculture (USDA), National Animal Research Center and the United States Department of Health and Human Services (DHHS), Centers for Disease Control and Prevention (CDC) and National Institutes of Health (NIH). These governmental organizations subsequently developed several national biosafety guidelines that preceded the first edition of BMBL.

In 1974, the CDC published *Classification of Etiologic Agents on the Basis of Hazard*.[10] This report introduced the concept for establishing ascending levels of containment that correspond to risks associated with handling infectious microorganisms that present similar hazardous characteristics. Human pathogens were grouped into four classes according to mode of transmission and the severity of disease they caused. A fifth class included non-indigenous animal pathogens whose entry into the United States was restricted by USDA policy.

The NIH published *National Cancer Institute Safety Standards for Research Involving Oncogenic Viruses* in 1974.[11] These guidelines established three levels of containment based on an assessment of the hypothetical risk of cancer in humans from exposure to animal oncogenic viruses or a suspected human oncogenic virus isolate from man.[12,13] In 1976 NIH first published the *NIH Guidelines for Research Involving Recombinant DNA Molecules* (*NIH Guidelines*).[14] The *NIH Guidelines* described in detail the microbiological practices, equipment, and facility safeguards that correspond to four ascending levels of physical containment and established criteria for assigning experiments to a containment level based on an assessment of potential hazards of this emerging technology. The evolution of these guidelines set the foundation for developing a code of practice for biosafety in microbiological and biomedical laboratories. Led by the CDC and NIH, a broad collaborative initiative involving scientists, laboratory directors, occupational physicians, epidemiologists, public health officials and health and safety professionals developed the first edition of BMBL in 1984. The BMBL provided the technical content not previously available in biosafety guidelines by adding summary statements conveying guidance pertinent to infectious microorganisms that had caused LAIs. The fifth edition of BMBL is also the product of a broad collaborative initiative committed to perpetuate the value of this national biosafety code of practice.

Risk Criteria for Establishing Ascending Levels of Containment

The primary risk criteria used to define the four ascending levels of containment, referred to as biosafety levels 1 through 4, are infectivity, severity of disease, transmissibility, and the nature of the work being conducted. Another important risk factor for agents that cause moderate to severe disease is the origin of the agent, whether indigenous or exotic. Each level of containment describes the microbiological practices, safety equipment and facility safeguards for the corresponding level of risk associated with handling a particular agent. The basic practices and equipment are appropriate for protocols common to most research and clinical laboratories. The facility safeguards help protect non-laboratory occupants of the building and the public health and environment.

Biosafety level 1 (BSL-1) is the basic level of protection and is appropriate for agents that are not known to cause disease in normal, healthy humans. Biosafety level 2 (BSL-2) is appropriate for handling moderate-risk agents that cause human disease of varying severity by ingestion or through percutaneous or mucous membrane exposure. Biosafety level 3 (BSL-3) is appropriate for agents with a known potential for aerosol transmission, for agents that may cause serious and potentially lethal infections and that are indigenous or exotic in origin. Exotic agents that pose a high individual risk of life-threatening disease by infectious aerosols and for which no treatment is available are restricted to high containment laboratories that meet biosafety level 4 (BSL-4) standards.

It is important to emphasize that the causative incident for most LAIs is unknown.[7,8] Less obvious exposures such as the inhalation of infectious aerosols or direct contact of the broken skin or mucous membranes with droplets containing an infectious microorganism or surfaces contaminated by droplets may possibly explain the incident responsible for a number of LAIs. Most manipulations of liquid suspensions of microorganisms produce aerosols and droplets. Small-particle aerosols have respirable size particles that may contain one or several microorganisms. These small particles stay airborne and easily disperse throughout the laboratory. When inhaled, the human lung will retain those particles. Larger particle droplets rapidly fall out of the air, contaminating gloves, the immediate work area, and the mucous membranes of unprotected workers. A procedure's potential to release microorganisms into the air as aerosols and droplets is the most important operational risk factor that supports the need for containment equipment and facility safeguards.

Agent Summary Statements

The fifth edition, as in all previous editions, includes agent summary statements that describe the hazards, recommended precautions, and levels of containment appropriate for handling specific human and zoonotic pathogens in the laboratory and in facilities that house laboratory vertebrate animals. Agent summary statements are included for agents that meet one or more of the following three criteria:

1. the agent is a proven hazard to laboratory personnel working with infectious materials;
2. the agent has a high potential for causing LAIs even though no documented cases exist; and
3. the agent causes grave disease or presents a significant public health hazard.

Scientists, clinicians, and biosafety professionals prepared the statements by assessing the risks of handling the agents using standard protocols followed in many laboratories. **No one should conclude that the absence of an agent summary statement for a human pathogen means that the agent is safe to handle at BSL-1, or without a risk assessment to determine the appropriate level of containment.** Laboratory directors should also conduct independent risk assessments before beginning work with an agent or procedure new to the laboratory, even though an agent summary statement is available. There may be situations where a laboratory director should consider modifying the precautionary measures or recommended practices, equipment, and facility safeguards described in an agent summary statement. In addition, laboratory directors should seek guidance when conducting risk assessments. Knowledgeable colleagues; institutional biosafety committees; biosafety officers; and public health, biosafety, and scientific associations are excellent resources.

The agent summary statements in the fourth edition BMBL were reviewed in the course of preparing the fifth edition of BMBL. There are new and updated agent summary statements including those for agents now classified as Select Agents. For example, there is an updated section on arboviruses and related zoonotic viruses including new agent summary statements. There are also substantive revisions to the Influenza Agent Summary Statement that address non-contemporary human influenza strains and recommend safeguards for research involving reverse genetics of the 1918 influenza strain.

The fifth edition also includes a revised section on risk assessment that gives more emphasis on the importance of this process in selecting the appropriate practices and level of containment. That section intentionally follows this introduction because risk assessment represents the foundation—a code of practice for safe handling of infectious agents in microbiological and biomedical laboratories.

Biosecurity

Today, the nation is facing a new challenge in safeguarding the public health from potential domestic or international terrorism involving the use of dangerous biological agents or toxins. Existing standards and practices may require adaptation to ensure protection from such hostile actions. In addition, recent federal regulations mandate increased security within the microbiological and biomedical community in order to protect biological pathogens and toxins from theft, loss, or misuse. The fifth edition of BMBL includes an important new section on biosecurity—the discipline addressing the security of microbiological agents and toxins and the threats posed to human and animal health, the environment, and the economy by deliberate misuse or release. A careful review of the biosecurity concepts and guidelines introduced in this new section is essential for all laboratory workers.

Using BMBL

BMBL is both a code of practice and an authoritative reference. Knowledge sufficient to work safely with hazardous microorganisms requires a careful review of the entire BMBL. This will offer the reader an understanding of the biosafety principles that serve as the basis for the concepts and recommendations included in this reference. Reading only selected sections will not adequately prepare even an experienced laboratory worker to handle potentially infectious agents safely.

The recommended practices, safety equipment, and facility safeguards described in the first edition of BMBL and expanded in the fifth edition are advisory in most circumstances. The intent was and is to establish a voluntary code of practice, one that all members of a laboratory community will together embrace to safeguard themselves and their colleagues, and to protect the public health and environment.

Looking Ahead

Laboratory-associated infections from exposure to biological agents known to cause disease are infrequent. It is critical that the microbiological and biomedical community continue its resolve to remain vigilant and not to become complacent. The LAIs reported in the last 25 years demonstrate that accidents and unrecognized exposures continue to occur. The absence of clear evidence of the means of transmission in most documented LAI should motivate persons at risk to be alert to all potential routes of exposure. The accidental release of microbial aerosols is a probable cause of many LAI[15], which demonstrates the importance of worker training and the ability to recognize potential hazards and correct unsafe habits. Attention to and proficient use of work practices, safety equipment and engineering controls are also essential.

The nation's response to recent world events brings with it a heightened concern for a potential increase in LAIs. In 2003, the United States federal government awarded significant funding for the construction of National Biocontainment Laboratories (NBL) and Regional Biocontainment Laboratories (RBL). The NBLs will house BSL-2, 3, and 4 laboratories; the RBLs will house BSL-2 and 3 laboratories. In addition, construction of new containment facilities by private and public institutions is underway nationwide. The expansion of biocontainment laboratories nationwide dramatically increases the need for training in microbiological practices and biosafety principles.

Understanding the principles of biosafety and adherence to the microbiological practices, containment and facility safeguards described in BMBL will contribute to a safer and healthier working environment for laboratory staff and adjacent personnel, and the community.

References

1. Richardson JH, Barkley WE, editors. Biosafety in microbiological and biomedical laboratories. 1st ed. Washington, DC. 1984.
2. Sulkin SE, Pike RM. Survey of laboratory-acquired infections. Am J Pub Hlth. 1951;41:769-81.
3. Pike RM, Sulkin SE, Schulze ML. Continuing importance of laboratory-acquired infections. Am J Pub Hlth. 1965;55:190-99.
4. Pike RM. Laboratory-associated infections: summary and analysis of 3921 cases. Health Lab Sci. 1976;13:105-14.
5. Pike RM. Past and present hazards of working with infectious agents. Arch Pathol Lab Med. 1978;102:333-36.
6. Pike RM. Laboratory-associated infections: incidence, fatalities, causes, and prevention. Annu Rev Microbiol. 1979;33:41-66.
7. Harding AL, Byers KB. Epidemiology of laboratory-associated infections. In: Fleming DO, Hunt DL, editors. Biological safety: principles and practices. 3rd ed. Washington, DC: ASM Press; 2000:35-54.
8. Staskiewicz J, Lewis CM, Colville J, et al. Outbreak of *Brucella melitensis* among microbiology laboratory workers in a community hospital. J Clin Microbiol. 1991;29:287-90.
9. Wedum AG. Laboratory safety in research with infectious diseases. Public Health Rep. 1964;79:619-33.
10. Classification of etiological agents on the basis of hazard. 4th ed. Atlanta, Centers for Disease Control (US); 1974.
11. National Cancer Institute safety standards for research involving oncogenic viruses. Bethesda: The National Institutes of Health (US), National Cancer Institute, Office of Research Safety; 1974, DHEW Publication No. (NIH) 75-790.
12. Wedum AG. History and epidemiology of laboratory-acquired infections (in relation to the cancer research program). JABSA. 1997;2:12-29.

13. West DL, Twardzik DR, McKinney RW, et al. Identification, analysis, and control of biohazards in viral cancer research. In: Fuscaldo AA, Erlick BJ, Hindman B, editors. Laboratory safety theory and practice. New York: Academic Press; 1980. p. 167-223.
14. NIH guidelines for research involving recombinant DNA molecules. Bethesda: The National Institutes of Health (US), Office of Biotechnology Activities; 2002, April.
15. Rusnak JM, Kortepeter MG, Hawley RJ, et al. Risk of occupationally acquired illnesses from biological threat agents in unvaccinated laboratory workers. Biosecur Bioterror. 2004;2:281-93.

Section II—Biological Risk Assessment

Risk assessment is an important responsibility for directors and principal investigators of microbiological and biomedical laboratories. Institutional biosafety committees (IBC), animal care and use committees, biological safety professionals, and laboratory animal veterinarians share in this responsibility. Risk assessment is a process used to identify the hazardous characteristics of a known infectious or potentially infectious agent or material, the activities that can result in a person's exposure to an agent, the likelihood that such exposure will cause a LAI, and the probable consequences of such an infection. The information identified by risk assessment will provide a guide for the selection of appropriate biosafety levels and microbiological practices, safety equipment, and facility safeguards that can prevent LAIs.

Laboratory directors and principal investigators should use risk assessment to alert their staffs to the hazards of working with infectious agents and to the need for developing proficiency in the use of selected safe practices and containment equipment. Successful control of hazards in the laboratory also protects persons not directly associated with the laboratory, such as other occupants of the same building, and the public.

Risk assessment requires careful judgment. Adverse consequences are more likely to occur if the risks are underestimated. By contrast, imposition of safeguards more rigorous than actually needed may result in additional expense and burden for the laboratory, with little safety enhancement. Unnecessary burden may result in circumvention of required safeguards. However, where there is insufficient information to make a clear determination of risk, it is prudent to consider the need for additional safeguards until more data are available.

The primary factors to consider in risk assessment and selection of precautions fall into two broad categories: agent hazards and laboratory procedure hazards. In addition, the capability of the laboratory staff to control hazards must be considered. This capability will depend on the training, technical proficiency, and good habits of all members of the laboratory, and the operational integrity of containment equipment and facility safeguards.

The agent summary statements contained in BMBL identify the primary agent and procedure hazards for specific pathogens and recommend precautions for their control. The guest editors and contributors of this and previous editions of BMBL based their recommendations on an assessment of the risks associated with the handling of pathogens using generally routine generic laboratory procedures. A review of the summary statement for a specific pathogen is a helpful starting point for assessment of the risks of working with that agent and those for a similar agent.

Hazardous Characteristics of an Agent

The principal hazardous characteristics of an agent are: its capability to infect and cause disease in a susceptible human or animal host, its virulence as measured by the severity of disease, and the availability of preventive measures and effective treatments for the disease. The World Health Organization (WHO) has recommended an agent risk group classification for laboratory use that describes four general risk groups based on these principal characteristics and the route of transmission of the natural disease.[1] The four groups address the risk to both the laboratory worker and the community. The *NIH Guidelines* established a comparable classification and assigned human etiological agents into four risk groups on the basis of hazard.[2] The descriptions of the WHO and NIH risk group classifications are presented in Table 1. They correlate with but do not equate to biosafety levels. A risk assessment will determine the degree of correlation between an agent's risk group classification and biosafety level. See Section 3 for a further discussion of the differences and relatedness of risk groups and biosafety levels.

Table 1: Classification of Infectious Microorganisms by Risk Group

Risk Group Classification	NIH Guidelines for Research involving Recombinant DNA Molecules 2002[2]	World Health Organization Laboratory Biosafety Manual 3rd Edition 2004[1]
Risk Group 1	Agents not associated with disease in healthy adult humans.	(No or low individual and community risk) A microorganism unlikely to cause human or animal disease.
Risk Group 2	Agents associated with human disease that is rarely serious and for which preventive or therapeutic interventions are *often* available.	(Moderate individual risk; low community risk) A pathogen that can cause human or animal disease but is unlikely to be a serious hazard to laboratory workers, the community, livestock or the environment. Laboratory exposures may cause serious infection, but effective treatment and preventive measures are available and the risk of spread of infection is limited.
Risk Group 3	Agents associated with serious or lethal human disease for which preventive or therapeutic interventions may be available (high individual risk but low community risk).	(High individual risk; low community risk) A pathogen that usually causes serious human or animal disease but does not ordinarily spread from one infected individual to another. Effective treatment and preventive measures are available.
Risk Group 4	Agents likely to cause serious or lethal human disease for which preventive or therapeutic interventions are not usually available (high individual risk and high community risk).	(High individual and community risk) A pathogen that usually causes serious human or animal disease and can be readily transmitted from one individual to another, directly or indirectly. Effective treatment and preventive measures are not usually available.[3]

Other hazardous characteristics of an agent include probable routes of transmission of laboratory infection, infective dose, stability in the environment, host range, and its endemic nature. In addition, reports of LAIs are a clear indicator of hazard and often are sources of information helpful for identifying agent and procedural hazards, and the precautions for their control. The absence of a report does not indicate minimal risk. Reports seldom provide incidence data, making comparative judgments on risks among agents difficult. The number of infections reported for a single agent may be an indication of the frequency of use as well as risk. Nevertheless, reporting of LAIs by laboratory directors in the scientific and medical literature is encouraged. Reviews of such reports and analyses of LAIs identified through extensive surveys are a valuable resource for risk assessment and reinforcement of the biosafety principles. The summary statements in BMBL include specific references to reports on LAIs.

The predominant probable routes of transmission in the laboratory are: 1) direct skin, eye or mucosal membrane exposure to an agent; 2) parenteral inoculation by a syringe needle or other contaminated sharp, or by bites from infected animals and arthropod vectors; 3) ingestion of liquid suspension of an infectious agent, or by contaminated hand to mouth exposure; and 4) inhalation of infectious aerosols. An awareness of the routes of transmission for the natural human disease is helpful in identifying probable routes of transmission in the laboratory and the potential for any risk to the public health. For example, transmission of infectious agents can occur by direct contact with discharges from respiratory mucous membranes of infected persons, which would be a clear indication that a laboratory worker is at risk of infection from mucosal membrane exposure to droplets generated while handling that agent. The American Public Health Association publication *Control of Communicable Diseases Manual* is an excellent reference for identifying both natural and often noted laboratory modes of transmission.[3] However, it is important to remember that the nature and severity of disease caused by a laboratory infection and the probable laboratory route of transmission of the infectious agent may differ from the route of transmission and severity associated with the naturally-acquired disease.[4]

An agent capable of transmitting disease through respiratory exposure to infectious aerosols is a serious laboratory hazard, both for the person handling the agent and for other laboratory occupants. This hazard requires special caution because infectious aerosols may not be a recognized route of transmission for the natural disease. Infective dose and agent stability are particularly important in establishing the risk of airborne transmission of disease. For example, the reports of multiple infections in laboratories associated with the use of *Coxiella burnetii* are explained by its low inhalation infective dose, which is estimated to be ten inhaled infectious particles, and its resistance to environmental stresses that enables the agent to survive outside of a living host or culture media long enough to become an aerosol hazard.[5]

When work involves the use of laboratory animals, the hazardous characteristics of zoonotic agents require careful consideration in risk assessment. Evidence that experimental animals can shed zoonotic agents and other infectious agents under study in saliva, urine, or feces is an important indicator of hazard. The death of a primate center laboratory worker from Cercopithecine herpes virus 1 (CHV-1, also known as Monkey B virus) infection following an ocular splash exposure to biologic material from a rhesus macaque emphasizes the seriousness of this hazard.[6] Lack of awareness for this potential hazard can make laboratory staff vulnerable to an unexpected outbreak involving multiple infections.[7] Experiments that demonstrate transmission of disease from an infected animal to a normal animal housed in the same cage are reliable indicators of hazard. Experiments that do not demonstrate transmission, however, do not rule out hazard. For example, experimental animals infected with *Francisella tularensis*, *Coxiella burnetii*, *Coccidioides immitis*, or *Chlamydia psittaci*—agents that have caused many LAIs—rarely infect cagemates.[8]

The origin of the agent is also important in risk assessment. Non-indigenous agents are of special concern because of their potential to introduce risk of transmission, or spread of human and animal or infectious diseases from foreign countries into the United States. Importation of etiological agents of human disease requires a permit from the CDC. Importation of many etiological agents of livestock, poultry and other animal diseases requires a permit from the USDA's Animal and Plant Health Inspection Service (APHIS). For additional details, see Appendix C.

Genetically modified agent hazards. The identification and assessment of hazardous characteristics of genetically modified agents involve consideration of the same factors used in risk assessment of the wild-type organism. It is particularly important to address the possibility that the genetic modification could increase an agent's pathogenicity or affect its susceptibility to antibiotics or other effective treatments. The risk assessment can be difficult or incomplete, because important information may not be available for a newly engineered agent. Several investigators have reported that they observed unanticipated enhanced virulence in recent studies with engineered agents.[9-12] These observations give reason to remain alert to the possibility that experimental alteration of virulence genes may lead to increased risk. It also suggests that risk assessment is a continuing process that requires updating as research progresses.

The *NIH Guidelines* are the key reference in assessing risk and establishing an appropriate biosafety level for work involving recombinant DNA molecules.[2] The purpose of the *NIH Guidelines* is to promote the safe conduct of research involving recombinant DNA. The guidelines specify appropriate practices and procedures for research involving constructing and handling both recombinant DNA molecules and organisms and viruses that contain recombinant DNA. They define recombinant DNA as a molecule constructed outside of a living cell with the capability to replicate in a living cell. The *NIH Guidelines* explicitly address experiments that involve introduction of recombinant DNA into Risk Groups 2, 3, and 4 agents, and experiments in which the DNA from Risk Groups 2, 3, and 4 agents is cloned into nonpathogenic prokaryotic or lower eukaryotic host-vector systems. Compliance with the *NIH Guidelines* is mandatory for investigators conducting recombinant DNA research funded by the NIH or performed at, or sponsored by, any public or private entity that receives any NIH funding for recombinant DNA research. Many other institutions have adopted these guidelines as the best current practice.

The *NIH Guidelines* were first published in 1976 and are revised on an ongoing basis in response to scientific and policy developments. The guidelines outline the roles and responsibilities of various entities affiliated with recombinant DNA research, including institutions, investigators, and the NIH. Recombinant DNA research subject to the *NIH Guidelines* may require: 1) approval by the NIH Director, review by the NIH Recombinant DNA Advisory Committee (RAC), and approval by the IBC; or 2) review by the NIH Office of Biotechnology Activities (OBA) and approval by the IBC; or 3) review by the RAC and approvals by the IBC and Institutional Review Board; or 4) approval by the IBC prior to initiation of the research; or 5) notification of the IBC simultaneous with initiation of the work. It is important to note that review by an IBC is required for all non-exempt experiments as defined by the *NIH Guidelines*.

The *NIH Guidelines* were the first documents to formulate the concept of an IBC as the responsible entity for biosafety issues stemming from recombinant DNA research. The *NIH Guidelines* outline the membership, procedures, and functions of an IBC. The institution is ultimately responsible for the effectiveness of the IBC, and may define additional roles and responsibilities for the IBC apart from those specified in the *NIH Guidelines*. See Appendix J for more information about the *NIH Guidelines* and OBA.

Cell cultures. Workers who handle or manipulate human or animal cells and tissues are at risk for possible exposure to potentially infectious latent and adventitious agents that may be present in those cells and tissues. This risk is well understood and illustrated by the reactivation of herpes viruses from latency,[13,14] the inadvertent transmission of disease to organ recipients,[15,16] and the persistence of human immunodeficiency virus (HIV), HBV, and hepatitis C virus (HCV) within infected individuals in the U.S. population.[17] There also is evidence of accidental transplantation of human tumor cells to healthy recipients which indicates that these cells are potentially hazardous to laboratory workers who handle them.[18] In addition, human and animal cell lines that are not well characterized or are obtained from secondary sources may introduce an infectious hazard to the laboratory. For example, the handling of nude mice inoculated with a tumor cell line unknowingly infected with lymphocytic choriomeningitis virus resulted in multiple LAIs.[19] The potential for human cell lines to harbor a bloodborne pathogen led the Occupational Health and Safety Administration (OSHA) to interpret that the occupational exposure to bloodborne pathogens final rule would include primary human cell lines and explants.[17]

Hazardous Characteristics of Laboratory Procedures

Investigations of LAIs have identified five principal routes of laboratory transmission. These are parenteral inoculations with syringe needles or other contaminated sharps, spills and splashes onto skin and mucous membranes, ingestion through mouth pipetting, animal bites and scratches, and inhalation exposures to infectious aerosols. The first four routes of laboratory transmission are easy to detect, but account for less than 20 percent of all reported LAIs.[20] Most reports of such infections do not include information sufficient to identify the route of transmission of infection. Work has shown that the probable sources of infection—animal or ectoparasite, clinical specimen, agent, and aerosol—are apparent in approximately 50 percent of cases.[21]

Aerosols are a serious hazard because they are ubiquitous in laboratory procedures, are usually undetected, and are extremely pervasive, placing the laboratory worker carrying out the procedure and other persons in the laboratory at risk of infection. There is general agreement among biosafety professionals, laboratory directors and principal investigators who have investigated LAIs that an aerosol generated by procedures and operations is the probable source of many LAIs, particularly in cases involving workers whose only known risk factor was that they worked with an agent or in an area where that work was done.

Procedures that impart energy to a microbial suspension will produce aerosols. Procedures and equipment used routinely for handling infectious agents in laboratories, such as pipetting, blenders, non-self contained centrifuges, sonicators and vortex mixers are proven sources of aerosols. These procedures and equipment generate respirable-size particles that remain airborne for protracted periods. When inhaled, these particles are retained in the lungs creating an exposure hazard for the person performing the operation, coworkers in the laboratory, and a potential hazard for persons occupying adjacent spaces open to air flow from the laboratory. A number of investigators have determined the aerosol output of common laboratory procedures. In addition, investigators have proposed a model for estimating inhalation dosage from a laboratory aerosol source. Parameters that characterize aerosol hazards include an agent's inhalation infective dose, its viability in an aerosol, aerosol concentration, and particle size.[22,23,24]

Procedures and equipment that generate respirable size particles also generate larger size droplets that can contain multiple copies of an infectious agent. The larger size droplets settle out of the air rapidly, contaminating the gloved hands and work surface and possibly the mucous membranes of the persons performing the procedure. An evaluation of the release of both respirable particles and droplets from laboratory operations determined that the respirable component is relatively small and does not vary widely; in contrast hand and surface contamination is substantial and varies widely.[25] The potential risk from exposure to droplet contamination requires as much attention in a risk assessment as the respirable component of aerosols.

Technique can significantly impact aerosol output and dose. The worker who is careful and proficient will minimize the generation of aerosols. A careless and hurried worker will substantially increase the aerosol hazard. For example, the hurried worker may operate a sonic homogenizer with maximum aeration whereas the careful worker will consistently operate the device to assure minimal aeration. Experiments show that the aerosol burden with maximal aeration is approximately 200 times greater than aerosol burden with minimal aeration.[22] Similar results were shown for pipetting with bubbles and with minimal bubbles. Containment and good laboratory practices also reduce this risk.

Potential Hazards Associated with Work Practices, Safety Equipment and Facility Safeguards

Workers are the first line of defense for protecting themselves, others in the laboratory, and the public from exposure to hazardous agents. Protection depends on the conscientious and proficient use of good microbiological practices and the correct use of safety equipment. A risk assessment should identify any potential deficiencies in the practices of the laboratory workers. Carelessness is the most serious concern, because it can compromise any safeguards of the laboratory and increase the risk for coworkers. Training, experience, knowledge of the agent and procedure hazards, good habits, caution, attentiveness, and concern for the health of coworkers are prerequisites for a laboratory staff in order to reduce the inherent risks that attend work with hazardous agents. Not all workers who join a laboratory staff will have these prerequisite traits even though they may possess excellent scientific credentials. Laboratory directors or principal investigators should train and retrain new staff to the point where aseptic techniques and safety precautions become second nature.[26]

There may be hazards that require specialized personal protective equipment in addition to safety glasses, laboratory gowns, and gloves. For example, a procedure that presents a splash hazard may require the use of a mask and a face shield to provide adequate protection. Inadequate training in the proper use of personal protective equipment may reduce its effectiveness, provide a false sense of security, and could increase the risk to the laboratory worker. For example, a respirator may impart a risk to the wearer independent of the agents being manipulated.

Safety equipment such as biological safety cabinets (BSC), centrifuge safety cups, and sealed rotors are used to provide a high degree of protection for the laboratory worker from exposure to microbial aerosols and droplets. Safety equipment that is not working properly is hazardous, especially when the user is unaware of the malfunction. Poor location, room air currents, decreased airflow, leaking filters, raised sashes, crowded work surfaces, and poor user technique compromise the containment capability of a BSC. The safety characteristics of modern centrifuges are only effective if the equipment is operated properly. Training in the correct use of equipment, proper procedure, routine inspections and potential malfunctions, and periodic re-certification of equipment, as needed, is essential.

Facility safeguards help prevent the accidental release of an agent from the laboratory. Their use is particularly important at BSL-3 and BSL-4 because the agents assigned to those levels can transmit disease by the inhalation route or can cause life-threatening disease. For example, one facility safeguard is directional airflow. This safeguard helps to prevent aerosol transmission from a laboratory into other areas of the building. Directional airflow is dependent on the operational integrity of the laboratory's heating, ventilation, and air conditioning (HVAC) system. HVAC systems require careful monitoring and periodic maintenance to sustain operational integrity. Loss of directional airflow compromises safe laboratory operation. BSL-4 containment facilities provide more complex safeguards that require significant expertise to design and operate.

Consideration of facility safeguards is an integral part of the risk assessments. A biological safety professional, building and facilities staff, and the IBC should help assess the facility's capability to provide appropriate protection for the planned work, and recommend changes as necessary. Risk assessment may support the need to include additional facility safeguards in the construction of new or renovation of old BSL-3 facilities.

An Approach to Assess Risks and Select Appropriate Safeguards

Biological risk assessment is a subjective process requiring consideration of many hazardous characteristics of agents and procedures, with judgments based often on incomplete information. There is no standard approach for conducting a biological risk assessment, but some structure can be helpful in guiding the process. This section describes a five-step approach that gives structure to the risk assessment process.

First, identify agent hazards and perform an initial assessment of risk. Consider the principal hazardous characteristics of the agent, which include its capability to infect and cause disease in a susceptible human host, severity of disease, and the availability of preventive measures and effective treatments.

Several excellent resources provide information and guidance for making an initial risk assessment. The BMBL provides agent summary statements for some agents that are associated with LAIs or are of increased public concern. Agent summary statements also identify known and suspected routes of transmission of laboratory infection and, when available, information on infective dose, host range, agent stability in the environment, protective immunizations, and attenuated strains of the agent.

A thorough examination of the agent hazards is necessary when the intended use of an agent does not correspond with the general conditions described in the summary statement or when an agent summary statement is not available. Although a summary statement for one agent may provide helpful information for assessing the risk of a similar agent, it should not serve as the primary resource for making the risk determination for that agent. Refer to other resources for guidance in identifying the agent hazards.

The *Control of Communicable Diseases Manual* provides information on communicable diseases including concise summaries on severity, mode of transmission, and the susceptibility and resistance of humans to disease.[3] In addition, it is always helpful to seek guidance from colleagues with experience in handling the agent and from biological safety professionals.

Often there is not sufficient information to make an appropriate assessment of risk. For example, the hazard of an unknown agent that may be present in a diagnostic specimen will be unknown until after completing agent identification and typing procedures. It would be prudent in this case to assume the specimen contains an agent presenting the hazardous classification that correlates with BSL-2, unless additional information suggests the presence of an agent of higher risk. Identification of agent hazards associated with newly emergent pathogens also requires judgments based on incomplete information. Consult interim biosafety guidelines prepared by the CDC and the WHO for risk assessment guidance. When assessing the hazards of a newly attenuated pathogen, experimental data should support a judgment that the attenuated pathogen is less hazardous than the wild-type parent pathogen before making any reduction in the containment recommended for that pathogen.

Make a preliminary determination of the biosafety level that best correlates with the initial risk assessment based on the identification and evaluation of the agent hazards. Remember that aerosol and droplet routes of agent transmission also are important considerations in specification of safety equipment and facility design that result in a given BSL level.

Second, identify laboratory procedure hazards. The principal laboratory procedure hazards are agent concentration, suspension volume, equipment and procedures that generate small particle aerosols and larger airborne particles (droplets), and use of sharps. Procedures involving animals can present a number of hazards such as bites and scratches, exposure to zoonotic agents, and the handling of experimentally generated infectious aerosols.

The complexity of a laboratory procedure can also present a hazard. The agent summary statement provides information on the primary laboratory hazards associated with typically routine procedures used in handling an agent. In proposed laboratory procedures where the procedure hazards differ from the general conditions of the agent summary statement or where an agent summary statement is not available, the risk assessment should identify specific hazards associated with the procedures.

Third, make a determination of the appropriate biosafety level and select additional precautions indicated by the risk assessment. The selection of the appropriate biosafety level and the selection of any additional laboratory precautions require a comprehensive understanding of the practices, safety equipment, and facility safeguards described in Sections III, IV, and V of this publication.

There will be situations where the intended use of an agent requires greater precautions than those described in the agent's Summary Statement. These situations will require the careful selection of additional precautions. An obvious example would be a procedure for exposing animals to experimentally generated infectious aerosols.

It is unlikely that a risk assessment would indicate a need to alter the recommended facility safeguards specified for the selected biosafety level. If this does occur, however, it is important that a biological safety professional validate this judgment independently before augmenting any facility secondary barrier.

It is also important to recognize that individuals in the laboratory may differ in their susceptibility to disease. Pre-existing diseases, medications, compromised immunity, and pregnancy or breast-feeding that may increase exposure to infants to certain agents, are some of the conditions that may increase the risk of an individual for acquiring a LAI. Consultation with an occupational physician knowledgeable in infectious diseases is advisable in these circumstances.

Fourth, evaluate the proficiencies of staff regarding safe practices and the integrity of safety equipment. The protection of laboratory workers, other persons associated with the laboratory, and the public will depend ultimately on the laboratory workers themselves. In conducting a risk assessment, the laboratory director or principal investigator should ensure that laboratory workers have acquired the technical proficiency in the use of microbiological practices and safety equipment required for the safe handling of the agent, and have developed good habits that sustain excellence in the performance of those practices. An evaluation of a person's training, experience in handling infectious agents, proficiency in the use of sterile techniques and BSCs, ability to respond to emergencies, and willingness to accept responsibility for protecting one's self and others is important insurance that a laboratory worker is capable of working safely.

The laboratory director or principal investigator should also ensure that the necessary safety equipment is available and operating properly. For example, a BSC that is not certified represents a potentially serious hazard to the laboratory worker using it and to others in the laboratory. The director should have all equipment deficiencies corrected before starting work with an agent.

Fifth, review the risk assessment with a biosafety professional, subject matter expert, and the IBC. A review of the risk assessment and selected safeguards by knowledgeable individuals is always beneficial and sometimes required by regulatory or funding agencies, as is the case with the *NIH Guidelines*.[2] Review of potentially high risk protocols by the local IBC should become standard practice. Adopting this step voluntarily will promote the use of safe practices in work with hazardous agents in microbiological and biomedical laboratories.

Conclusion

Risk assessment is the basis for the safeguards developed by the CDC, the NIH, and the microbiological and biomedical community to protect the health of laboratory workers and the public from the risks associated with the use of hazardous biological agents in laboratories. Experience shows that these established safe practices, equipment, and facility safeguards work.

New knowledge and experience may justify altering these safeguards. Risk assessment, however, must be the basis for recommended change. Assessments conducted by laboratory directors and principal investigators for the use of emergent agents and the conduct of novel experiments will contribute to our understanding of the risks these endeavors may present and the means for their control. Those risk assessments will likely mirror progress in science and technology and serve as the basis for future revisions of BMBL.

References

1. World Health Organization. Laboratory biosafety manual. 3rd ed. Geneva; 2004.
2. The National Institutes of Health (US), Office of Biotechnology Activities. NIH guidelines for research involving recombinant DNA molecules. Bethesda; 2002, April.
3. American Public Health Association. Control of communicable diseases manual. 18th ed. DL Heymann, editor. Washington, DC; 2005.
4. Lennette EH, Koprowski H. Human infection with Venezuelan equine encephalomyelitis virus. JAMA. 1943;123:1088-95.
5. Tigertt WD, Benenson AS, Gochenour WS. Airborne Q fever. Bacteriol Rev. 1961;25:285-93.
6. Centers for Disease Control and Prevention. Fatal Cercopithecine herpesvirus 1 (B virus) infection following a mucocutaneous exposure and interim recommendations for worker protection. MMWR Morb Mortal Wkly Rep. 1998;47:1073-6,1083.
7. Centers for Disease Control and Prevention. Laboratory management of agents associated with hantavirus pulmonary syndrome: interim biosafety guidelines. MMWR Recomm Rep. 1994;43:1-7.
8. Wedum AG, Barkley WE, Hellman A. Handling of infectious agents. J Am Vet Med Assoc. 1972;161:1557-67.

9. Jackson RJ, Ramsay AJ, Christensen CD, et al. Expression of mouse interleukin-4 by a recombinant ectromelia virus suppresses cytolytic lymphocyte responses and overcomes genetic resistance to mousepox. J Virol. 2001;75:1205-10.

10. Shimono N, Morici L, Casali N, et al. Hypervirulent mutant of *Mycobacterium tuberculosis* resulting from disruption of the mce1 operon. Proc Natl Acad Sci U S A. 2003;100:15918-23.

11. Cunningham ML, Titus RG, Turco SJ, et al. Regulation of differentiation to the infective stage of the protozoan parasite Leishmania major by tetrahydrobiopterin. Science. 2001;292:285-7.

12. Kobasa D, Takada A, Shinya K, et al. Enhanced virulence of influenza A viruses with the haemagglutinin of the 1918 pandemic virus. Nature. 2004;431:703-7.

13. Efstathiou S, Preston CM. Towards an understanding of the molecular basis of herpes simplex virus latency. Virus Res. 2005;111:108-19.

14. Oxman MN, Levin MJ, Johnson GR, et al. A vaccine to prevent herpes zoster and postherpetic neuralgia in older adults. N Engl J Med. 2005;352:2271-84.

15. Centers for Disease Control and Prevention. Update: Investigation of rabies infections in organ donor and transplant recipients—Alabama, Arkansas, Oklahoma, and Texas. MMWR Morb Mortal Wkly Rep. 2004;53:615-16.

16. Centers for Disease Control and Prevention. Lymphocytic choriomeningitis virus infection in organ transplant recipients—Massachusetts, Rhode Island, 2005. MMWR Morb Mortal Wkly Rep. 2005;54:537-39.

17. US Department of Labor. Occupational exposure to bloodborne pathogens. Final Rule. Standard interpretations: applicability of 1910.1030 to established human cell lines. June 24, 1994.

18. Gartner HV, Seidl C, Luckenbach C, et al. Genetic analysis of a sarcoma accidentally transplanted from a patient to a surgeon. N Engl J Med. 1996;335:1494-7.

19. Dykewicz CA, Dato VM, Fisher-Hoch SP, et al. Lymphocytic choriomeningitis outbreak associated with nude mice in a research institute. JAMA. 1992;267:1349-53.

20. Pike RM. Laboratory-associated infections: incidence, fatalities, causes, and prevention. Annu Rev Microbiol. 1979;33:41-66.

21. Harding AL, Byers KB. Epidemiology of laboratory-associated infections. In: Fleming DO, Hunt DL, editors. Biological safety: principles and practices. 3rd ed. Washington, DC: ASM Press; 2000:35-54.

22. Dimmick RL, Fogl WF, Chatigny MA. Potential for accidental microbial aerosol transmission in the biology laboratory. In: Hellman A, Oxman MN, Pollack R, editors. Biohazards in biological research. Proceedings of a conference held at the Asilomar conference center; 1973 Jan 22-24; Pacific Grove, CA. New York: Cold Spring Harbor Laboratory; 1973. p. 246-66.

23. Kenny MT, Sable FL. Particle size distribution of *Serratia marcescens* aerosols created during common laboratory procedures and simulated laboratory accidents. Appl Microbiol. 1968;16:1146-50.
24. Chatigny MA, Barkley WE, Vogl WF. Aerosol biohazard in microbiological laboratories and how it is affected by air conditioning systems. ASHRAE Transactions. 1974;80(Pt 1):463-469.
25. Chatigny MA, Hatch MT, Wolochow H, et al. Studies on release and survival of biological substances used in recombinant DNA laboratory procedures. National Institutes of Health Recombinant DNA Technical Bulletin. 1979.
26. Lennette EH. Panel V common sense in the laboratory: recommendations and priorities. Biohazards in biological research. Proceedings of a conference held at the Asilomar conference center; 1973 Jan 22-24; Pacific Grove, CA. New York: Cold Spring Harbor Laboratory; 1973. p. 353.

Section III—Principles of Biosafety

A fundamental objective of any biosafety program is the containment of potentially harmful biological agents. The term "containment" is used in describing safe methods, facilities and equipment for managing infectious materials in the laboratory environment where they are being handled or maintained. The purpose of containment is to reduce or eliminate exposure of laboratory workers, other persons, and the outside environment to potentially hazardous agents. The use of vaccines may provide an increased level of personal protection. The risk assessment of the work to be done with a specific agent will determine the appropriate combination of these elements.

Laboratory Practices and Technique

The most important element of containment is strict adherence to standard microbiological practices and techniques. Persons working with infectious agents or potentially infected materials must be aware of potential hazards, and must be trained and proficient in the practices and techniques required for handling such material safely. The director or person in charge of the laboratory is responsible for providing or arranging the appropriate training of personnel.

Each laboratory should develop or adopt a biosafety or operations manual that identifies the hazards that will or may be encountered, and that specifies practices and procedures designed to minimize or eliminate exposures to these hazards. Personnel should be advised of special hazards and should be required to read and follow the required practices and procedures. A scientist, trained and knowledgeable in appropriate laboratory techniques, safety procedures, and hazards associated with handling infectious agents must be responsible for the conduct of work with any infectious agents or materials. This individual should consult with biosafety or other health and safety professionals with regard to risk assessment.

When standard laboratory practices are not sufficient to control the hazards associated with a particular agent or laboratory procedure, additional measures may be needed. The laboratory director is responsible for selecting additional safety practices, which must be in keeping with the hazards associated with the agent or procedure.

Appropriate facility design and engineering features, safety equipment, and management practices must supplement laboratory personnel, safety practices, and techniques.

Safety Equipment (Primary Barriers and Personal Protective Equipment)

Safety equipment includes BSCs, enclosed containers, and other engineering controls designed to remove or minimize exposures to hazardous biological materials. The BSC is the principal device used to provide containment of infectious droplets or aerosols generated by many microbiological procedures. Three types of BSCs (Class I, II, III) used in microbiological laboratories are described and illustrated in Appendix A. Open-fronted Class I and Class II BSCs are primary barriers that offer significant levels of protection to laboratory personnel and to the environment when used with good microbiological techniques. The Class II biological safety cabinet also provides protection from external contamination of the materials (e.g., cell cultures, microbiological stocks) being manipulated inside the cabinet. The gas-tight Class III biological safety cabinet provides the highest attainable level of protection to personnel and the environment.

An example of another primary barrier is the safety centrifuge cup, an enclosed container designed to prevent aerosols from being released during centrifugation. To minimize aerosol hazards, containment controls such as BSCs or centrifuge cups must be used when handling infectious agents.

Safety equipment also may include items for personal protection, such as gloves, coats, gowns, shoe covers, boots, respirators, face shields, safety glasses, or goggles. Personal protective equipment is often used in combination with BSCs and other devices that contain the agents, animals, or materials being handled. In some situations in which it is impractical to work in BSCs, personal protective equipment may form the primary barrier between personnel and the infectious materials. Examples include certain animal studies, animal necropsy, agent production activities, and activities relating to maintenance, service, or support of the laboratory facility.

Facility Design and Construction (Secondary Barriers)

The design and construction of the facility contributes to the laboratory workers' protection, provides a barrier to protect persons outside the laboratory, and protects persons or animals in the community from infectious agents that may be accidentally released from the laboratory. Laboratory directors are responsible for providing facilities commensurate with the laboratory's function and the recommended biosafety level for the agents being manipulated.

The recommended secondary barrier(s) will depend on the risk of transmission of specific agents. For example, the exposure risks for most laboratory work in BSL-1 and BSL-2 facilities will be direct contact with the agents, or inadvertent contact exposures through contaminated work environments. Secondary barriers in these laboratories may include separation of the laboratory work area from public access, availability of a decontamination facility (e.g., autoclave), and hand washing facilities.

When the risk of infection by exposure to an infectious aerosol is present, higher levels of primary containment and multiple secondary barriers may become necessary to prevent infectious agents from escaping into the environment. Such design features include specialized ventilation systems to ensure directional airflow, air treatment systems to decontaminate or remove agents from exhaust air, controlled access zones, airlocks at laboratory entrances, or separate buildings or modules to isolate the laboratory. Design engineers for laboratories may refer to specific ventilation recommendations as found in the ASHRAE Laboratory Design Guide published by the American Society of Heating, Refrigerating, and Air-Conditioning Engineers (ASHRAE).[1]

Biosafety Levels

Four BSLs are described in Section 4, which consist of combinations of laboratory practices and techniques, safety equipment, and laboratory facilities. Each combination is specifically appropriate for the operations performed, the documented or suspected routes of transmission of the infectious agents, and the laboratory function or activity. The BSLs described in this manual should be differentiated from Risk Groups, as described in the *NIH Guidelines* and the World Health Organization Laboratory Biosafety Manual. Risk groups are the result of a classification of microbiological agents based on their association with, and resulting severity of, disease in humans. The risk group of an agent should be one factor considered in association with mode of transmission, procedural protocols, experience of staff, and other factors in determining the BSL in which the work will be conducted.

The recommended biosafety level(s) for the organisms in Section VIII (Agent Summary Statements) represent those conditions under which the agent ordinarily can be safely handled. Of course, not all of the organisms capable of causing disease are included in Section VIII and an institution must be prepared to perform risk assessments for these agents using the best available information. Detailed information regarding the conduct of biological risk assessments can be found in Section II. The laboratory director is specifically and primarily responsible for assessing the risks and applying the appropriate biosafety levels. The institution's Biological Safety Officer (BSO) and IBC can be of great assistance in performing and reviewing the required risk assessment. At one point, under the *NIH Guidelines*, BSOs were required only when large-scale research or production of organisms containing recombinant DNA molecules was performed or when work with recombinant DNA molecules was conducted at BSL-3 or above. IBCs were required only when an institution was performing non-exempt recombinant DNA experiments. Today, however, it is strongly suggested that an institution conducting research or otherwise working with pathogenic agents have a BSO and properly constituted and functioning IBC. The responsibilities of each now extend beyond those described in the *NIH Guidelines* and depend on the size and complexity of the program.

Generally, work with known agents should be conducted at the biosafety level recommended in Section VIII. When information is available to suggest that virulence, pathogenicity, antibiotic resistance patterns, vaccine and treatment availability, or other factors are significantly altered, more (or less) stringent practices may be specified. Often an increased volume or a high concentration of agent may require additional containment practices.

Biosafety Level 1 practices, safety equipment, and facility design and construction are appropriate for undergraduate and secondary educational training and teaching laboratories, and for other laboratories in which work is done with defined and characterized strains of viable microorganisms not known to consistently cause disease in healthy adult humans. *Bacillus subtilis, Nigeria gruberi*, infectious canine hepatitis virus, and exempt organisms under the *NIH Guidelines* are representative of microorganisms meeting these criteria. Many agents not ordinarily associated with disease processes in humans are, however, opportunistic pathogens and may cause infection in the young, the aged, and immunodeficient or immunosuppressed individuals. Vaccine strains that have undergone multiple *in vivo* passages should not be considered avirulent simply because they are vaccine strains.

BSL-1 represents a basic level of containment that relies on standard microbiological practices with no special primary or secondary barriers recommended, other than a sink for hand washing.

Biosafety Level 2 practices, equipment, and facility design and construction are applicable to clinical, diagnostic, teaching, and other laboratories in which work is done with the broad spectrum of indigenous moderate-risk agents that are present in the community and associated with human disease of varying severity. With good microbiological techniques, these agents can be used safely in activities conducted on the open bench, provided the potential for producing splashes or aerosols is low. Hepatitis B virus, HIV, the *Salmonella*, and *Toxoplasma* are representative of microorganisms assigned to this containment level. BSL-2 is appropriate when work is done with any human-derived blood, body fluids, tissues, or primary human cell lines where the presence of an infectious agent may be unknown. (Laboratory personnel working with human-derived materials should refer to the OSHA Bloodborne Pathogen Standard[2] for specific required precautions).

Primary hazards to personnel working with these agents relate to accidental percutaneous or mucous membrane exposures, or ingestion of infectious materials. Extreme caution should be taken with contaminated needles or sharp instruments. Even though organisms routinely manipulated at BSL-2 are not known to be transmissible by the aerosol route, procedures with aerosol or high splash potential that may increase the risk of such personnel exposure must be conducted in primary containment equipment, or in devices such as a BSC or safety centrifuge cups. Personal protective equipment should be used as appropriate, such as splash shields, face protection, gowns, and gloves.

Secondary barriers, such as hand washing sinks and waste decontamination facilities, must be available to reduce potential environmental contamination.

Biosafety Level 3 practices, safety equipment, and facility design and construction are applicable to clinical, diagnostic, teaching, research, or production facilities in which work is done with indigenous or exotic agents with a potential for respiratory transmission, and which may cause serious and potentially lethal infection. *Mycobacterium tuberculosis*, St. Louis encephalitis virus, and *Coxiella burnetii* are representative of the microorganisms assigned to this level. Primary hazards to personnel working with these agents relate to autoinoculation, ingestion, and exposure to infectious aerosols. At BSL-3, more emphasis is placed on primary and secondary barriers to protect personnel in contiguous areas, the community, and the environment from exposure to potentially infectious aerosols. For example, all laboratory manipulations should be performed in a BSC or other enclosed equipment, such as a gas-tight aerosol generation chamber. Secondary barriers for this level include controlled access to the laboratory and ventilation requirements that minimize the release of infectious aerosols from the laboratory.

Biosafety Level 4 practices, safety equipment, and facility design and construction are applicable for work with dangerous and exotic agents that pose a high individual risk of life-threatening disease, which may be transmitted via the aerosol route and for which there is no available vaccine or therapy. Agents with a close or identical antigenic relationship to BSL-4 agents also should be handled at this level. When sufficient data are obtained, work with these agents may continue at this level or at a lower level. Viruses such as Marburg or Congo-Crimean hemorrhagic fever are manipulated at BSL-4.

The primary hazards to personnel working with BSL-4 agents are respiratory exposure to infectious aerosols, mucous membrane or broken skin exposure to infectious droplets, and autoinoculation. All manipulations of potentially infectious diagnostic materials, isolates, and naturally or experimentally infected animals, pose a high risk of exposure and infection to laboratory personnel, the community, and the environment.

The laboratory worker's complete isolation from aerosolized infectious materials is accomplished primarily by working in a Class III BSC or in a full-body, air-supplied positive-pressure personnel suit. The BSL-4 facility itself is generally a separate building or completely isolated zone with complex, specialized ventilation requirements and waste management systems to prevent release of viable agents to the environment.

The laboratory director is specifically and primarily responsible for the safe operation of the laboratory. His/her knowledge and judgment are critical in assessing risks and appropriately applying these recommendations. The recommended biosafety level represents those conditions under which the agent can ordinarily be safely handled. Special characteristics of the agents used, the training and experience of personnel, procedures being conducted and the nature or function of the laboratory may further influence the director in applying these recommendations.

Animal Facilities

Four standard biosafety levels are also described for activities involving infectious disease work with commonly used experimental animals. These four combinations of practices, safety equipment, and facilities are designated Animal Biosafety Levels 1, 2, 3, and 4, and provide increasing levels of protection to personnel and the environment.

One additional biosafety level, designated BSL-3-Agriculture (or BSL 3-Ag) addresses activities involving large or loose-housed animals and/or studies involving agents designated as High Consequence Pathogens by the USDA. BSL 3-Ag laboratories are designed so that the laboratory facility itself acts as a primary barrier to prevent release of infectious agents into the environment. More information on the design and operation of BSL 3-Ag facilities and USDA High Consequence Pathogens is provided in Appendix D.

Clinical Laboratories

Clinical laboratories, especially those in health care facilities, receive clinical specimens with requests for a variety of diagnostic and clinical support services. Typically, the infectious nature of clinical material is unknown, and specimens are often submitted with a broad request for microbiological examination for multiple agents (e.g., sputa submitted for "routine," acid-fast, and fungal cultures). It is the responsibility of the laboratory director to establish standard procedures in the laboratory that realistically address the issue of the infective hazard of clinical specimens.

Except in extraordinary circumstances (e.g., suspected hemorrhagic fever), the initial processing of clinical specimens and serological identification of isolates can be done safely at BSL-2, the recommended level for work with bloodborne pathogens such as HBV and HIV. The containment elements described in BSL-2 are consistent with the OSHA standard, "Occupational Exposure to Bloodborne Pathogens.[2,3] This requires the use of specific precautions with all clinical specimens of blood or other potentially infectious material (Universal or Standard* Precautions).[4,5] Additionally, other recommendations specific for clinical laboratories may be obtained from the Clinical Laboratory Standards Institute (formerly known as the National Committee for Clinical Laboratory Standards).[6]

BSL-2 recommendations and OSHA requirements focus on the prevention of percutaneous and mucous membrane exposures to clinical material. Primary barriers such as BSCs (Class I or II) should be used when performing procedures that might cause splashing, spraying, or splattering of droplets. Biological safety cabinets also should be used for the initial processing of clinical specimens when the nature of the test requested or other information suggests the likely presence of an agent readily transmissible by infectious aerosols (e.g., *M. tuberculosis*), or when the use of a BSC (Class II) is indicated to protect the integrity of the specimen.

The segregation of clinical laboratory functions and limited or restricted access to such areas is the responsibility of the laboratory director. It is also the director's responsibility to establish standard, written procedures that address the potential hazards and the required precautions to be implemented.

Importation and Interstate Shipment of Certain Biomedical Materials

The importation of etiologic agents and vectors of human diseases is subject to the requirements of the Public Health Service Foreign Quarantine regulations. Companion regulations of the Public Health Service and the Department of Transportation specify packaging, labeling, and shipping requirements for etiologic agents and diagnostic specimens shipped in interstate commerce. (See Appendix C.)

The USDA regulates the importation and interstate shipment of animal pathogens and prohibits the importation, possession, or use of certain exotic animal disease agents that pose a serious disease threat to domestic livestock and poultry. (See Appendix F.)

Select Agents

In recent years, with the passing of federal legislation regulating the possession, use, and transfer of agents with high adverse public health and/or agricultural consequences (DHHS and USDA Select Agents), much greater emphasis has been placed in the emerging field of biosecurity. Biosecurity and select agent issues are covered in detail in Section 6 and Appendix F of this document. In contrast with biosafety, a field dedicated to the protection of workers and the environment from exposures to infectious materials, the field of biosecurity prevents loss of valuable research materials and limits access to infectious materials by individuals who would use them for harmful purposes. Nevertheless, adequate containment of biological materials is a fundamental program component for both biosafety and biosecurity.

References

1. McIntosh IBD, Morgan CB, Dorgan CE. ASHRAE laboratory design guide. Atlanta: American Society of Heating, Refrigerating and Air-Conditioning Engineers, Inc; 2001.
2. U.S. Department of Labor. Occupational exposure to bloodborne pathogens. Final Rule. 29 CFR. 1910.1030 (1991).
3. Richmond JY. HIV biosafety: guidelines and regulations. In: Schochetman G, George JR, editors. AIDS testing. 2nd ed. New York: Springer-Verlag; 1994. p. 346-60.
4. Centers for Disease Control and Prevention. Update: universal precautions for prevention of transmission of human immunodeficiency virus, hepatitis

B virus and other bloodborne pathogens in health-care settings. MMWR Morb Mortal Wkly Rep. 1988;37:377-82,387-8.
5.   Garner JS. Guideline for isolation precautions in hospitals. Infect Control Hosp Epidemiol. 1996;17:53-80.
6.   National Committee for Clinical Laboratory Standards. Protection of laboratory workers from occupationally-acquired infections; approved guideline, 3rd ed.

*   In 1996 the United States Hospital Infection Control Practices Advisory Committee introduced a new set of guidelines, "Standard Precautions," to synthesize the major features of Universal Precautions (blood and body fluid) with Body Substance Isolation Precautions (designed to reduce the risk of transmission of pathogens from moist body substances).[6] Standard Precautions apply to 1) blood; 2) all body fluids, secretions, and excretions except sweat, regardless of whether or not they contain visible blood; 3) non-intact skin; and 4) mucous membranes. For additional information on Standard Precautions, see reference 6 or the CDC Web site: *www.cdc.gov.*

Section IV—Laboratory Biosafety Level Criteria

The essential elements of the four biosafety levels for activities involving infectious microorganisms and laboratory animals are summarized in Table 2 of this section and discussed in Section 2. The levels are designated in ascending order, by degree of protection provided to personnel, the environment, and the community. Standard microbiological practices are common to all laboratories. Special microbiological practices enhance worker safety, environmental protection, and address the risk of handling agents requiring increasing levels of containment.

Biosafety Level 1

Biosafety Level 1 is suitable for work involving well-characterized agents not known to consistently cause disease in immunocompetent adult humans, and present minimal potential hazard to laboratory personnel and the environment. BSL-1 laboratories are not necessarily separated from the general traffic patterns in the building. Work is typically conducted on open bench tops using standard microbiological practices. Special containment equipment or facility design is not required, but may be used as determined by appropriate risk assessment. Laboratory personnel must have specific training in the procedures conducted in the laboratory and must be supervised by a scientist with training in microbiology or a related science.

The following standard practices, safety equipment, and facility requirements apply to BSL-1.

A. Standard Microbiological Practices

1. The laboratory supervisor must enforce the institutional policies that control access to the laboratory.

2. Persons must wash their hands after working with potentially hazardous materials and before leaving the laboratory.

3. Eating, drinking, smoking, handling contact lenses, applying cosmetics, and storing food for human consumption must not be permitted in laboratory areas. Food must be stored outside the laboratory area in cabinets or refrigerators designated and used for this purpose.

4. Mouth pipetting is prohibited; mechanical pipetting devices must be used.

5. Policies for the safe handling of sharps, such as needles, scalpels, pipettes, and broken glassware must be developed and implemented. Whenever practical, laboratory supervisors should adopt improved engineering and work practice controls that reduce risk of sharps injuries.Precautions, including those listed below, must always be taken with sharp items. These include:

a. Careful management of needles and other sharps are of primary importance. Needles must not be bent, sheared, broken, recapped, removed from disposable syringes, or otherwise manipulated by hand before disposal.

b. Used disposable needles and syringes must be carefully placed in conveniently located puncture-resistant containers used for sharps disposal.

c. Non-disposable sharps must be placed in a hard walled container for transport to a processing area for decontamination, preferably by autoclaving.

d. Broken glassware must not be handled directly. Instead, it must be removed using a brush and dustpan, tongs, or forceps. Plastic ware should be substituted for glassware whenever possible.

6. Perform all procedures to minimize the creation of splashes and/or aerosols.

7. Decontaminate work surfaces after completion of work and after any spill or splash of potentially infectious material with appropriate disinfectant.

8. Decontaminate all cultures, stocks, and other potentially infectious materials before disposal using an effective method. Depending on where the decontamination will be performed, the following methods should be used prior to transport.

a. Materials to be decontaminated outside of the immediate laboratory must be placed in a durable, leak proof container and secured for transport.

b. Materials to be removed from the facility for decontamination must be packed in accordance with applicable local, state, and federal regulations.

9. A sign incorporating the universal biohazard symbol must be posted at the entrance to the laboratory when infectious agents are present. The sign may include the name of the agent(s) in use, and the name and phone number of the laboratory supervisor or other responsible personnel. Agent information should be posted in accordance with the institutional policy.

10. An effective integrated pest management program is required. (See Appendix G.)

11. The laboratory supervisor must ensure that laboratory personnel receive appropriate training regarding their duties, the necessary precautions to prevent exposures, and exposure evaluation procedures. Personnel must receive annual updates or additional training when procedural or policy changes occur. Personal health status may impact an individual's susceptibility to infection, ability to receive immunizations or prophylactic interventions. Therefore, all laboratory personnel and particularly women of childbearing age should be provided with information regarding immune competence and conditions that may predispose them to infection. Individuals having these conditions should be encouraged to self-identify to the institution's healthcare provider for appropriate counseling and guidance.

B. *Special Practices*

None required.

C. *Safety Equipment (Primary Barriers and Personal Protective Equipment)*

1. Special containment devices or equipment, such as BSCs, are not generally required.

2. Protective laboratory coats, gowns, or uniforms are recommended to prevent contamination of personal clothing.

3. Wear protective eyewear when conducting procedures that have the potential to create splashes of microorganisms or other hazardous materials. Persons who wear contact lenses in laboratories should also wear eye protection.

4. Gloves must be worn to protect hands from exposure to hazardous materials. Glove selection should be based on an appropriate risk assessment. Alternatives to latex gloves should be available. Wash hands prior to leaving the laboratory. In addition, BSL-1 workers should:

    a. Change gloves when contaminated, glove integrity is compromised, or when otherwise necessary.

    b. Remove gloves and wash hands when work with hazardous materials has been completed and before leaving the laboratory.

    c. Do not wash or reuse disposable gloves. Dispose of used gloves with other contaminated laboratory waste. Hand washing protocols must be rigorously followed.

D. Laboratory Facilities (Secondary Barriers)

1. Laboratories should have doors for access control.

2. Laboratories must have a sink for hand washing.

3. The laboratory should be designed so that it can be easily cleaned. Carpets and rugs in laboratories are not appropriate.

4. Laboratory furniture must be capable of supporting anticipated loads and uses. Spaces between benches, cabinets, and equipment should be accessible for cleaning.

   a. Bench tops must be impervious to water and resistant to heat, organic solvents, acids, alkalis, and other chemicals.

   b. Chairs used in laboratory work must be covered with a non-porous material that can be easily cleaned and decontaminated with appropriate disinfectant.

5. Laboratories windows that open to the exterior should be fitted with screens.

Biosafety Level 2

Biosafety Level 2 builds upon BSL-1. BSL-2 is suitable for work involving agents that pose moderate hazards to personnel and the environment. It differs from BSL-1 in that: 1) laboratory personnel have specific training in handling pathogenic agents and are supervised by scientists competent in handling infectious agents and associated procedures; 2) access to the laboratory is restricted when work is being conducted; and 3) all procedures in which infectious aerosols or splashes may be created are conducted in BSCs or other physical containment equipment.

The following standard and special practices, safety equipment, and facility requirements apply to BSL-2.

A. Standard Microbiological Practices

1. The laboratory supervisor must enforce the institutional policies that control access to the laboratory.

2. Persons must wash their hands after working with potentially hazardous materials and before leaving the laboratory.

3. Eating, drinking, smoking, handling contact lenses, applying cosmetics, and storing food for human consumption must not be permitted in laboratory areas. Food must be stored outside the laboratory area in cabinets or refrigerators designated and used for this purpose.

4.   Mouth pipetting is prohibited; mechanical pipetting devices must be used.

5.   Policies for the safe handling of sharps, such as needles, scalpels, pipettes, and broken glassware must be developed and implemented. Whenever practical, laboratory supervisors should adopt improved engineering and work practice controls that reduce risk of sharps injuries. Precautions, including those listed below, must always be taken with sharp items. These include:

   a.   Careful management of needles and other sharps are of primary importance. Needles must not be bent, sheared, broken, recapped, removed from disposable syringes, or otherwise manipulated by hand before disposal.

   b.   Used disposable needles and syringes must be carefully placed in conveniently located puncture-resistant containers used for sharps disposal.

   c.   Non-disposable sharps must be placed in a hard walled container for transport to a processing area for decontamination, preferably by autoclaving.

   d.   Broken glassware must not be handled directly. Instead, it must be removed using a brush and dustpan, tongs, or forceps. Plastic ware should be substituted for glassware whenever possible.

6.   Perform all procedures to minimize the creation of splashes and/or aerosols.

7.   Decontaminate work surfaces after completion of work and after any spill or splash of potentially infectious material with appropriate disinfectant.

8.   Decontaminate all cultures, stocks, and other potentially infectious materials before disposal using an effective method. Depending on where the decontamination will be performed, the following methods should be used prior to transport:

   a.   Materials to be decontaminated outside of the immediate laboratory must be placed in a durable, leak proof container and secured for transport.

   b.   Materials to be removed from the facility for decontamination must be packed in accordance with applicable local, state, and federal regulations.

9.   A sign incorporating the universal biohazard symbol must be posted at the entrance to the laboratory when infectious agents are present. Posted information must include: the laboratory's biosafety level, the

supervisor's name (or other responsible personnel), telephone number, and required procedures for entering and exiting the laboratory. Agent information should be posted in accordance with the institutional policy.

10. An effective integrated pest management program is required. (See Appendix G.)

11. The laboratory supervisor must ensure that laboratory personnel receive appropriate training regarding their duties, the necessary precautions to prevent exposures, and exposure evaluation procedures. Personnel must receive annual updates or additional training when procedural or policy changes occur. Personal health status may impact an individual's susceptibility to infection, ability to receive immunizations or prophylactic interventions. Therefore, all laboratory personnel and particularly women of childbearing age should be provided with information regarding immune competence and conditions that may predispose them to infection. Individuals having these conditions should be encouraged to self-identify to the institution's healthcare provider for appropriate counseling and guidance.

B. Special Practices

1. All persons entering the laboratory must be advised of the potential hazards and meet specific entry/exit requirements.

2. Laboratory personnel must be provided medical surveillance, as appropriate, and offered available immunizations for agents handled or potentially present in the laboratory.

3. Each institution should consider the need for collection and storage of serum samples from at-risk personnel.

4. A laboratory-specific biosafety manual must be prepared and adopted as policy. The biosafety manual must be available and accessible.

5. The laboratory supervisor must ensure that laboratory personnel demonstrate proficiency in standard and special microbiological practices before working with BSL-2 agents.

6. Potentially infectious materials must be placed in a durable, leak proof container during collection, handling, processing, storage, or transport within a facility.

7. Laboratory equipment should be routinely decontaminated, as well as, after spills, splashes, or other potential contamination.

a. Spills involving infectious materials must be contained, decontaminated, and cleaned up by staff properly trained and equipped to work with infectious material.

b. Equipment must be decontaminated before repair, maintenance, or removal from the laboratory.

8. Incidents that may result in exposure to infectious materials must be immediately evaluated and treated according to procedures described in the laboratory biosafety manual. All such incidents must be reported to the laboratory supervisor. Medical evaluation, surveillance, and treatment should be provided and appropriate records maintained.

9. Animal and plants not associated with the work being performed must not be permitted in the laboratory.

10. All procedures involving the manipulation of infectious materials that may generate an aerosol should be conducted within a BSC or other physical containment devices.

C. *Safety Equipment (Primary Barriers and Personal Protective Equipment)*

1. Properly maintained BSCs, other appropriate personal protective equipment, or other physical containment devices must be used whenever:

a. Procedures with a potential for creating infectious aerosols or splashes are conducted. These may include pipetting, centrifuging, grinding, blending, shaking, mixing, sonicating, opening containers of infectious materials, inoculating animals intranasally, and harvesting infected tissues from animals or eggs.

b. High concentrations or large volumes of infectious agents are used. Such materials may be centrifuged in the open laboratory using sealed rotor heads or centrifuge safety cups.

2. Protective laboratory coats, gowns, smocks, or uniforms designated for laboratory use must be worn while working with hazardous materials. Remove protective clothing before leaving for non-laboratory areas, e.g., cafeteria, library, and administrative offices). Dispose of protective clothing appropriately, or deposit it for laundering by the institution. It is recommended that laboratory clothing not be taken home.

3. Eye and face protection (goggles, mask, face shield or other splatter guard) is used for anticipated splashes or sprays of infectious or other hazardous materials when the microorganisms must be handled outside the BSC or containment device. Eye and face protection must be disposed of with other contaminated laboratory waste or

decontaminated before reuse. Persons who wear contact lenses in laboratories should also wear eye protection.

4. Gloves must be worn to protect hands from exposure to hazardous materials. Glove selection should be based on an appropriate risk assessment. Alternatives to latex gloves should be available. Gloves must not be worn outside the laboratory. In addition, BSL-2 laboratory workers should:

    a. Change gloves when contaminated, glove integrity is compromised, or when otherwise necessary.

    b. Remove gloves and wash hands when work with hazardous materials has been completed and before leaving the laboratory.

    c. Do not wash or reuse disposable gloves. Dispose of used gloves with other contaminated laboratory waste. Hand washing protocols must be rigorously followed.

5. Eye, face and respiratory protection should be used in rooms containing infected animals as determined by the risk assessment.

D. Laboratory Facilities (Secondary Barriers)

1. Laboratory doors should be self-closing and have locks in accordance with the institutional policies.

2. Laboratories must have a sink for hand washing. The sink may be manually, hands-free, or automatically operated. It should be located near the exit door.

3. The laboratory should be designed so that it can be easily cleaned and decontaminated. Carpets and rugs in laboratories are not permitted.

4. Laboratory furniture must be capable of supporting anticipated loads and uses. Spaces between benches, cabinets, and equipment should be accessible for cleaning.

    a. Bench tops must be impervious to water and resistant to heat, organic solvents, acids, alkalis, and other chemicals.

    b. Chairs used in laboratory work must be covered with a non-porous material that can be easily cleaned and decontaminated with appropriate disinfectant.

5. Laboratory windows that open to the exterior are not recommended. However, if a laboratory does have windows that open to the exterior, they must be fitted with screens.

6. BSCs must be installed so that fluctuations of the room air supply and exhaust do not interfere with proper operations. BSCs should be located away from doors, windows that can be opened, heavily traveled laboratory areas, and other possible airflow disruptions.

7. Vacuum lines should be protected with liquid disinfectant traps.

8. An eyewash station must be readily available.

9. There are no specific requirements for ventilation systems. However, planning of new facilities should consider mechanical ventilation systems that provide an inward flow of air without recirculation to spaces outside of the laboratory.

10. HEPA filtered exhaust air from a Class II BSC can be safely recirculation back into the laboratory environment if the cabinet is tested and certified at least annually and operated according to manufacturer's recommendations. BSCs can also be connected to the laboratory exhaust system by either a thimble (canopy) connection or directly exhausted to the outside through a hard connection. Provisions to assure proper safety cabinet performance and air system operation must be verified.

11. A method for decontaminating all laboratory wastes should be available in the facility (e.g., autoclave, chemical disinfection, incineration, or other validated decontamination method).

Biosafety Level 3

Biosafety Level 3 is applicable to clinical, diagnostic, teaching, research, or production facilities where work is performed with indigenous or exotic agents that may cause serious or potentially lethal disease through the inhalation route of exposure. Laboratory personnel must receive specific training in handling pathogenic and potentially lethal agents, and must be supervised by scientists competent in handling infectious agents and associated procedures.

All procedures involving the manipulation of infectious materials must be conducted within BSCs or other physical containment devices.

A BSL-3 laboratory has special engineering and design features.

The following standard and special safety practices, equipment, and facility requirements apply to BSL-3.

A. Standard Microbiological Practices

1. The laboratory supervisor must enforce the institutional policies that control access to the laboratory.

2. Persons must wash their hands after working with potentially hazardous materials and before leaving the laboratory.

3. Eating, drinking, smoking, handling contact lenses, applying cosmetics, and storing food for human consumption must not be permitted in laboratory areas. Food must be stored outside the laboratory area in cabinets or refrigerators designated and used for this purpose.

4. Mouth pipetting is prohibited; mechanical pipetting devices must be used.

5. Policies for the safe handling of sharps, such as needles, scalpels, pipettes, and broken glassware must be developed and implemented. Whenever practical, laboratory supervisors should adopt improved engineering and work practice controls that reduce risk of sharps injuries.

    Precautions, including those listed below, must always be taken with sharp items. These include:

    a. Careful management of needles and other sharps are of primary importance. Needles must not be bent, sheared, broken, recapped, removed from disposable syringes, or otherwise manipulated by hand before disposal.

    b. Used disposable needles and syringes must be carefully placed in conveniently located puncture-resistant containers used for sharps disposal.

    c. Non-disposable sharps must be placed in a hard walled container for transport to a processing area for decontamination, preferably by autoclaving.

    d. Broken glassware must not be handled directly. Instead, it must be removed using a brush and dustpan, tongs, or forceps. Plastic ware should be substituted for glassware whenever possible.

6. Perform all procedures to minimize the creation of splashes and/or aerosols.

7. Decontaminate work surfaces after completion of work and after any spill or splash of potentially infectious material with appropriate disinfectant.

8. Decontaminate all cultures, stocks, and other potentially infectious materials before disposal using an effective method. A method for decontaminating all laboratory wastes should be available in the facility, preferably within the laboratory (e.g., autoclave, chemical disinfection, incineration, or other validated decontamination method). Depending on where the decontamination will be performed, the following methods should be used prior to transport:

a. Materials to be decontaminated outside of the immediate laboratory must be placed in a durable, leak proof container and secured for transport.

b. Materials to be removed from the facility for decontamination must be packed in accordance with applicable local, state, and federal regulations.

9. A sign incorporating the universal biohazard symbol must be posted at the entrance to the laboratory when infectious agents are present. Posted information must include the laboratory's biosafety level, the supervisor's name (or other responsible personnel), telephone number, and required procedures for entering and exiting the laboratory. Agent information should be posted in accordance with the institutional policy.

10. An effective integrated pest management program is required. (See Appendix G.)

11. The laboratory supervisor must ensure that laboratory personnel receive appropriate training regarding their duties, the necessary precautions to prevent exposures, and exposure evaluation procedures. Personnel must receive annual updates or additional training when procedural or policy changes occur. Personal health status may impact an individual's susceptibility to infection, ability to receive immunizations or prophylactic interventions. Therefore, all laboratory personnel and particularly women of childbearing age should be provided with information regarding immune competence and conditions that may predispose them to infection. Individuals having these conditions should be encouraged to self-identify to the institution's healthcare provider for appropriate counseling and guidance.

B. Special Practices

1. All persons entering the laboratory must be advised of the potential hazards and meet specific entry/exit requirements.

2. Laboratory personnel must be provided medical surveillance and offered appropriate immunizations for agents handled or potentially present in the laboratory.

3. Each institution should consider the need for collection and storage of serum samples from at-risk personnel.

4. A laboratory-specific biosafety manual must be prepared and adopted as policy. The biosafety manual must be available and accessible.

5. The laboratory supervisor must ensure that laboratory personnel demonstrate proficiency in standard and special microbiological practices before working with BSL-3 agents.

6. Potentially infectious materials must be placed in a durable, leak proof container during collection, handling, processing, storage, or transport within a facility.

7. Laboratory equipment should be routinely decontaminated, as well as, after spills, splashes, or other potential contamination.

    a. Spills involving infectious materials must be contained, decontaminated, and cleaned up by staff properly trained and equipped to work with infectious material.

    b. Equipment must be decontaminated before repair, maintenance, or removal from the laboratory.

8. Incidents that may result in exposure to infectious materials must be immediately evaluated and treated according to procedures described in the laboratory biosafety manual. All such incidents must be reported to the laboratory supervisor. Medical evaluation, surveillance, and treatment should be provided and appropriate records maintained.

9. Animals and plants not associated with the work being performed must not be permitted in the laboratory.

10. All procedures involving the manipulation of infectious materials must be conducted within a BSC, or other physical containment devices. No work with open vessels is conducted on the bench. When a procedure cannot be performed within a BSC, a combination of personal protective equipment and other containment devices, such as a centrifuge safety cup or sealed rotor must be used.

C. *Safety Equipment (Primary Barriers and Personal Protective Equipment)*

1. All procedures involving the manipulation of infectious materials must be conducted within a BSC (preferably Class II or Class III), or other physical containment devices.

2. Workers in the laboratory where protective laboratory clothing with a solid-front, such as tie-back or wrap-around gowns, scrub suits, or coveralls. Protective clothing is not worn outside of the laboratory. Reusable clothing is decontaminated before being laundered. Clothing is changed when contaminated.

3. Eye and face protection (goggles, mask, face shield or other splash guard) is used for anticipated splashes or sprays of infectious or other hazardous materials. Eye and face protection must be disposed of with other contaminated laboratory waste or decontaminated before reuse. Persons who wear contact lenses in laboratories must also wear eye protection.

4. Gloves must be worn to protect hands from exposure to hazardous materials. Glove selection should be based on an appropriate risk assessment. Alternatives to latex gloves should be available. Gloves must not be worn outside the laboratory. In addition, BSL-3 laboratory workers:

    a. Changes gloves when contaminated, glove integrity is compromised, or when otherwise necessary. Wear two pairs of gloves when appropriate.

    b. Remove gloves and wash hands when work with hazardous materials has been completed and before leaving the laboratory.

    c. Do not wash or reuse disposable gloves. Dispose of used gloves with other contaminated laboratory waste. Hand washing protocols must be rigorously followed.

5. Eye, face, and respiratory protection must be used in rooms containing infected animals.

D. Laboratory Facilities (Secondary Barriers)

1. Laboratory doors must be self-closing and have locks in accordance with the institutional policies. The laboratory must be separated from areas that are open to unrestricted traffic flow within the building. Laboratory access is restricted. Access to the laboratory is through two self-closing doors. A clothing change room (anteroom) may be included in the passageway between the two self-closing doors.

2. Laboratories must have a sink for hand washing. The sink must be hands-free or automatically operated. It should be located near the exit door. If the laboratory is segregated into different laboratories, a sink must also be available for hand washing in each zone. Additional sinks may be required as determined by the risk assessment.

3. The laboratory must be designed so that it can be easily cleaned and decontaminated. Carpets and rugs are not permitted. Seams, floors, walls, and ceiling surfaces should be sealed. Spaces around doors and ventilation openings should be capable of being sealed to facilitate space decontamination.

a. Floors must be slip resistant, impervious to liquids, and resistant to chemicals. Consideration should be given to the installation of seamless, sealed, resilient or poured floors, with integral cove bases.

b. Walls should be constructed to produce a sealed smooth finish that can be easily cleaned and decontaminated.

c. Ceilings should be constructed, sealed, and finished in the same general manner as walls.

Decontamination of the entire laboratory should be considered when there has been gross contamination of the space, significant changes in laboratory usage, for major renovations, or maintenance shut downs. Selection of the appropriate materials and methods used to decontaminate the laboratory must be based on the risk assessment.

4. Laboratory furniture must be capable of supporting anticipated loads and uses. Spaces between benches, cabinets, and equipment must be accessible for cleaning.

a. Bench tops must be impervious to water and resistant to heat, organic solvents, acids, alkalis, and other chemicals.

b. Chairs used in laboratory work must be covered with a non-porous material that can be easily cleaned and decontaminated with appropriate disinfectant.

5. All windows in the laboratory must be sealed.

6. BSCs must be installed so that fluctuations of the room air supply and exhaust do not interfere with proper operations. BSCs should be located away from doors, heavily traveled laboratory areas, and other possible airflow disruptions.

7. Vacuum lines must be protected with HEPA filters, or their equivalent. Filters must be replaced as needed. Liquid disinfectant traps may be required.

8. An eyewash station must be readily available in the laboratory.

9. A ducted air ventilation system is required. This system must provide sustained directional airflow by drawing air into the laboratory from "clean" areas toward "potentially contaminated" areas. The laboratory shall be designed such that under failure conditions the airflow will not be reversed.

a. Laboratory personnel must be able to verify directional airflow. A visual monitoring device, which confirms directional airflow, must be provided at the laboratory entry. Audible alarms should be considered to notify personnel of air flow disruption.

b. The laboratory exhaust air must not re-circulate to any other area of the building.

c. The laboratory building exhaust air should be dispersed away from occupied areas and from building air intake locations or the exhaust air must be HEPA filtered.

HEPA filter housings should have gas-tight isolation dampers, decontamination ports, and/or bag-in/bag-out (with appropriate decontamination procedures) capability. The HEPA filter housing should allow for leak testing of each filter and assembly. The filters and the housing should be certified at least annually.

10. HEPA filtered exhaust air from a Class II BSC can be safely re-circulated into the laboratory environment if the cabinet is tested and certified at least annually and operated according to manufacturer's recommendations. BSCs can also be connected to the laboratory exhaust system by either a thimble (canopy) connection or directly exhausted to the outside through a hard connection. Provisions to assure proper safety cabinet performance and air system operation must be verified. BSCs should be certified at least annually to assure correct performance. Class III BSCs must be directly (hard) connected up through the second exhaust HEPA filter of the cabinet. Supply air must be provided in such a manner that prevents positive pressurization of the cabinet.

11. A method for decontaminating all laboratory wastes should be available in the facility, preferably within the laboratory (e.g., autoclave, chemical disinfection, or other validated decontamination method).

12. Equipment that may produce infectious aerosols must be contained in primary barrier devices that exhaust air through HEPA filtration or other equivalent technology before being discharged into the laboratory. These HEPA filters should be tested and/or replaced at least annually.

13. Facility design consideration should be given to means of decontaminating large pieces of equipment before removal from the laboratory.

14. Enhanced environmental and personal protection may be required by the agent summary statement, risk assessment, or applicable local, state, or federal regulations. These laboratory enhancements may include, for example, one or more of the following: an anteroom for clean storage of equipment and supplies with dress-in, shower-out capabilities; gas tight dampers to facilitate laboratory isolation; final HEPA filtration of the laboratory exhaust air; laboratory effluent decontamination; and advanced access control devices, such as biometrics.

15. The BSL-3 facility design, operational parameters, and procedures must be verified and documented prior to operation. Facilities must be re-verified and documented at least annually.

Biosafety Level 4

Biosafety Level 4 is required for work with dangerous and exotic agents that pose a high individual risk of aerosol-transmitted laboratory infections and life-threatening disease that is frequently fatal, for which there are no vaccines or treatments, or a related agent with unknown risk of transmission. Agents with a close or identical antigenic relationship to agents requiring BSL-4 containment must be handled at this level until sufficient data are obtained either to confirm continued work at this level, or re-designate the level. Laboratory staff must have specific and thorough training in handling extremely hazardous infectious agents. Laboratory staff must understand the primary and secondary containment functions of standard and special practices, containment equipment, and laboratory design characteristics. All laboratory staff and supervisors must be competent in handling agents and procedures requiring BSL-4 containment. The laboratory supervisor in accordance with institutional policies controls access to the laboratory.

There are two models for BSL-4 laboratories:

1. A *Cabinet Laboratory*—Manipulation of agents must be performed in a Class III BSC; and
2. A *Suit Laboratory*—Personnel must wear a positive pressure supplied air protective suit.

BSL-4 cabinet and suit laboratories have special engineering and design features to prevent microorganisms from being disseminated into the environment.

The following standard and special safety practices, equipment, and facilities apply to BSL-4.

A. Standard Microbiological Practices

1. The laboratory supervisor must enforce the institutional policies that control access to the laboratory.

2. Eating, drinking, smoking, handling contact lenses, applying cosmetics, and storing food for human consumption must not be permitted in laboratory areas. Food must be stored outside the laboratory area in cabinets or refrigerators designated and used for this purpose.

3. Mechanical pipetting devices must be used.

4. Policies for the safe handling of sharps, such as needles, scalpels, pipettes, and broken glassware must be developed and implemented.

Precautions, including those listed below, must be taken with any sharp items. These include:

a. Broken glassware must not be handled directly. Instead, it must be removed using a brush and dustpan, tongs, or forceps. Plastic ware should be substituted for glassware whenever possible.

b. Use of needles and syringes or other sharp instruments should be restricted in the laboratory, except when there is no practical alternative.

c. Used needles must not be bent, sheared, broken, recapped, removed from disposable syringes, or otherwise manipulated by hand before disposal or decontamination. Used disposable needles must be carefully placed in puncture-resistant containers used for sharps disposal, located as close to the point of use as possible.

d. Whenever practical, laboratory supervisors should adopt improved engineering and work practice controls that reduce risk of sharps injuries.

5. Perform all procedures to minimize the creation of splashes and/or aerosols.

6. Decontaminate work surfaces with appropriate disinfectant after completion of work and after any spill or splash of potentially infectious material.

7. Decontaminate all wastes before removal from the laboratory by an effective and validated method.

8. A sign incorporating the universal biohazard symbol must be posted at the entrance to the laboratory when infectious agents are present. Posted information must include the laboratory's biosafety level, the supervisor's name (or other responsible personnel), telephone number, and required procedures for entering and exiting the laboratory. Agent information should be posted in accordance with the institutional policy.

9. An effective integrated pest management program is required. (See Appendix G.)

10. The laboratory supervisor must ensure that laboratory personnel receive appropriate training regarding their duties, the necessary precautions to prevent exposures, and exposure evaluation procedures. Personnel must receive annual updates or additional training when procedural or policy changes occur. Personal health status may impact an individual's susceptibility to infection, ability to receive immunizations or prophylactic interventions. Therefore, all laboratory personnel and particularly women of childbearing age should be provided with information

regarding immune competence and conditions that may predispose them to infection. Individuals having these conditions should be encouraged to self-identify to the institution's healthcare provider for appropriate counseling and guidance.

B. Special Practices

1. All persons entering the laboratory must be advised of the potential hazards and meet specific entry requirements in accordance with institutional policies.

    Only persons whose presence in the facility or individual laboratory rooms is required for scientific or support purposes are authorized to enter.

    Entry into the facility must be limited by means of secure, locked doors. A logbook, or other means of documenting the date and time of all persons entering and leaving the laboratory must be maintained.

    While the laboratory is operational, personnel must enter and exit the laboratory through the clothing change and shower rooms except during emergencies. All personal clothing must be removed in the outer clothing change room. All persons entering the laboratory must use laboratory clothing, including undergarments, pants, shirts, jumpsuits, shoes, and gloves (as appropriate). All persons leaving the laboratory must take a personal body shower. Used laboratory clothing must not be removed from the inner change room through the personal shower. These items must be treated as contaminated materials and decontaminated before laundering.

    After the laboratory has been completely decontaminated and all infectious agents are secured, necessary staff may enter and exit without following the clothing change and shower requirements described above.

2. Laboratory personnel and support staff must be provided appropriate occupational medical services including medical surveillance and available immunizations for agents handled or potentially present in the laboratory. A system must be established for reporting and documenting laboratory accidents, exposures, employee absenteeism and for the medical surveillance of potential laboratory-associated illnesses. An essential adjunct to such an occupational medical services system is the availability of a facility for the isolation and medical care of personnel with potential or known laboratory-acquired infections.

3. Each institution should consider the need for collection and storage of serum samples from at-risk personnel.

4. A laboratory-specific biosafety manual must be prepared. The biosafety manual must be available, accessible, and followed.

5. The laboratory supervisor is responsible for ensuring that laboratory personnel:

    a. Demonstrate high proficiency in standard and special microbiological practices, and techniques for working with agents requiring BSL-4 containment.

    b. Receive appropriate training in the practices and operations specific to the laboratory facility.

    c. Receive annual updates and additional training when procedural or policy changes occur.

6. Removal of biological materials that are to remain in a viable or intact state from the laboratory must be transferred to a non-breakable, sealed primary container and then enclosed in a non-breakable, sealed secondary container. These materials must be transferred through a disinfectant dunk tank, fumigation chamber, or decontamination shower. Once removed, packaged viable material must not be opened outside BSL-4 containment unless inactivated by a validated method.

7. Laboratory equipment musts be routinely decontaminated, as well as after spills, splashes, or other potential contamination.

    a. Spills involving infectious materials must be contained, decontaminated, and cleaned up by appropriate professional staff, or others properly trained and equipped to work with infectious material. A spill procedure must be developed and posted within the laboratory.

    b. Equipment must be decontaminated using an effective and validated method before repair, maintenance, or removal from the laboratory. The interior of the Class III cabinet as well as all contaminated plenums, fans and filters must be decontaminated using a validated gaseous or vapor method.

    c. Equipment or material that might be damaged by high temperatures or steam must be decontaminated using an effective and validated procedure such as a gaseous or vapor method in an airlock or chamber designed for this purpose.

8. Incidents that may result in exposure to infectious materials must be immediately evaluated and treated according to procedures described in the laboratory biosafety manual. All incidents must be reported to the laboratory supervisor, institutional management and appropriate

laboratory personnel as defined in the laboratory biosafety manual. Medical evaluation, surveillance, and treatment should be provided and appropriate records maintained.

9. Animals and plants not associated with the work being performed must not be permitted in the laboratory.

10. Supplies and materials that are not brought into the BSL-4 laboratory through the change room, must be brought in through a previously decontaminated double-door autoclave, fumigation chamber, or airlock. After securing the outer doors, personnel within the laboratory retrieve the materials by opening the interior doors of the autoclave, fumigation chamber, or airlock. These doors must be secured after materials are brought into the facility. The doors of the autoclave or fumigation chamber are interlocked in a manner that prevents opening of the outer door unless the autoclave or fumigation chamber has been operated through a decontamination cycle.

    Only necessary equipment and supplies should be stored inside the BSL-4 laboratory. All equipment and supplies taken inside the laboratory must be decontaminated before removal from the laboratory.

11. Daily inspections of essential containment and life support systems must be completed and documented before laboratory work is initiated to ensure that the laboratory is operating according to established parameters.

12. Practical and effective protocols for emergency situations must be established. These protocols must include plans for medical emergencies, facility malfunctions, fires, escape of animals within the laboratory, and other potential emergencies. Training in emergency response procedures must be provided to emergency response personnel and other responsible staff according to institutional policies.

C. *Safety Equipment (Primary Barriers and Personal Protective Equipment)*

Cabinet Laboratory

1. All manipulations of infectious materials within the laboratory must be conducted in the Class III biological safety cabinet.

    Double-door, pass through autoclaves must be provided for decontaminating materials passing out of the Class III BSC(s). The autoclave doors must be interlocked so that only one can be opened at any time and be automatically controlled so that the outside door to the autoclave can only be opened after the decontamination cycle has been completed.

The Class III cabinet must also have a pass-through dunk tank, fumigation chamber, or equivalent decontamination method so that materials and equipment that cannot be decontaminated in the autoclave can be safely removed from the cabinet. Containment must be maintained at all times.

The Class III cabinet must have a HEPA filter on the supply air intake and two HEPA filters in series on the exhaust outlet of the unit. There must be gas tight dampers on the supply and exhaust ducts of the cabinet to permit gas or vapor decontamination of the unit. Ports for injection of test medium must be present on all HEPA filter housings.

The interior of the Class III cabinet must be constructed with smooth finishes that can be easily cleaned and decontaminated. All sharp edges on cabinet finishes must be eliminated to reduce the potential for cuts and tears of gloves. Equipment to be placed in the Class III cabinet should also be free of sharp edges or other surfaces that may damage or puncture the cabinet gloves.

Class III cabinet gloves must be inspected for damage prior to use and changed if necessary. Gloves should be replaced annually during cabinet re-certification.

The cabinet should be designed to permit maintenance and repairs of cabinet mechanical systems (refrigeration, incubators, centrifuges, etc.) to be performed from the exterior of the cabinet whenever possible.

Manipulation of high concentrations or large volumes of infectious agents within the Class III cabinet should be performed using physical containment devices inside the cabinet whenever practical. Such materials should be centrifuged inside the cabinet using sealed rotor heads or centrifuge safety cups.

The Class III cabinet must be certified at least annually.

2. Workers in the laboratory must wear protective laboratory clothing with a solid-front, such as tie-back or wrap-around gowns, scrub suits, or coveralls. No personal clothing, jewelry, or other items except eyeglasses should be taken past the personal shower area. All protective clothing must be removed in the dirty side change room before showering. Reusable clothing must be autoclaved prior to removal from the laboratory for laundering.

3. Eye, face and respiratory protection should be used in rooms containing infected animals as determined by the risk assessment. Prescription eyeglasses must be decontaminated before removal through the personal body shower.

4.   Disposable gloves must be worn underneath cabinet gloves to protect the worker from exposure should a break or tear occur in a cabinet glove. Gloves must not be worn outside the laboratory. Alternatives to latex gloves should be available. Do not wash or reuse disposable gloves. Dispose of used gloves with other contaminated laboratory waste.

Suit Laboratory

1.   All procedures must be conducted by personnel wearing a one-piece positive pressure supplied air suit.

     All manipulations of infectious agents must be performed within a BSC or other primary barrier system.

     Equipment that may produce aerosols must be contained in primary barrier devices that exhaust air through HEPA filtration before being discharged into the laboratory. These HEPA filters should be tested annually and replaced as needed.

     HEPA filtered exhaust air from a Class II BSC can be safely re-circulated into the laboratory environment if the cabinet is tested and certified at least annually and operated according to manufacturer's specifications.

2.   Workers must wear laboratory clothing, such as scrub suits, before entering the room used for donning positive pressure suits. All laboratory clothing must be removed in the dirty side change room before entering the personal shower.

3.   Inner disposable gloves must be worn to protect against break or tears in the outer suit gloves. Disposable gloves must not be worn outside the change area. Alternatives to latex gloves should be available. Do not wash or reuse disposable gloves. Inner gloves must be removed and discarded in the inner change room prior to entering the personal shower. Dispose of used gloves with other contaminated waste.

4.   Decontamination of outer suit gloves is performed during laboratory operations to remove gross contamination and minimize further contamination of the laboratory.

D. *Laboratory Facilities (Secondary Barriers)*

Cabinet Laboratory

1.   The BSL-4 cabinet laboratory consists of either a separate building or a clearly demarcated and isolated zone within a building. Laboratory doors must have locks in accordance with the institutional policies.

Rooms in the facility must be arranged to ensure sequential passage through an inner (dirty) changing area, a personal shower and an outer (clean) change room upon exiting the room(s) containing the Class III BSC(s).

An automatically activated emergency power source must be provided at a minimum for the laboratory exhaust system, life support systems, alarms, lighting, entry and exit controls, BSCs, and door gaskets. Monitoring and control systems for air supply, exhaust, life support, alarms, entry and exit controls, and security systems should be on an uninterrupted power supply (UPS).

A double-door autoclave, dunk tank, fumigation chamber, or ventilated airlock must be provided at the containment barrier for the passage of materials, supplies, or equipment.

2. A hands-free sink must be provided near the door of the cabinet room(s) and the inner change room. A sink must be provided in the outer change room. All sinks in the room(s) containing the Class III BSC must be connected to the wastewater decontamination system.

3. Walls, floors, and ceilings of the laboratory must be constructed to form a sealed internal shell to facilitate fumigation and prohibit animal and insect intrusion. The internal surfaces of this shell must be resistant to chemicals used for cleaning and decontamination of the area. Floors must be monolithic, sealed and coved.

All penetrations in the internal shell of the laboratory and inner change room must be sealed.

Openings around doors into the cabinet room and inner change room must be minimized and capable of being sealed to facilitate decontamination.

Drains in the laboratory floor (if present) must be connected directly to the liquid waste decontamination system.

Services and plumbing that penetrate the laboratory walls, floors, or ceiling must be installed to ensure that no backflow from the laboratory occurs. These penetrations must be fitted with two (in series) backflow prevention devices. Consideration should be given to locating these devices outside of containment. Atmospheric venting systems must be provided with two HEPA filters in series and be sealed up to the second filter.

Decontamination of the entire cabinet must be performed using a validated gaseous or vapor method when there have been significant changes in cabinet usage, before major renovations or maintenance shut downs, and in other situations, as determined by risk assessment.

Selection of the appropriate materials and methods used for decontamination must be based on the risk assessment.

4. Laboratory furniture must be of simple construction, capable of supporting anticipated loading and uses. Spaces between benches, cabinets, and equipment must be accessible for cleaning and decontamination. Chairs and other furniture must be covered with a non-porous material that can be easily decontaminated.

5. Windows must be break-resistant and sealed.

6. If Class II BSCs are needed in the cabinet laboratory, they must be installed so that fluctuations of the room air supply and exhaust do not interfere with proper operations. Class II cabinets should be located away from doors, heavily traveled laboratory areas, and other possible airflow disruptions.

7. Central vacuum systems are not recommended. If, however, there is a central vacuum system, it must not serve areas outside the cabinet room. Two in-line HEPA filters must be placed near each use point. Filters must be installed to permit in-place decontamination and replacement.

8. An eyewash station must be readily available in the laboratory.

9. A dedicated non-recirculating ventilation system is provided. Only laboratories with the same HVAC requirements (i.e., other BSL-4 labs, ABSL-4, BSL-3-Ag labs) may share ventilation systems if gas-tight dampers and HEPA filters isolate each individual laboratory system.

   The supply and exhaust components of the ventilation system must be designed to maintain the laboratory at negative pressure to surrounding areas and provide differential pressure or directional airflow, as appropriate, between adjacent areas within the laboratory.

   Redundant supply fans are recommended. Redundant exhaust fans are required. Supply and exhaust fans must be interlocked to prevent positive pressurization of the laboratory.

   The ventilation system must be monitored and alarmed to indicate malfunction or deviation from design parameters. A visual monitoring device must be installed near the clean change room so proper differential pressures within the laboratory may be verified prior to entry.

   Supply air to and exhaust air from the cabinet room, inner change room, and fumigation/decontamination chambers must pass through HEPA filter(s). The air exhaust discharge must be located away from occupied spaces and building air intakes.

All HEPA filters should be located as near as practicable to the cabinet and laboratory in order to minimize the length of potentially contaminated ductwork. All HEPA filters must be tested and certified annually.

The HEPA filter housings should be designed to allow for *in situ* decontamination and validation of the filter prior to removal. The design of the HEPA filter housing must have gas-tight isolation dampers, decontamination ports, and ability to scan each filter assembly for leaks.

10. HEPA filtered exhaust air from a Class II BSC can be safely re-circulated into the laboratory environment if the cabinet is tested and certified at least annually and operated according to the manufacturer's recommendations. If BSC exhaust is to be recirculated to the outside, BSCs can also be connected to the laboratory exhaust system by either a thimble (canopy) connection or a hard ducted, direct connection ensuring that cabinet exhaust air passes through two (2) HEPA filters—including the HEPA in the BSC—prior to release outside. Provisions to assure proper safety cabinet performance and air system operation must be verified.

    Class III BSCs must be directly and independently exhausted through two HEPA filters in series. Supply air must be provided in such a manner that prevents positive pressurization of the cabinet.

11. Pass through dunk tanks, fumigation chambers, or equivalent decontamination methods must be provided so that materials and equipment that cannot be decontaminated in the autoclave can be safely removed from the cabinet room(s). Access to the exit side of the pass-through shall be limited to those individuals authorized to be in the BSL-4 laboratory.

12. Liquid effluents from cabinet room sinks, floor drains, autoclave chambers, and other sources within the cabinet room must be decontaminated by a proven method, preferably heat treatment, before being discharged to the sanitary sewer.

    Decontamination of all liquid wastes must be documented. The decontamination process for liquid wastes must be validated physically and biologically. Biological validation must be performed annually or more often if required by institutional policy.

    Effluents from showers and toilets may be discharged to the sanitary sewer without treatment.

13 A double-door, pass through autoclave(s) must be provided for decontaminating materials passing out of the cabinet laboratory. Autoclaves that open outside of the laboratory must be sealed to the interior wall. This bioseal must be durable and airtight and capable of

expansion and contraction. Positioning the bioseal so that the equipment can be accessed and maintained from outside the laboratory is strongly recommended. The autoclave doors must be interlocked so that only one can be opened at any time and be automatically controlled so that the outside door to the autoclave can only be opened after the decontamination cycle has been completed.

Gas and liquid discharge from the autoclave chamber must be decontaminated. When feasible, autoclave decontamination processes should be designed so that unfiltered air or steam exposed to infectious material cannot be released to the environment.

14. The BSL-4 facility design parameters and operational procedures must be documented. The facility must be tested to verify that the design and operational parameters have been met prior to operation. Facilities must also be re-verified annually. Verification criteria should be modified as necessary by operational experience.

15. Appropriate communication systems must be provided between the laboratory and the outside (e.g., voice, fax, and computer). Provisions for emergency communication and emergency access or egress must be developed and implemented.

Suit Laboratory

1. The BSL-4 suit laboratory consists of either a separate building or a clearly demarcated and isolated zone within a building. Laboratory doors must have locks in accordance with the institutional policies.

Rooms in the facility must be arranged to ensure exit by sequential passage through the chemical shower, inner (dirty) change room, personal shower, and outer (clean) changing area.

Entry into the BSL-4 laboratory must be through an airlock fitted with airtight doors. Personnel who enter this area must wear a positive pressure suit supplied with HEPA filtered breathing air. The breathing air systems must have redundant compressors, failure alarms and emergency backup.

A chemical shower must be provided to decontaminate the surface of the positive pressure suit before the worker leaves the laboratory. In the event of an emergency exit or failure of the chemical shower system, a method for decontaminating positive pressure suits, such as a gravity fed supply of chemical disinfectant, is needed.

An automatically activated emergency power source must be provided, at a minimum, for the laboratory exhaust system, life support systems, alarms, lighting, entry and exit controls, BSCs, and door gaskets.

Monitoring and control systems for air supply, exhaust, life support, alarms, entry and exit controls, and security systems should be on a UPS.

A double-door autoclave, dunk tank, or fumigation chamber must be provided at the containment barrier for the passage of materials, supplies, or equipment in or out of the laboratory.

2. Sinks inside the suit laboratory should be placed near procedure areas and be connected to the wastewater decontamination system.

3. Walls, floors, and ceilings of the laboratory must be constructed to form a sealed internal shell to facilitate fumigation and prohibit animal and insect intrusion. The internal surfaces of this shell must be resistant to chemicals used for cleaning and decontamination of the area. Floors must be monolithic, sealed and coved.

   All penetrations in the internal shell of the laboratory, suit storage room and the inner change room must be sealed.

   Drains, if present, in the laboratory floor must be connected directly to the liquid waste decontamination system. Sewer vents must have protection against insect and animal intrusion.

   Services and plumbing that penetrate the laboratory walls, floors, or ceiling must be installed to ensure that no backflow from the laboratory occurs. These penetrations must be fitted with two (in series) backflow prevention devices. Consideration should be given to locating these devices outside of containment. Atmospheric venting systems must be provided with two HEPA filters in series and be sealed up to the second filter.

4. Laboratory furniture must be of simple construction, capable of supporting anticipated loading and uses. Sharp edges and corners should be avoided. Spaces between benches, cabinets, and equipment must be accessible for cleaning and decontamination. Chairs and other furniture must be covered with a non-porous material that can be easily decontaminated.

5. Windows must be break-resistant and sealed.

6. BSCs and other primary containment barrier systems must be installed so that fluctuations of the room air supply and exhaust do not interfere with proper operations. BSCs should be located away from doors, heavily traveled laboratory areas, and other possible airflow disruptions.

7. Central vacuum systems are not recommended. If, however, there is a central vacuum system, it must not serve areas outside the BSL-4 laboratory. Two in-line HEPA filters must be placed near each use point. Filters must be installed to permit in-place decontamination and replacement.

8. An eyewash station must be readily available in the laboratory area for use during maintenance and repair activities.

9. A dedicated, non-recirculating ventilation system is provided. Only laboratories with the same HVAC requirements (i.e., other BSL-4 labs, ABSL-4, BSL-3 Ag labs) may share ventilation systems if gas-tight dampers and HEPA filters isolate each individual laboratory system.

   The supply and exhaust components of the ventilation system must be designed to maintain the laboratory at negative pressure to surrounding areas and provide differential pressure or directional airflow as appropriate between adjacent areas within the laboratory.

   Redundant supply fans are recommended. Redundant exhaust fans are required. Supply and exhaust fans must be interlocked to prevent positive pressurization of the laboratory.

   The ventilation system must be monitored and alarmed to indicate malfunction or deviation from design parameters. A visual monitoring device must be installed near the clean change room so proper differential pressures within the laboratory may be verified prior to entry.

   Supply air to the laboratory, including the decontamination shower, must pass through a HEPA filter. All exhaust air from the suit laboratory, decontamination shower and fumigation or decontamination chambers must pass through two HEPA filters, in series, before discharge to the outside. The exhaust air discharge must be located away from occupied spaces and air intakes.

   All HEPA filters must be located as near as practicable to the laboratory in order to minimize the length of potentially contaminated ductwork. All HEPA filters must be tested and certified annually.

   The HEPA filter housings must be designed to allow for *in situ* decontamination and validation of the filter prior to removal. The design of the HEPA filter housing must have gas-tight isolation dampers, decontamination ports, and ability to scan each filter assembly for leaks.

10. HEPA filtered exhaust air from a Class II BSC can be safely re-circulated back into the laboratory environment if the cabinet is tested and certified at least annually and operated according to the manufacturer's recommendations. Biological safety cabinets can also be connected to the laboratory exhaust system by either a thimble (canopy) connection or a direct (hard) connection. Provisions to assure proper safety cabinet performance and air system operation must be verified.

11. Pass through dunk tanks, fumigation chambers, or equivalent decontamination methods must be provided so that materials and equipment that cannot be decontaminated in the autoclave can be safely removed from the BSL-4 laboratory. Access to the exit side of the pass-through shall be limited to those individuals authorized to be in the BSL-4 laboratory.

12. Liquid effluents from chemical showers, sinks, floor drains, autoclave chambers, and other sources within the laboratory must be decontaminated by a proven method, preferably heat treatment, before being discharged to the sanitary sewer.

    Decontamination of all liquid wastes must be documented. The decontamination process for liquid wastes must be validated physically and biologically. Biological validation must be performed annually or more often if required by institutional policy.

    Effluents from personal body showers and toilets may be discharged to the sanitary sewer without treatment.

13. A double-door, pass through autoclave(s) must be provided for decontaminating materials passing out of the cabinet laboratory. Autoclaves that open outside of the laboratory must be sealed to the interior wall. This bioseal must be durable, airtight, and capable of expansion and contraction. Positioning the bioseal so that the equipment can be accessed and maintained from outside the laboratory is strongly recommended. The autoclave doors must be interlocked so that only one can be opened at any time and be automatically controlled so that the outside door to the autoclave can only be opened after the decontamination cycle has been completed.

    Gas and liquid discharge from the autoclave chamber must be decontaminated. When feasible, autoclave decontamination processes should be designed so that unfiltered air or steam exposed to infectious material cannot be released to the environment.

14. The BSL-4 facility design parameters and operational procedures must be documented. The facility must be tested to verify that the design and operational parameters have been met prior to operation. Facilities must also be re-verified annually. Verification criteria should be modified as necessary by operational experience.

15. Appropriate communication systems must be provided between the laboratory and the outside (e.g., voice, fax, and computer). Provisions for emergency communication and emergency access or egress must be developed and implemented.

Table 2. Summary of Recommended Biosafety Levels for Infectious Agents

BSL	Agents	Practices	Primary Barriers and Safety Equipment	Facilities (Secondary Barriers)
1	Not known to consistently cause diseases in healthy adults	Standard microbiological practices	■ No primary barriers required. ■ PPE: laboratory coats and gloves; eye, face protection, as needed	Laboratory bench and sink required
2	■ Agents associated with human disease ■ Routes of transmission include per-cutaneous injury, ingestion, mucous membrane exposure	BSL-1 practice plus: ■ Limited access ■ Biohazard warning signs ■ "Sharps" precautions ■ Biosafety manual defining any needed waste decontamination or medical surveillance policies	Primary barriers: ■ BSCs or other physical containment devices used for all manipulations of agents that cause splashes or aerosols of infectious materials ■ PPE: Laboratory coats, gloves, face and eye protection, as needed	BSL-1 plus: ■ Autoclave available
3	Indigenous or exotic agents that may cause serious or potentially lethal disease through the inhalation route of exposure	BSL-2 practice plus: ■ Controlled access ■ Decontamination of all waste ■ Decontamination of laboratory clothing before laundering	Primary barriers: ■ BSCs or other physical containment devices used for all open manipulations of agents ■ PPE: Protective laboratory clothing, gloves, face, eye and respiratory protection, as needed	BSL-2 plus: ■ Physical separation from access corridors ■ Self-closing, double-door access ■ Exhausted air not recirculated ■ Negative airflow into laboratory ■ Entry through airlock or anteroom ■ Hand washing sink near laboratory exit
4	■ Dangerous/exotic agents which post high individual risk of aerosol-trans-mitted laboratory infections that are frequently fatal, for which there are no vaccines or treatments ■ Agents with a close or identical anti-genic relationship to an agent requir-ing BSL-4 until data are available to redesignate the level ■ Related agents with unknown risk of transmission	BSL-3 practices plus: ■ Clothing change before entering ■ Shower on exit ■ All material decontaminated on exit from facility	Primary barriers: ■ All procedures conducted in Class III BSCs or Class I or II BSCs in com-bination with full-body, air-supplied. positive pressure suit	BSL-3 plus: ■ Separate building or isolated zone ■ Dedicated supply and exhaust, vacuum, and decontamination systems ■ Other requirements outlined in the text

Section V—Vertebrate Animal Biosafety Level Criteria for Vivarium Research Facilities

This guidance is provided for the use of experimentally infected animals housed in indoor research facilities (e.g., vivaria), and is also useful in the maintenance of laboratory animals that may naturally harbor zoonotic infectious agents. In both instances, the institutional management must provide facilities, staff, and established practices that reasonably ensure appropriate levels of environmental quality, safety, security and care for the laboratory animal. Laboratory animal facilities are a special type of laboratory. As a general principle, the biosafety level (facilities, practices, and operational requirements) recommended for working with infectious agents *in vivo* and *in vitro* are comparable.

The animal room can present unique problems. In the animal room, the activities of the animals themselves can present unique hazards not found in standard microbiological laboratories. Animals may generate aerosols, they may bite and scratch, and they may be infected with a zoonotic agent. The co-application of Biosafety Levels and the Animal Biosafety Levels are determined by a protocol-driven risk assessment.

These recommendations presuppose that laboratory animal facilities, operational practices, and quality of animal care meet applicable standards and regulations (e.g., *Guide for the Care and Use of Laboratory Animals*[1] and *Laboratory Animal Welfare Regulations*[2]) and that appropriate species have been selected for animal experiments. In addition, the organization must have an occupational health and safety program that addresses potential hazards associated with the conduct of laboratory animal research. The following publication by the Institute for Laboratory Animal Research (ILAR), *Occupational Health and Safety in the Care and Use of Research Animals*[3], is most helpful in this regard. Additional safety guidance on working with non-human primates is available in the ILAR publication, *Occupational Health and Safety in the Care and Use of Nonhuman Primates.*[4]

Facilities for laboratory animals used in studies of infectious or non-infectious disease should be physically separate from other activities such as animal production and quarantine, clinical laboratories, and especially from facilities providing patient care. Traffic flow that will minimize the risk of cross contamination should be incorporated into the facility design.

The recommendations detailed below describe four combinations of practices, safety equipment, and facilities for experiments with animals involved in infectious disease research and other studies that may require containment. These four combinations, designated Animal Biosafety Levels (ABSL) 1-4, provide increasing levels of protection to personnel and to the environment, and are recommended as minimal standards for activities involving infected laboratory animals. The four ABSLs describe animal facilities and practices applicable to work with animals infected with agents assigned to Biosafety Levels 1-4, respectively. Investigators that are inexperienced in conducting these types of experiments should seek help in designing their experiments from individuals who are experienced in this special work.

In addition to the animal biosafety levels described in this section, the USDA has developed facility parameters and work practices for handling agents of agriculture significance. Appendix D includes a discussion on Animal Biosafety Level 3 Agriculture (BSL-3-Ag). USDA requirements are unique to agriculture because of the necessity to protect the environment from pathogens of economic or environmental impact. Appendix D also describes some of the enhancements beyond BSL/ABSL-3 that may be required by USDA-APHIS when working in the laboratory or vivarium with certain veterinary agents of concern.

Facility standards and practices for invertebrate vectors and hosts are not specifically addressed in this section. The reader is referred to Appendix E for more information on the Arthropod Containment Guidelines.

Animal Biosafety Level 1

Animal Biosafety Level 1 is suitable for work in animals involving well-characterized agents that are not known to cause disease in immunocompetent adult humans, and present minimal potential hazard to personnel and the environment.

ABSL-1 facilities should be separated from the general traffic patterns of the building and restricted as appropriate. Special containment equipment or facility design may be required as determined by appropriate risk assessment. (See Section 2, Biological Risk Assessment.)

Personnel must have specific training in animal facility procedures and must be supervised by an individual with adequate knowledge of potential hazards and experimental animal procedures.

The following standard practices, safety equipment, and facility requirements apply to ABSL-1.

A. Standard Microbiological Practices

1. The animal facility director establishes and enforces policies, procedures, and protocols for institutional policies and emergencies.

   Each institute must assure that worker safety and health concerns are addressed as part of the animal protocol review.

   Prior to beginning a study animal protocols must also be reviewed and approved by the Institutional Animal Care and Use Committee (IACUC)[5] and the Institutional Biosafety Committee.

2. A safety manual specific to the animal facility is prepared or adopted in consultation with the animal facility director and appropriate safety professionals. The safety manual must be available and accessible. Personnel are advised of potential hazards and are required to read and follow instructions on practices and procedures.

3. The supervisor must ensure that animal care, laboratory and support personnel receive appropriate training regarding their duties, animal husbandry procedures, potential hazards, manipulations of infectious agents, necessary precautions to prevent exposures, and hazard/ exposure evaluation procedures (physical hazards, splashes, aerosolization, etc.). Personnel must receive annual updates and additional training when procedures or policies change. Records are maintained for all hazard evaluations, employee training sessions and staff attendance.

4. An appropriate medical surveillance program is in place, as determined by risk assessment. The need for an animal allergy prevention program should be considered.

   Facility supervisors should ensure that medical staff is informed of potential occupational hazards within the animal facility, to include those associated with research, animal husbandry duties, animal care and manipulations.

   Personal health status may impact an individual's susceptibility to infection, ability to receive immunizations or prophylactic interventions. Therefore, all personnel and particularly women of childbearing age should be provided information regarding immune competence and conditions that may predispose them to infection. Individuals having these conditions should be encouraged to self-identify to the institution's healthcare provider for appropriate counseling and guidance.

   Personnel using respirators must be enrolled in an appropriately constituted respiratory protection program.

5. A sign incorporating safety information must be posted at the entrance to the areas where infectious materials and/or animals are housed or are manipulated. The sign must include the animal biosafety level, general occupational health requirements, personal protective equipment requirements, the supervisor's name (or other responsible personnel), telephone number, and required procedures for entering and exiting the animal areas. Identification of specific infectious agents is recommended when more than one agent is being used within an animal room.

   Security-sensitive agent information should be posted in accordance with the institutional policy.

Advance consideration should be given to emergency and disaster recovery plans, as a contingency for man-made or natural disasters.[1,3,4]

6.   Access to the animal room is limited. Only those persons required for program or support purposes are authorized to enter the facility.

   All persons including facility personnel, service workers, and visitors are advised of the potential hazards (natural or research pathogens, allergens, etc.) and are instructed on the appropriate safeguards.

7.   Protective laboratory coats, gowns, or uniforms are recommended to prevent contamination of personal clothing.

   Gloves are worn to prevent skin contact with contaminated, infectious and hazardous materials, and when handling animals.

   Gloves and personal protective equipment should be removed in a manner that minimizes transfer of infectious materials outside of the areas where infectious materials and/or animals are housed or are manipulated.

   Persons must wash their hands after removing gloves, and before leaving the areas where infectious materials and/or animals are housed or are manipulated.

   Eye and face and respiratory protection should be used in rooms containing infected animals, as dictated by the risk assessment.

8.   Eating, drinking, smoking, handling contact lenses, applying cosmetics, and storing food for human consumption must not be permitted in laboratory areas. Food must be stored outside of the laboratory in cabinets or refrigerators designed and used for this purpose.

9.   All procedures are carefully performed to minimize the creation of aerosols or splatters of infectious materials and waste.

10.   Mouth pipetting is prohibited. Mechanical pipetting devices must be used.

11.   Policies for the safe handling of sharps, such as needles, scalpels, pipettes, and broken glassware must be developed and implemented.

   When applicable, laboratory supervisors should adopt improved engineering and work practice controls that reduce the risk of sharps injuries. Precautions, including those listed below, must always be taken with sharp items. These include:

   a.   Use of needles and syringes or other sharp instruments in the animal facility is limited to situations where there is no alternative for such procedures as parenteral injection, blood collection, or aspiration of fluids from laboratory animals and diaphragm bottles.

b. Disposable needles must not be bent, sheared, broken, recapped, removed from disposable syringes, or otherwise manipulated by hand before disposal. Used disposable needles must be carefully placed in puncture-resistant containers used for sharps disposal. Sharps containers should be located as close to the work site as possible.

c. Non-disposable sharps must be placed in a hard-walled container for transport to a processing area for decontamination, preferably by autoclaving.

d. Broken glassware must not be handled directly. Instead, it must be removed using a brush and dustpan, tongs, or forceps. Plastic ware should be substituted for glassware whenever possible.

e. Equipment containing sharp edges and corners should be avoided.

12. Equipment and work surfaces are routinely decontaminated with an appropriate disinfectant after work with an infectious agent, and after any spills, splashes, or other overt contamination.

13. Animals and plants not associated with the work being performed must not be permitted in the areas where infectious materials and/ or animals are housed or are manipulated.

14. An effective integrated pest management program is required. (See Appendix G.)

15. All wastes from the animal room (including animal tissues, carcasses, and bedding) are transported from the animal room in leak-proof, covered containers for appropriate disposal in compliance with applicable institutional, local and state requirements.

Decontaminate all potentially infectious materials before disposal using an effective method.

B. Special Practices

None required.

C. Safety Equipment (Primary Barriers and Personal Protective Equipment)

1. A risk assessment should determine the appropriate type of personal protective equipment to be utilized.

2. Special containment devices or equipment may not be required as determined by appropriate risk assessment.

3. Protective laboratory coats, gowns, or uniforms may be required to prevent contamination of personal clothing.

   Protective outer clothing is not worn outside areas where infectious materials and/or animals are housed or manipulated. Gowns and uniforms are not worn outside the facility.

4. Protective eyewear is worn when conducting procedures that have the potential to create splashes of microorganisms or other hazardous materials. Persons who wear contact lenses should also wear eye protection when entering areas with potentially high concentrations or airborne particulates.

   Persons having contact with NHPs must assess risk of mucous membrane exposure and wear protective equipment (e.g., masks, goggles, face shields, etc.) as appropriate for the task to be performed.

5. Gloves are worn to protect hands from exposure to hazardous materials.

   A risk assessment should be performed to identify the appropriate glove for the task and alternatives to latex gloves should be available.

   Change gloves when contaminated, glove integrity is compromised, or when otherwise necessary.

   Gloves must not be worn outside the animal rooms.

   Gloves and personal protective equipment should be removed in a manner that prevents transfer of infectious materials.

   Do not wash or reuse disposable gloves. Dispose of used gloves with other contaminated waste.

6. Persons must wash their hands after handling animals and before leaving the areas where infectious materials and/or animals are housed or are manipulated. Hand washing should occur after the removal of gloves.

D. Laboratory Facilities (Secondary Barriers)

1. The animal facility is separated from areas that are open to unrestricted personnel traffic within the building. External facility doors are self-closing and self-locking.

   Access to the animal facility is restricted.

Doors to areas where infectious materials and/or animals are housed, open inward, are self-closing, are kept closed when experimental animals are present, and should never be propped open. Doors to cubicles inside an animal room may open outward or slide horizontally or vertically.

2. The animal facility must have a sink for hand washing.

    Sink traps are filled with water, and/or appropriate liquid to prevent the migration of vermin and gases.

3. The animal facility is designed, constructed, and maintained to facilitate cleaning and housekeeping. The interior surfaces (walls, floors and ceilings) are water resistant. Floors must be slip resistant, impervious to liquids, and resistant to chemicals.

    It is recommended that penetrations in floors, walls and ceiling surfaces be sealed, including openings around ducts, doors and doorframes, to facilitate pest control and proper cleaning.

4. Cabinets and bench tops must be impervious to water and resistant to heat, organic solvents, acids, alkalis, and other chemicals. Spaces between benches, cabinets, and equipment should be accessible for cleaning.

    Chairs used in animal area must be covered with a non-porous material that can be easily cleaned and decontaminated. Furniture must be capable of supporting anticipated loads and uses. Sharp edges and corners should be avoided.

5. External windows are not recommended; if present windows must be resistant to breakage. Where possible, windows should be sealed. If the animal facility has windows that open, they are fitted with fly screens. The presence of windows may impact facility security and therefore should be assessed by security personnel.

6. Ventilation should be provided in accordance with the *Guide for Care and Use of Laboratory Animals*.[1] No recirculation of exhaust air may occur. It is recommended that animal rooms have inward directional airflow.

    Ventilation system design should consider the heat and high moisture load produced during the cleaning of animal rooms and the cage wash process.

7. Internal facility appurtenances, such as light fixtures, air ducts, and utility pipes, are arranged to minimize horizontal surface areas to facilitate cleaning and minimize the accumulation of debris or fomites.

8. If floor drains are provided, the traps are filled with water, and/or appropriate disinfectant to prevent the migration of vermin and gases.

9. Cages are washed manually or preferably in a mechanical cage washer. The mechanical cage washer should have a final rinse temperature of at least 180°F. If manual cage washing is utilized, ensure that appropriate disinfectants are selected.

10. Illumination is adequate for all activities, avoiding reflections and glare that could impede vision.

11. Emergency eyewash and shower are readily available; location is determined by risk assessment.

Animal Biosafety Level 2

Animal Biosafety Level 2 builds upon the practices, procedures, containment equipment, and facility requirements of ABSL-1. ABSL-2 is suitable for work involving laboratory animals infected with agents associated with human disease and pose moderate hazards to personnel and the environment. It also addresses hazards from ingestion as well as from percutaneous and mucous membrane exposure.

ABSL-2 requires that: 1) access to the animal facility is restricted; 2) personnel must have specific training in animal facility procedures, the handling of infected animals and the manipulation of pathogenic agents; 3) personnel must be supervised by individuals with adequate knowledge of potential hazards, microbiological agents, animal manipulations and husbandry procedures; and 4) BSCs or other physical containment equipment is used when procedures involve the manipulation of infectious materials, or where aerosols or splashes may be created.

Appropriate personal protective equipment must be utilized to reduce exposure to infectious agents, animals, and contaminated equipment. Implementation of employee occupational health programs should be considered.

The following standard and special practices, safety equipment, and facility requirements apply to ABSL-2:

A. *Standard Microbiological Practices*

1. The animal facility director establishes and enforces policies, procedures, and protocols for institutional policies and emergencies.

   Each organization must assure that worker safety and health concerns are addressed as part of the animal protocol review.

   Prior to beginning a study, animal protocols must also be reviewed and approved by the IACUC[5] and the Institutional Biosafety Committee.

2. A safety manual specific to the animal facility is prepared or adopted in consultation with the animal facility director and appropriate safety professionals.

The safety manual must be available and accessible. Personnel are advised of potential hazards, and are required to read and follow instructions on practices and procedures.

Consideration should be given to specific biohazards unique to the animal species and protocol in use.

3. The supervisor must ensure that animal care, laboratory, and support personnel receive appropriate training regarding their duties, animal husbandry procedure, potential hazards, manipulations of infectious agents, necessary precautions to prevent hazard or exposures, and hazard/exposure evaluation procedures (physical hazards, splashes, aerosolization, etc.). Personnel must receive annual updates or additional training when procedures or policies change. Records are maintained for all hazard evaluations, employee training sessions and staff attendance.

4. An appropriate medical surveillance program is in place, as determined by risk assessment. The need for an animal allergy prevention program should be considered.

   Facility supervisors should ensure that medical staff is informed of potential occupational hazards within the animal facility, to include those associated with research, animal husbandry duties, animal care and manipulations.

   Personal health status may impact an individual's susceptibility to infection, ability to receive immunizations or prophylactic interventions. Therefore, all personnel and particularly women of childbearing age should be provided information regarding immune competence and conditions that may predispose them to infection. Individuals having these conditions should be encouraged to self-identify to the institution's healthcare provider for appropriate counseling and guidance.

   Personnel using respirators must be enrolled in an appropriately constituted respiratory protection program.

5. A sign incorporating the universal biohazard symbol must be posted at the entrance to areas where infectious materials and/ or animals are housed or are manipulated when infectious agents are present. The sign must include the animal biosafety level, general occupational health requirements, personal protective equipment requirements, the supervisor's name (or names of other responsible personnel), telephone number, and required procedures for entering and exiting the animal areas. Identification of all infectious agents is necessary when more than one agent is being used within an animal room.

Security-sensitive agent information and occupational health requirements should be posted in accordance with the institutional policy.

Advance consideration should be given to emergency and disaster recovery plans, as a contingency for man-made or natural disasters.[1,3,4]

6. Access to the animal room is limited. Only those persons required for program or support purposes are authorized to enter the animal facility and the areas where infectious materials and/or animals are housed or manipulated.

   All persons including facility personnel, service workers, and visitors are advised of the potential hazards (physical, naturally occurring, or research pathogens, allergens, etc.) and are instructed on the appropriate safeguards.

7. Protective laboratory coats, gowns, or uniforms are recommended to prevent contamination of personal clothing.

   Gloves are worn to prevent skin contact with contaminated, infectious and hazardous materials and when handling animals.

   Gloves and personal protective equipment should be removed in a manner that prevents transfer of infectious materials outside of the areas where infectious materials and/or animals are housed or are manipulated.

   Persons must wash their hands after removing gloves, and before leaving the areas where infectious materials and/or animals are housed or are manipulated.

   Eye, face and respiratory protection should be used in rooms containing infected animals, as dictated by the risk assessment.

8. Eating, drinking, smoking, handling contact lenses, applying cosmetics, and storing food for human consumption must not be permitted in laboratory areas. Food must be stored outside of the laboratory in cabinets or refrigerators designated and used for this purpose.

9. All procedures are carefully performed to minimize the creation of aerosols or splatters of infectious materials and waste.

10. Mouth pipetting is prohibited. Mechanical pipetting devices must be used.

11. Policies for the safe handling of sharps, such as needles, scalpels, pipettes, and broken glassware must be developed and implemented. When applicable, laboratory supervisors should adopt improved engineering and work practice controls that reduce the risk of sharps injuries. Precautions must always be taken with sharp items. These include:

a. The use of needles and syringes or other sharp instruments in the animal facility is limited to situations where there is no alternative such as parenteral injection, blood collection, or aspiration of fluids from laboratory animals and diaphragm bottles.

b. Disposable needles must not be bent, sheared, broken, recapped, removed from disposable syringes, or otherwise manipulated by hand before disposal. Used, disposable needles must be carefully placed in puncture-resistant containers used for sharps disposal. Sharps containers should be located as close to the work site as possible.

c. Non-disposable sharps must be placed in a hard-walled container for transport to a processing area for decontamination, preferably by autoclaving.

d. Broken glassware must not be handled directly; it should be removed using a brush and dustpan, tongs, or forceps. Plastic ware should be substituted for glassware whenever possible.

e. Use of equipment with sharp edges and corners should be avoided.

12. Equipment and work surfaces are routinely decontaminated with an appropriate disinfectant after work with an infectious agent, and after any spills, splashes, or other overt contamination.

13. Animals and plants not associated with the work being performed must not be permitted in the areas where infectious materials and/ or animals are housed or manipulated.

14. An effective integrated pest management program is required. (See Appendix G.)

15. All wastes from the animal room (including animal tissues, carcasses, and bedding) are transported from the animal room in leak-proof containers for appropriate disposal in compliance with applicable institutional, local and state requirements.

Decontaminate all potentially infectious materials before disposal using an effective method.

B. Special Practices

1. Animal care staff, laboratory and routine support personnel must be provided a medical surveillance program as dictated by the risk assessment and administered appropriate immunizations for agents handled or potentially present, before entry into animal rooms.

When appropriate, a base line serum sample should be stored.

2. Procedures involving a high potential for generating aerosols should be conducted within a biosafety cabinet or other physical containment device. When a procedure cannot be performed within a biosafety cabinet, a combination of personal protective equipment and other containment devices must be used.

    Restraint devices and practices that reduce the risk of exposure during animal manipulations (e.g., physical restraint devices, chemical restraint medications) should be used whenever possible.

3. Decontamination by an appropriate method (e.g. autoclave, chemical disinfection, or other approved decontamination methods) is necessary for all potentially infectious materials and animal waste before movement outside the areas where infectious materials and/or animals are housed or are manipulated. This includes potentially infectious animal tissues, carcasses, contaminated bedding, unused feed, sharps, and other refuse.

    A method for decontaminating routine husbandry equipment, sensitive electronic and medical equipment should be identified and implemented.

    Materials to be decontaminated outside of the immediate areas where infectious materials and/or animals are housed or are manipulated must be placed in a durable, leak proof, covered container and secured for transport. The outer surface of the container is disinfected prior to moving materials. The transport container must have a universal biohazard label.

    Develop and implement an appropriate waste disposal program in compliance with applicable institutional, local and state requirements. Autoclaving of content prior to incineration is recommended.

4. Equipment, cages, and racks should be handled in a manner that minimizes contamination of other areas.

    Equipment must be decontaminated before repair, maintenance, or removal from the areas where infectious materials and/or animals are housed or are manipulated.

5. Spills involving infectious materials must be contained, decontaminated, and cleaned up by staff properly trained and equipped to work with infectious material.

6. Incidents that may result in exposure to infectious materials must be immediately evaluated and treated according to procedures described in the safety manual. All such incidents must be reported to the animal facility supervisor or personnel designated by the institution. Medical evaluation, surveillance, and treatment should be provided as appropriate and records maintained.

C. Safety Equipment (Primary Barriers and Personal Protective Equipment)

1. Properly maintained BSCs, personal protective equipment (e.g., gloves, lab coats, face shields, respirators, etc.) and/or other physical containment devices or equipment, are used whenever conducting procedures with a potential for creating aerosols, splashes, or other potential exposures to hazardous materials. These include necropsy of infected animals, harvesting of tissues or fluids from infected animals or eggs, and intranasal inoculation of animals.

    When indicated by risk assessment, animals are housed in primary biosafety containment equipment appropriate for the animal species, such as solid wall and bottom cages covered with filter bonnets for rodents or other equivalent primary containment systems for larger animal cages.

2. A risk assessment should determine the appropriate type of personal protective equipment to be utilized.

    Scrub suits and uniforms are removed before leaving the animal facility. Reusable clothing is appropriately contained and decontaminated before being laundered. Laboratory and protective clothing should never be taken home.

    Gowns, uniforms, laboratory coats and personal protective equipment are worn while in the areas where infectious materials and/or animals are housed or manipulated and removed prior to exiting. Disposable personal protective equipment and other contaminated waste are appropriately contained and decontaminated prior to disposal.

3. Eye and face protection (mask, goggles, face shield or other splatter guard) are used for manipulations or activities that may result in splashes or sprays from infectious or other hazardous materials and when the animal or microorganisms must be handled outside the BSC or containment device. Eye and face protection must be disposed of with other contaminated laboratory waste or decontaminated before reuse. Persons who wear contact lenses should also wear eye protection when entering areas with potentially high concentrations or airborne particulates.

    Persons having contact with NHPs should assess risk of mucous membrane exposure and wear protective equipment (e.g., masks, goggles, face shields) appropriate for the task to be performed. Respiratory protection is worn based upon risk assessment.

4. Gloves are worn to protect hands from exposure to hazardous materials. A risk assessment should be performed to identify the appropriate glove for the task and alternatives to latex gloves should be available.

Gloves are changed when contaminated, glove integrity is compromised, or when otherwise necessary.

Gloves must not be worn outside the animal rooms.

Gloves and personal protective equipment should be removed in a manner that prevents transfer of infectious materials.

Do not wash or reuse disposable gloves. Dispose of used gloves with other contaminated waste.

Persons must wash their hands after handling animals and before leaving the areas where infectious materials and/or animals are housed or are manipulated. Hand washing should occur after the removal of gloves.

D. Laboratory Facilities (Secondary Barriers)

1. The animal facility is separated from areas that are open to unrestricted personnel traffic within the building. External facility doors are self-closing and self-locking.

    Doors to areas where infectious materials and/or animals are housed, open inward, are self-closing, are kept closed when experimental animals are present, and should never be propped open. Doors to cubicles inside an animal room may open outward or slide horizontally or vertically.

2. A hand-washing sink is located at the exit of the areas where infectious materials and/or animals are housed or are manipulated. Additional sinks for hand washing should be located in other appropriate locations within the facility.

    If the animal facility has segregated areas where infectious materials and/or animals are housed or manipulated, a sink must also be available for hand washing at the exit from each segregated area.

    Sink traps are filled with water, and/or appropriate disinfectant to prevent the migration of vermin and gases.

3. The animal facility is designed, constructed, and maintained to facilitate cleaning and housekeeping. The interior surfaces (walls, floors and ceilings) are water resistant.

    Penetrations in floors, walls and ceiling surfaces are sealed, including openings around ducts, doors and doorframes, to facilitate pest control and proper cleaning.

    Floors must be slip-resistant, impervious to liquids, and resistant to chemicals.

4. Cabinets and bench tops must be impervious to water and resistant to heat, organic solvents, acids, alkalis, and other chemicals. Spaces between benches, cabinets, and equipment should be accessible for cleaning.

   Furniture should be minimized. Chairs used in animal area must be covered with a non-porous material that can be easily cleaned and decontaminated. Furniture must be capable of supporting anticipated loads and uses. Sharp edges and corners should be avoided.

5. External windows are not recommended; if present, windows must be sealed and resistant to breakage. The presence of windows may impact facility security and therefore should be assessed by security personnel.

6. Ventilation should be provided in accordance with the *Guide for Care and Use of Laboratory Animals*.[1] The direction of airflow into the animal facility is inward; animal rooms maintain inward directional airflow compared to adjoining hallways. A ducted exhaust air ventilation system is provided. Exhaust air is discharged to the outside without being recirculated to other rooms.

   Ventilation system design should consider the heat and high moisture load produced during the cleaning of animal rooms and the cage wash process.

7. Internal facility appurtenances, such as light fixtures, air ducts, and utility pipes, are arranged to minimize horizontal surface areas, to facilitate cleaning and minimize the accumulation of debris or fomites.

8. Floor drains must be maintained and filled with water, and/or appropriate disinfectant to prevent the migration of vermin and gases.

9. Cages should be autoclaved or otherwise decontaminated prior to washing. Mechanical cage washer should have a final rinse temperature of at least 180°F. The cage wash area should be designed to accommodate the use of high-pressure spray systems, humidity, strong chemical disinfectants and 180°F water temperatures during the cage/equipment cleaning process.

10. Illumination is adequate for all activities, avoiding reflections and glare that could impede vision.

11. If BSCs are present, they must be installed so that fluctuations of the room air supply and exhaust do not interfere with proper operations. BSCs should be located away from doors, heavily traveled laboratory areas, and other possible airflow disruptions.

HEPA filtered exhaust air from a Class II BSC can be safely re-circulated back into the laboratory environment if the cabinet is tested and certified at least annually and operated according to manufacturer's recommendations. BSCs can also be connected to the laboratory exhaust system by either a thimble (canopy) connection or directly to the outside through an independent, hard connection. Provisions to assure proper safety cabinet performance and air system operation must be verified. BSCs should be recertified at least once a year to ensure correct performance.

All BSCs should be used according to manufacturer's specifications to protect the worker and avoid creating a hazardous environment from volatile chemicals and gases.

12. If vacuum service (i.e., central or local) is provided, each service connection should be fitted with liquid disinfectant traps and an in-line HEPA filter placed as near as practicable to each use point or service cock. Filters are installed to permit in-place decontamination and replacement.

13. An autoclave should be present in the animal facility to facilitate decontamination of infectious materials and waste.

14. Emergency eyewash and shower are readily available; location is determined by risk assessment.

Animal Biosafety Level 3

Animal Biosafety Level 3 involves practices suitable for work with laboratory animals infected with indigenous or exotic agents, agents that present a potential for aerosol transmission, and agents causing serious or potentially lethal disease. ABSL-3 builds upon the standard practices, procedures, containment equipment, and facility requirements of ABSL-2.

The ABSL-3 laboratory has special engineering and design features.

ABSL-3 requires that: 1) access to the animal facility is restricted; 2) personnel must have specific training in animal facility procedures, the handling of infected animals, and the manipulation of potentially lethal agents; 3) personnel must be supervised by individuals with adequate knowledge of potential hazards, microbiological agents, animal manipulations, and husbandry procedures; and 4) procedures involving the manipulation of infectious materials, or where aerosols or splashes may be created, must be conducted in BSCs or by use of other physical containment equipment.

Appropriate personal protective equipment must be utilized to reduce exposure to infectious agents, animals, and contaminated equipment. Employee occupational health programs must be implemented.

The following standard and special safety practices, safety equipment, and facility requirements apply to ABSL-3.

A. Standard Microbiological Practices

1. The animal facility director establishes and enforces policies, procedures, and protocols for institutional policies and emergencies.

   Each institute must assure that worker safety and health concerns are addressed as part of the animal protocol review.

   Prior to beginning a study, animal protocols must be reviewed and approved by the IACUC[5] and the Institutional Biosafety Committee.

2. A safety manual specific to the animal facility is prepared or adopted in consultation with the animal facility director and appropriate safety professionals.

   The safety manual must be available and accessible. Personnel are advised of potential and special hazards, and are required to read and follow instructions on practices and procedures.

   Consideration must be given to specific biohazards unique to the animal species and protocol in use.

3. The supervisor must ensure that animal care, laboratory and support personnel receive appropriate training regarding their duties, animal husbandry procedures, potential hazards, manipulations of infectious agents, necessary precautions to prevent hazard or exposures, and hazard/exposure evaluation procedures (physical hazards, splashes, aerosolization, etc.). Personnel must receive annual updates or additional training when procedures or policies change. Records are maintained for all hazard evaluations, employee training sessions and staff attendance.

4. An appropriate medical surveillance program is in place, as determined by risk assessment. The need for an animal allergy prevention program should be considered.

   Facility supervisors should ensure that medical staff is informed of potential occupational hazards within the animal facility, to include those associated with the research, animal husbandry duties, animal care, and manipulations.

   Personal health status may impact an individual's susceptibility to infection, ability to receive immunizations or prophylactic interventions. Therefore, all personnel and particularly women of childbearing age should be provided information regarding immune competence and conditions that may predispose them to infection. Individuals having these conditions should be encouraged to self-identify to the institution's healthcare provider for appropriate counseling and guidance.

Personnel using respirators must be enrolled in an appropriately constituted respiratory protection program.

5. A sign incorporating the universal biohazard symbol must be posted at the entrance to areas where infectious materials and/or animals are housed or are manipulated. The sign must include the animal biosafety level, general occupational health requirements, personal protective equipment requirements, the supervisor's name (or other responsible personnel), telephone number, and required procedures for entering and exiting the animal areas. Identification of specific infectious agents is recommended when more than one agent is used within an animal room.

    Security-sensitive agent information and occupational health requirements should be posted in accordance with the institutional policy.

    Advance consideration should be given to emergency and disaster recovery plans, as a contingency for man-made or natural disasters.[1,3,4]

6. Access to the animal room is limited to the fewest number of individuals possible. Only those persons required for program or support purposes are authorized to enter the animal facility and the areas where infectious materials and/or animals are housed or are manipulated.

    All persons, including facility personnel, service workers, and visitors, are advised of the potential hazards (natural or research pathogens, allergens, etc.) and are instructed on the appropriate safeguards.

7. Protective laboratory coats, gowns, or uniforms are recommended to prevent contamination of personal clothing.

    Gloves are worn to prevent skin contact with contaminated, infectious/hazardous materials and when handling animals. Double-glove practices should be used when dictated by risk assessment.

    Gloves and personal protective equipment should be removed in a manner that prevents transfer of infectious materials outside of the areas where infectious materials and/or animals are housed or are manipulated.

    Persons must wash their hands after removing gloves and before leaving the areas where infectious materials and/or animals are housed or are manipulated.

    Eye, face and respiratory protection should be used in rooms containing infected animals, as dictated by the risk assessment.

8. Eating, drinking, smoking, handling contact lenses, applying cosmetics, and storing food for human consumption must not be permitted in laboratory areas. Food must be stored outside the laboratory area in cabinets or refrigerators designated and used for this purpose.

9. All procedures are carefully performed to minimize the creation of aerosols or splatters of infectious materials and waste.

10. Mouth pipetting is prohibited. Mechanical pipetting devices must be used.

11. Policies for the safe handling of sharps, such as needles, scalpels, pipettes, and broken glassware must be developed and implemented.

    When applicable, laboratory supervisors should adopt improved engineering and work practice controls that reduce the risk of sharps injuries. Precautions must always be taken with sharp items. These include:

    a. Use of needles and syringes or other sharp instruments in the animal facility is limited to situations where there is no alternative such as parenteral injection, blood collection, or aspiration of fluids from laboratory animals and diaphragm bottles.

    b. Disposable needles must not be bent, sheared, broken, recapped, removed from disposable syringes, or otherwise manipulated by hand before disposal. Used, disposable needles must be carefully placed in puncture-resistant containers used for sharps disposal. Sharps containers should be located as close to the work site as possible.

    c. Non-disposable sharps must be placed in a hard-walled container for transport to a processing area for decontamination, preferably by autoclaving.

    d. Broken glassware must not be handled directly; it should be removed using a brush and dustpan, tongs, or forceps. Plastic ware should be substituted for glassware whenever possible.

    e. Use of equipment with sharp edges and corners should be avoided.

12. Equipment and work surfaces are routinely decontaminated with an appropriate disinfectant after work with an infectious agent, and after any spills, splashes, or other overt contamination.

13. Animals and plants not associated with the work being performed must not be permitted in the areas where infectious materials and/ or animals are housed or are manipulated.

14. An effective integrated pest management program is required. (See Appendix G.)

15. All wastes from the animal room (including animal tissues, carcasses, and bedding) are transported from the animal room in leak-proof containers for appropriate disposal in compliance with applicable institutional, local and state requirements.

    Decontaminate all potentially infectious materials before disposal using an effective method.

B. *Special Practices*

1. Animal care staff, laboratory and routine support personnel must be provided a medical surveillance program as dictated by the risk assessment and administered appropriate immunizations for agents handled or potentially present, before entry into animal rooms.

    When appropriate, a base line serum sample should be stored.

2. All procedures involving the manipulation of infectious materials, handling of infected animals or the generation of aerosols must be conducted within BSCs or other physical containment devices when practical.

    When a procedure cannot be performed within a biosafety cabinet, a combination of personal protective equipment and other containment devices must be used.

    Restraint devices and practices are used to reduce the risk of exposure during animal manipulations (e.g., physical restraint devices, chemical restraint medications).

3. The risk of infectious aerosols from infected animals or their bedding also can be reduced if animals are housed in containment caging systems, such as solid wall and bottom cages covered with filter bonnets, open cages placed in inward flow ventilated enclosures, HEPA-filter isolators and caging systems, or other equivalent primary containment systems.

4. Actively ventilated caging systems must be designed to prevent the escape of microorganisms from the cage. Exhaust plenums for these systems should be sealed to prevent escape of microorganisms if the ventilation system becomes static, and the exhaust must be HEPA filtered. Safety mechanisms should be in place that prevent the cages and exhaust plenums from becoming positive to the surrounding area should the exhaust fan fail. The system should also be alarmed to indicate operational malfunctions.

5. A method for decontaminating all infectious materials must be available within the facility, preferably within the areas where infectious materials

and/or animals are housed or are manipulated (e.g., autoclave, chemical disinfection, or other approved decontamination methods).

Consideration must be given to means for decontaminating routine husbandry equipment, sensitive electronic and medical equipment.

Decontaminate all potential infectious materials (including animal tissues, carcasses, contaminated bedding, unused feed, sharps, and other refuse) by an appropriate method before removal from the areas where infectious materials and/or animals are housed or manipulated.

It is recommended that animal bedding and waste be decontaminated prior to manipulation and before removal from the areas where infectious materials and/or animals are housed or are manipulated, preferably within the caging system.

Develop and implement an appropriate waste disposal program in compliance with applicable institutional, local and state requirements.

6. Equipment, cages, and racks should be handled in a manner that minimizes contamination of other areas.

   Equipment must be decontaminated before repair, maintenance, or removal from the areas where infectious materials and/or animals are housed or are manipulated.

   Spills involving infectious materials must be contained, decontaminated, and cleaned up by staff properly trained and equipped to work with infectious material.

7. Incidents that may result in exposure to infectious materials must be immediately evaluated and treated according to procedures described in the safety manual. All such incidents must be reported to the animal facility supervisor or personnel designated by the institution. Medical evaluation, surveillance, and treatment should be provided as appropriate and records maintained.

C. *Safety Equipment (Primary Barriers and Personal Protective Equipment)*

1. Properly maintained BSCs and other physical containment devices or equipment should be used for all manipulations for infectious materials and when possible, animals. These manipulations include necropsy, harvesting of tissues or fluids from infected animals or eggs, and intranasal inoculation of animals.

   The risk of infectious aerosols from infected animals or bedding can be reduced by primary barrier systems. These systems may include solid wall and bottom cages covered with filter bonnets, ventilated cage rack systems, or for larger cages placed in inward flow ventilated enclosures or other equivalent systems or devices.

2. A risk assessment should determine the appropriate type of personal protective equipment to be utilized.

    Personnel within the animal facility where protective clothing, such as uniforms or scrub suits. Reusable clothing is appropriately contained and decontaminated before being laundered. Laboratory and protective clothing should never be taken home. Disposable personal protective equipment such as non-woven olefin cover-all suits, wrap-around or solid-front gowns should be worn over this clothing, before entering the areas where infectious materials and/or animals are housed or manipulated. Front-button laboratory coats are unsuitable.

    Disposable personal protective equipment must be removed when leaving the areas where infectious materials and/or animals are housed or are manipulated. Scrub suits and uniforms are removed before leaving the animal facility.

    Disposable personal protective equipment and other contaminated waste are appropriately contained and decontaminated prior to disposal.

3. All personnel entering areas where infectious materials and/or animals are housed or manipulated wear appropriate eye, face and respiratory protection. To prevent cross contamination, boots, shoe covers, or other protective footwear, are used where indicated.

    Eye and face protection must be disposed of with other contaminated laboratory waste or decontaminated before reuse. Persons who wear contact lenses should also wear eye protection when entering areas with potentially high concentrations or airborne particulates.

4. Gloves are worn to protect hands from exposure to hazardous materials.

    A risk assessment should be performed to identify the appropriate glove for the task and alternatives to latex gloves should be available.

    Procedures may require the use of wearing two pairs of gloves (double-glove).

    Gloves are changed when contaminated, glove integrity is compromised, or when otherwise necessary.

    Gloves must not be worn outside the animal rooms.

Gloves and personal protective equipment should be removed in a manner that prevents transfer of infectious materials.

Do not wash or reuse disposable gloves. Dispose of used gloves with other contaminated waste.

Persons must wash their hands after handling animals and before leaving the areas where infectious materials and/or animals are housed or are manipulated. Hand washing should occur after the removal of gloves.

D. Laboratory Facilities (Secondary Barriers)

1. The animal facility is separated from areas that are open to unrestricted personnel traffic within the building. External facility doors are self-closing and self-locking.

   Access to the animal facility is restricted.

   Doors to areas where infectious materials and/or animals are housed, open inward, are self-closing, are kept closed when experimental animals are present, and should never be propped open.

   Entry into the containment area is via a double-door entry, which constitutes an anteroom/airlock and a change room. Showers may be considered based on risk assessment. An additional double-door access anteroom or double-doored autoclave may be provided for movement of supplies and wastes into and out of the facility.

2. A hand-washing sink is located at the exit of the areas where infectious materials and/or animals are housed or are manipulated. Additional sinks for hand washing should be located in other appropriate locations within the facility. The sink should be hands-free or automatically operated.

   If the animal facility has multiple segregated areas where infectious materials and/or animals are housed or are manipulated, a sink must also be available for hand washing at the exit from each segregated area.

   Sink traps are filled with water, and/or appropriate liquid to prevent the migration of vermin and gases.

3. The animal facility is designed, constructed, and maintained to facilitate cleaning, decontamination and housekeeping. The interior surfaces (walls, floors and ceilings) are water resistant.

   Penetrations in floors, walls and ceiling surfaces are sealed, including openings around ducts and doorframes, to facilitate pest control, proper cleaning and decontamination. Walls, floors and ceilings should form a sealed and sanitizable surface.

Floors must be slip resistant, impervious to liquids, and resistant to chemicals. Flooring is seamless, sealed resilient or poured floors, with integral cove bases.

Decontamination of an entire animal room should be considered when there has been gross contamination of the space, significant changes in usage, for major renovations, or maintenance shut downs. Selection of the appropriate materials and methods used to decontaminate the animal room must be based on the risk assessment.

4. Cabinets and bench tops must be impervious to water and resistant to heat, organic solvents, acids, alkalis, and other chemicals. Spaces between benches, cabinets, and equipment should be accessible for cleaning.

    Furniture should be minimized. Chairs used in animal areas must be covered with a non-porous material that can be easily cleaned and decontaminated. Furniture must be capable of supporting anticipated loads and uses. Equipment and furnishings with sharp edges and corners should be avoided.

5. External windows are not recommended; if present, all windows must be sealed and must be resistant to breakage. The presence of windows may impact facility security and therefore should be assessed by security personnel.

6. Ventilation of the facility should be provided in accordance with the *Guide for Care and Use of Laboratory Animals*.[1] The direction of airflow into the animal facility is inward; animal rooms maintain inward directional airflow compared to adjoining hallways. A ducted exhaust air ventilation system is provided. Exhaust air is discharged to the outside without being recirculated to other rooms. This system creates directional airflow, which draws air into the animal room from "clean" areas and toward "contaminated" areas.

    Ventilation system design should consider the heat and high moisture load produced during the cleaning of animal rooms and the cage wash process. HEPA filtration and other treatments of the exhaust air may not be required, but should be considered based on site requirements, specific agent manipulations and use conditions. The exhaust must be dispersed away from occupied areas and air intakes.

    Personnel must verify that the direction of the airflow (into the animal areas) is proper. It is recommended that a visual monitoring device that indicates directional inward airflow be provided at the animal room entry. The ABSL-3 animal facility shall be designed such that under failure conditions the airflow will not be reversed. Alarms should be considered to notify personnel of ventilation and HVAC system failure.

7. Internal facility appurtenances, such as light fixtures, air ducts, and utility pipes, are arranged to minimize horizontal surface areas, to facilitate cleaning and minimize the accumulation of debris or fomites.

8. Floor drains must be maintained and filled with water, and/or appropriate disinfectant to prevent the migration of vermin and gases.

9. Cages are washed in a mechanical cage washer. The mechanical cage washer has a final rinse temperature of at least 180°F. Cages should be autoclaved or otherwise decontaminated prior to removal from ABSL-3 space. The cage wash facility should be designed and constructed to accommodate high-pressure spray systems, humidity, strong chemical disinfectants and 180°F water temperatures during the cage cleaning process.

10. Illumination is adequate for all activities, avoiding reflections and glare that could impede vision.

11. BSCs (Class II, Class III) must be installed so that fluctuations of the room air supply and exhaust do not interfere with proper operations. Class II BSCs should be located away from doors, heavily traveled laboratory areas, and other possible airflow disruptions.

    HEPA filtered exhaust air from a Class II BSC can be safely re-circulated into the laboratory environment if the cabinet is tested and certified at least annually and operated according to manufacturer's recommendations. BSCs can also be connected to the laboratory exhaust system by either a thimble (canopy) connection or exhausted directly to the outside through a direct (hard) connection. Provisions to assure proper safety cabinet performance and air system operation must be verified. BSCs should be certified at least annually to assure correct performance.

    Class III BSCs must supply air in such a manner that prevents positive pressurization of the cabinet or the laboratory room.

    All BSCs should be used according to manufacturers' specifications.

    When applicable, equipment that may produce infectious aerosols must be contained in devices that exhaust air through HEPA filtration or other equivalent technology before being discharged into the animal facility. These HEPA filters should be tested and/or replaced at least annually.

12. An autoclave is available which is convenient to the animal rooms where the biohazard is contained. The autoclave is utilized to decontaminate

infectious materials and waste before moving it to the other areas of the facility. If not convenient to areas where infectious materials and/or animals are housed or are manipulated, special practices should be developed for transport of infectious materials to designated alternate location/s within the facility.

13. Emergency eyewash and shower are readily available; location is determined by risk assessment.

14. The ABSL-3 facility design and operational procedures must be documented. The facility must be tested to verify that the design and operational parameters have been met prior to use. Facilities should be re-verified at least annually against these procedures as modified by operational experience.

15. Additional environmental protection (e.g., personnel showers, HEPA filtration of exhaust air, containment of other piped services, and the provision of effluent decontamination) should be considered if recommended by the agent summary statement, as determined by risk assessment of the site conditions, or other applicable federal, state or local regulations.

Animal Biosafety Level 4

Animal Biosafety Level 4 is required for work with animals infected with dangerous and exotic agents that pose a high individual risk of aerosol-transmitted laboratory infections and life-threatening disease that is frequently fatal, for which there are no vaccines or treatments; or a related agent with unknown risk of transmission. Agents with a close or identical antigenic relationship to agents requiring ABSL-4 containment must be handled at this level until sufficient data are obtained either to confirm continued work at this level, or to re-designate the level. Animal care staff must have specific and thorough training in handling extremely hazardous, infectious agents and infected animals. Animal care staff must understand the primary and secondary containment functions of standard and special practices, containment equipment, and laboratory design characteristics. All animal care staff and supervisors must be competent in handling animals, agents and procedures requiring ABSL-4 containment. The animal facility director and/or laboratory supervisor control access to the animal facility within the ABSL-4 laboratory in accordance with institutional policies.

There are two models for ABSL-4 laboratories:

1. A *Cabinet Laboratory*—All handling of agents, infected animals and housing of infected animals must be performed in Class III BSCs (see Appendix A); and
2. A *Suit Laboratory*—Personnel must wear a positive pressure protective suit (see Appendix A); infected animals must be housed in ventilated enclosures with inward directional airflow and HEPA filtered exhaust;

and infected animals should be handled within a primary barrier system, such as a Class II BSC or other equivalent containment system.

ABSL-4 builds upon the standard practices, procedures, containment equipment, and facility requirements of ABSL-3. However, ABSL-4 cabinet and suit laboratories have special engineering and design features to prevent microorganisms from being disseminated into the environment and personnel. The ABSL-4 cabinet laboratory is distinctly different from an ABSL-3 laboratory containing a Class III BSC. The following standard and special safety practices, equipment, and facilities apply to ABSL-4.

A. Standard Microbiological Practices

1. The animal facility directors must establish and enforce policies, procedures, and protocols for biosafety, biosecurity and emergencies within the ABSL-4 laboratory.

   The animal facility director and/or designated institutional officials are responsible for enforcing the policies that control access to the ABSL-4 facility. Laboratory personnel and support staff must be provided appropriate occupational medical service including medical surveillance and available immunizations for agents handled or potentially present in the laboratory.[3] A system must be established for reporting and documenting laboratory accidents, exposures, employee absenteeism and for the medical surveillance of potential laboratory-associated illnesses. An essential adjunct to such an occupational medical services system is the availability of a facility for the isolation and medical care of personnel with potential or known laboratory-acquired infections. Facility supervisors should ensure that medical staff are informed of potential occupational hazards within the animal facility including those associated with the research, animal husbandry duties, animal care, and manipulations.

   An ABSL-4 laboratory specific, biosafety manual must be prepared in consultation with the animal facility director, the laboratory supervisor, and the biosafety advisor. The biosafety manual must be available and accessible. Personnel are advised of special hazards, and are required to read and follow instructions on practices and procedures.

   Prior to beginning a study, appropriate policies and procedures for animal welfare during the conduct of research, must be developed and approved by the IACUC. The biosafety official, the IBC and/or other applicable committees, are responsible for review of protocols and polices to prevent hazardous exposures to personnel who manipulate and care for animals.

2. A complete clothing change is required in the ABSL-4 operation. Personnel within the animal facility where protective clothing, such as uniforms or scrub suits.

All persons leaving the BSL-4/ABSL-4 laboratory are required to take a personal body shower.

3. Eating, drinking, smoking, handling contact lenses, applying cosmetics, and storing food for human consumption must not be permitted in laboratory areas. Food must be stored outside the laboratory area in cabinets or refrigerators designated and used for this purpose.

4. Mechanical pipetting devices must be used.

5. Policies for the safe handling of sharps, such as needles, scalpels, pipettes, and broken glassware must be developed and implemented.

   When applicable, laboratory supervisors should adopt improved engineering and work practice controls that reduce the risk of sharps injuries. Precautions, including those listed below, must always be taken with sharp items. These include:

   a. Use of needles and syringes or other sharp instruments are limited for use in the animal facility is limited to situations where there is no alternative such as parenteral injection, blood collection, or aspiration of fluids from laboratory animals and diaphragm bottles.

   b. Disposable needles must not be bent, sheared, broken, recapped, removed from disposable syringes, or otherwise manipulated by hand before disposal. Used disposable needles must be carefully placed in puncture-resistant containers used for sharps disposal and placed as close to the work site as possible.

   c. Non-disposable sharps must be placed in a hard walled container for transport to a processing area for decontamination, preferably by autoclaving.

   d. Broken glassware must not be handled directly. Instead, it must be removed using a brush and dustpan, tongs, or forceps. Plastic ware should be substituted for glassware whenever possible.

   e. Equipment containing sharp edges and corners should be avoided.

6. Perform all procedures to minimize the creation of splashes and/or aerosols.

   Procedures involving the manipulation of infectious materials must be conducted within biological safety cabinets, or other physical containment devices. When procedures cannot be performed in a BSC, alternate containment equipment should be used.

7. Decontaminate work surfaces after completion of work and after any spill or splash of potentially infectious material with appropriate disinfectant.

Incidents that may result in exposure to infectious materials must be immediately evaluated and treated according to procedures described in the laboratory biosafety manual. All incidents must be reported to the animal facility director, laboratory supervisor, institutional management and appropriate facility safety personnel. Medical evaluation, surveillance, and treatment must be provided and appropriate records maintained.

8. Decontaminate all wastes (including animal tissues, carcasses, and contaminated bedding) and other materials before removal from the ABSL-4 laboratory by an effective and validated method. Laboratory clothing should be decontaminated before laundering.

Supplies and materials needed in the facility must be brought in through a double-door autoclave, fumigation chamber, or airlock. Supplies and materials that are not brought into the ABSL-4 laboratory through the change room must be brought in through a previously decontaminated double-door autoclave, fumigation chamber, or airlock. Containment should be maintained at all times. After securing the outer doors, personnel within the areas where infectious materials and/or animals are housed or are manipulated retrieve the materials by opening the interior doors of the autoclave, fumigation chamber, or airlock. These doors must be secured after materials are brought into the facility.

Only necessary equipment and supplies should be taken inside the ABSL-4 laboratory. All equipment and supplies taken inside the laboratory must be decontaminated before removal. Consideration should be given to means for decontaminating routine husbandry equipment and sensitive electronic and medical equipment.

The doors of the autoclave and fumigation chamber are interlocked in a manner that prevents opening of the outer door unless the autoclave has been operated through a decontamination cycle or the fumigation chamber has been decontaminated.

9. A sign incorporating the universal biohazard symbol must be posted at the entrance to the laboratory and the animal room/s when infectious agents are present. The sign must include the animal biosafety level, general occupational health requirements, personal protective equipment requirements, the supervisor's name (or other responsible personnel), telephone number, and required procedures for entering and exiting the animal areas. Identification of specific infectious agents is recommended when more than one agent is being used within an animal room.

Security sensitive agent information and occupational health requirements should be posted in accordance with the institutional policy.

Advance consideration must be given to emergency and disaster recovery plans, as a contingency for man-made or natural disasters.[1,3,4]

10. An effective integrated pest management program is required. (See Appendix G.)

11. The laboratory supervisor must ensure that laboratory personnel receive appropriate training regarding their duties, the necessary precautions to prevent exposures, and exposure evaluation procedures. Personnel must receive annual updates or additional training when procedural or policy changes occur. Personal health status may impact an individual's susceptibility to infection, ability to receive immunizations or prophylactic interventions. Therefore, all laboratory personnel and particularly women of childbearing age should be provided with information regarding immune competence and conditions that may predispose them to infection. Individuals having these conditions should be encouraged to self-identify to the institution's healthcare provider for appropriate counseling and guidance.

12. Animals and plants not associated with the work being performed must not be permitted in the areas where infectious materials and/ or animals are housed or are manipulated.

B. Special Practices

1. All persons entering the ABSL-4 laboratory must be advised of the potential hazards and meet specific entry/exit requirements.

   Only persons whose presence in the laboratory or individual animal rooms is required for scientific or support purposes are authorized to enter.

   Entry into the facility must be limited by means of secure, locked doors. A logbook, or other means of documenting the date and time of all persons entering and leaving the ABSL-4 laboratory must be maintained.

   While the laboratory is operational, personnel must enter and exit the laboratory through the clothing change and shower rooms except during emergencies. All personal clothing must be removed in the outer clothing change room. All personnel entering the laboratory must use laboratory clothing, including undergarments, pants, shirts, jumpsuits, shoes, and gloves.

   All persons leaving the ABSL-4 laboratory are required to take a personal body shower. Used laboratory clothing must not be removed from the inner change room through the personal shower. These items must be treated as contaminated materials and decontaminated before laundering or disposal.

After the laboratory has been completely decontaminated by validated method, necessary staff may enter and exit the laboratory without following the clothing change and shower requirements described above.

Personal health status may impact an individual's susceptibility to infection, ability to receive immunizations or prophylactic interventions. Therefore, all laboratory personnel and particularly women of childbearing age should be provided with information regarding immune competence and conditions that may predispose them to infection. Individuals having these conditions should be encouraged to self-identify to the institution's healthcare provider for appropriate counseling and guidance.

2. Animal facility personnel and support staff must be provided occupational medical services, including medical surveillance and available immunizations for agents handled or potentially present in the laboratory. A system must be established for reporting and documenting laboratory accidents, exposures, employee absenteeism and for the medical surveillance of potential laboratory-acquired illnesses. An essential adjunct to an occupational medical system is the availability of a facility for the isolation and medical care of personnel with potential or known laboratory-acquired illnesses.

3. Each institution must establish policies and procedures describing the collection and storage of serum samples from at-risk personnel.

4. The animal facility supervisor is responsible for ensuring that animal personnel:

   a. Receive appropriate training in the practices and operations specific to the animal facility, such as animal husbandry procedures, potential hazards present, manipulations of infectious agents, and necessary precautions to prevent potential exposures.

   b. Demonstrate high proficiency in standard and special microbiological practices, and techniques before entering the ABSL-4 facility or working with agents requiring ABSL-4 containment.

   c. Receive annual updates and additional training when procedure or policy changes occur. Records are maintained for all hazard evaluations and employee training.

5. Removal of biological materials that are to remain in a viable or intact state from the ABSL-4 laboratory must be transferred to a non-breakable,

sealed primary container and then enclosed in a non-breakable, sealed secondary container. These materials must be transferred through a disinfectant dunk tank, fumigation chamber, or decontamination shower. Once removed, packaged viable material must not be opened outside ABSL-4 containment unless inactivated by a validated method.

6. Laboratory equipment must be routinely decontaminated, as well as after spills, splashes, or other potential contamination. Equipment, cages, and racks should be handled in manner that minimizes contamination of other areas. Cages are autoclaved or thoroughly decontaminated before they are cleaned and washed.

   a. All equipment and contaminated materials must be decontaminated before removal from the animal facility. Equipment must be decontaminated using an effective and validated method before repair, maintenance, or removal from the animal facility.

   b. Equipment or material that might be damaged by high temperatures or steam must be decontaminated using an effective and validated procedure such as a gaseous or vapor method in an airlock or chamber designed for this purpose.

   c. Spills involving infectious materials must be contained, decontaminated, and cleaned up by staff properly trained and equipped to work with infectious material. A spill procedure must be developed and posted within the laboratory. Spills and accidents of potentially infectious materials must be immediately reported to the animal facility and laboratory supervisors or personnel designated by the institution.

7. The doors of the autoclave and fumigation chamber are interlocked in a manner that prevents opening of the outer door unless the autoclave/ decontamination chamber has been operated through a decontamination cycle or the fumigation chamber has been decontaminated.

8. Daily inspections of essential containment and life support systems must be completed before laboratory work is initiated to ensure that the laboratory and animal facilities are operating according to established parameters.

9. Practical and effective protocols for emergencies must be established. These protocols must include plans for medical emergencies, facility malfunctions, fires, escape of animals within the ABSL-4 laboratory, and other potential emergencies. Training in emergency response procedures must be provided to emergency response personnel according to institutional policies.

10. Based on site-specific risk assessment, personnel assigned to work with infected animals may be required to work in pairs. Procedures to reduce possible worker exposure must be instituted, such as use of squeeze cages, working only with anesthetized animals, or other appropriate practices.

C. *Safety Equipment (Primary Barriers and Personal Protective Equipment)*

Cabinet Laboratory

1. All manipulations of infectious animals and materials within the laboratory must be conducted in the Class III BSC. Double-door, pass through autoclaves must be provided for decontaminating materials passing out of the Class III BSC(s). The autoclave doors must be interlocked so that only one can be opened at any time and be automatically controlled so that the outside door to the autoclave can only be opened after the decontamination cycle has been completed.

   The Class III cabinet must also have a pass-through dunk tank, fumigation chamber, or equivalent decontamination method so that materials and equipment that cannot be decontaminated in the autoclave can be safely removed from the cabinet. Containment must be maintained at all times.

   The Class III cabinet must have a HEPA filter on the supply air intake and two HEPA filters in series on the exhaust outlet of the unit. There must be gas-tight dampers on the supply and exhaust ducts of the cabinet to permit gas or vapor decontamination of the unit. Ports for injection of test medium must be present on all HEPA filter housings.

   The interior of the Class III cabinet must be constructed with smooth finishes that can be easily cleaned and decontaminated. All sharp edges on cabinet finishes must be eliminated to reduce the potential for cuts and tears of gloves. Equipment to be placed in the Class III cabinet should also be free of sharp edges or other surfaces that may damage or puncture the cabinet gloves.

   Class III cabinet gloves must be inspected for leaks periodically and changed if necessary. Gloves should be replaced annually during cabinet re-certification.

   The cabinet should be designed to permit maintenance and repairs of cabinet mechanical systems (refrigeration, incubators, centrifuges, etc.) to be performed from the exterior of the cabinet whenever possible.

   Manipulation of high concentrations or large volumes of infectious agents within the Class III cabinet should be performed using physical containment devices inside the cabinet whenever practical. Such materials should be centrifuged inside the cabinet using sealed rotor heads or centrifuge safety cups.

The interior of the Class III cabinet as well as all contaminated plenums, fans and filters must be decontaminated using a validated gaseous or vapor method.

The Class III cabinet must be certified at least annually.

Restraint devices and practices that reduce the risk of exposure during animal manipulations must be used where practicable (e.g., physical restraint devices, chemical restraint medications, mesh or Kevlar gloves, etc.).

2. Workers must wear protective laboratory clothing such as solid-front or wrap-around gowns, scrub suits, or coveralls when in the laboratory. No personal clothing, jewelry, or other items except eyeglasses should be taken past the personal shower area. Upon exiting the laboratory, all protective clothing must be removed in the dirty side change room before showering. Reusable laboratory clothing must be autoclaved before being laundered.

3. Eye, face and respiratory protection should be used in rooms containing infected animals as determined by the risk assessment. Prescription eye glasses must be decontaminated before removal thought the personal body shower.

4. Gloves must be worn to protect against breaks or tears in the cabinet gloves. Gloves must not be worn outside the laboratory. Alternatives to latex gloves should be available. Do not wash or reuse disposable gloves. Dispose of used gloves with other contaminated waste.

Suit Laboratory

1. Infected animals should be housed in a primary containment system (such as open cages placed in ventilated enclosures, solid wall and bottom cages covered with filter bonnets and opened in laminar flow hoods, or other equivalent primary containment systems).

Personnel wearing a one-piece positive pressure suit ventilated with a life support system must conduct all procedures.

All manipulations of potentially infectious agents must be performed within a Class II BSC or other primary barrier system. Infected animals should be handled within a primary barrier system, such as a Class II BSC or other equivalent containment system.

Equipment that may produce aerosols must be contained in devices that exhaust air through HEPA filtration before being discharged into the laboratory. These HEPA filters should be tested annually and replaced as need.

HEPA filtered exhaust air from a Class II BSC can be safely re-circulated into the laboratory environment if the cabinet is tested and certified at least annually and operated according to manufacturer's recommendations.

2. Workers must wear protective laboratory clothing, such as scrub suits, before entering the room used for donning positive pressure suits. All protective clothing must be removed in the dirty side change room before entering the personal shower. Reusable laboratory clothing must be autoclaved before being laundered.

3. Inner gloves must be worn to protect against break or tears in the outer suit gloves. Disposable gloves must not be worn outside the change area. Alternatives to latex gloves should be available. Do not wash or reuse disposable gloves. Inner gloves must be removed and discarded in the inner change room prior to entering the personal shower. Dispose of used gloves with other contaminated waste.

4. Decontamination of outer suit gloves is performed during operations to remove gross contamination and minimize further contamination of the laboratory.

D. Laboratory Facilities (Secondary Barriers)

Cabinet Laboratory

1. The ABSL-4 cabinet laboratory consists of either a separate building or a clearly demarcated and isolated zone within a building. Laboratory doors must have locks in accordance with the institutional policies.

Rooms in the ABSL-4 facility must be arranged to ensure sequential passage through an inner (dirty) change area, personal shower and outer (clean) change room prior to exiting the room(s) containing the Class III BSC(s).

An automatically activated emergency power source must be provided at a minimum for the laboratory exhaust system, life support systems, alarms, lighting, entry and exit controls, BSCs, and door gaskets. Monitoring and control systems for air supply, exhaust, life support, alarms, entry and exit, and security systems should be on an uninterrupted power supply (UPS).

A double-door autoclave, dunk tank, fumigation chamber, or ventilated anteroom/airlock must be provided at the containment barrier for the passage of materials, supplies, or equipment.

2. A hands-free sink must be provided near the doors of the cabinet room(s) and the inner change rooms. A sink must be provided in the outer change room. All sinks in the room(s) containing the Class III BSC must be connected to the wastewater decontamination system.

3. Walls, floors, and ceilings of the laboratory must be constructed to form a sealed internal shell to facilitate fumigation and prohibit animal and insect intrusion. The internal surfaces of this shell must be resistant to liquids and chemicals used for cleaning and decontamination of the area. Floors must be monolithic, sealed and coved.

   All penetrations in the internal shell of the laboratory and inner change room must be sealed.

   Openings around doors into the cabinet room and inner change room must be minimized and capable of being sealed to facilitate decontamination.

   All drains in ABSL-4 laboratory area floor must be connected directly to the liquid waste decontamination system.

   Services and plumbing that penetrate the laboratory walls, floors or ceiling, must be installed to ensure that no backflow from the laboratory occurs. Services must be sealed and be provided with redundant backflow prevention. Consideration should be given to locating these devices outside of containment. Atmospheric venting systems must be provided with two HEPA filters in series and are sealed up to the second filter.

   Decontamination of the entire cabinet must be performed using a validated gaseous or vapor method when there have been significant changes in cabinet usage, before major renovations or maintenance shut downs, and in other situations, as determined by risk assessment. Selection of the appropriate materials and methods used for decontamination must be based on the risk assessment of the biological agents in use.

4. Laboratory furniture must be of simple construction, capable of supporting anticipated loading and uses. Spaces between benches, cabinets, and equipment must be accessible for cleaning and decontamination. Chairs and other furniture should be covered with a non-porous material that can be easily decontaminated.

5. Windows must be break-resistant and sealed.

6. If Class II BSCs are needed in the cabinet laboratory, they must be installed so that fluctuations of the room air supply and exhaust do not interfere with proper operations. Class II BSCs should be located away from doors, heavily traveled laboratory areas, and other possible airflow disruptions.

7. Central vacuum systems are not recommended. If, however, there is a central vacuum system, it must not serve areas outside the cabinet room. Two in-line HEPA filters must be placed near each use point. Filters must be installed to permit in-place decontamination and replacement.

8. An eyewash station must be readily available in the laboratory.

9. A dedicated non-recirculating ventilation system is provided. Only laboratories with the same HVAC requirements (i.e., other BSL-4 labs, ABSL-4, BSL-3-Ag labs) may share ventilation systems if gas-tight dampers and HEPA filters isolate each individual laboratory system.

   The supply and exhaust components of the ventilation system must be designed to maintain the ABSL-4 laboratory at negative pressure to surrounding areas and provide differential pressure/directional airflow between adjacent areas within the laboratory.

   Redundant supply fans are recommended. Redundant exhaust fans are required. Supply and exhaust fans must be interlocked to prevent positive pressurization of the laboratory.

   The ventilation system must be monitored and alarmed to indicate malfunction or deviation from design parameters. A visual monitoring device must be installed near the clean change room so proper differential pressures within the laboratory may be verified.

   Supply air to and exhaust air from the cabinet room and fumigation/decontamination chambers must pass through HEPA filter(s). The air exhaust discharge must be located away from occupied spaces and building air intakes.

   All HEPA filters should be located as near as practicable to the cabinet laboratory in order to minimize the length of potentially contaminated ductwork. All HEPA filters must be tested and certified annually.

   The HEPA filter housings should be designed to allow for *in situ* decontamination and validation of the filter prior to removal. The design of the HEPA filter housing must have gas-tight isolation dampers; decontamination ports, and ability to scan each filter assembly for leaks.

10. HEPA filtered exhaust air from a Class II BSC can be safely re-circulated into the laboratory environment if the cabinet is tested and certified at least annually and operated according to the manufacturer's recommendations. Biological safety cabinets can also be connected to the laboratory exhaust system by either a thimble (canopy) connection or a direct (hard) connection. Provisions to assure proper safety cabinet performance and air system operation must be verified.

    Class III BSCs must be directly and independently exhausted through two HEPA filters in series. Supply air must be provided in such a manner that prevents positive pressurization of the cabinet.

11. Pass through dunk tanks, fumigation chambers, or equivalent decontamination methods must be provided so that materials and equipment that cannot be decontaminated in the autoclave can be safely removed from the cabinet room(s). Access to the exit side of the pass though shall be limited to those individuals authorized to be in the ABSL-4 laboratory.

12. Liquid effluents from cabinet room sinks, floor drains, autoclave chambers, and other sources within the cabinet room must be decontaminated by a proven method, preferably heat treatment, before being discharged to the sanitary sewer.

    Decontamination of all liquid wastes must be documented. The decontamination process for liquid wastes must be validated physically and biologically. Biological validation must be performed annually or more often as required by institutional policy.

    Effluents from showers and toilets may be discharged to the sanitary sewer without treatment.

13. A double-door autoclave must be provided for decontaminating waste or other materials passing out of the cabinet room. Autoclaves that open outside of the laboratory must be sealed to the wall. This bioseal must be durable, airtight, and sealed to the wall. Positioning the bioseal so that the equipment can be accessed and maintained from outside the laboratory is recommended. The autoclave doors must be interlocked so that only one can be opened at any time and be automatically controlled so that the outside door can only be opened after the autoclave decontamination cycle has been completed.

    Gas and liquid discharge from the autoclave chamber must be decontaminated. When feasible, autoclave decontamination processes should be designed so that over-pressurization cannot release unfiltered air or steam exposed to infectious material to the environment.

14. The ABSL-4 facility design parameters and operational procedures must be documented. The facility must be tested to verify that the design and operational parameters have been met prior to operation. Facilities must also be re-verified annually. Verification criteria should be modified as necessary by operational experience.

15. Appropriate communication systems must be provided between the ABSL-4 laboratory and the outside (e.g., voice, fax, and computer). Provisions for emergency communication and access/egress must be considered.

Suit Laboratory

1. The ABSL-4 suit laboratory consists of either a separate building or a clearly demarcated and isolated zone within a building. Laboratory doors must have locks in accordance with the institutional policies.

   Entry to this laboratory must be through an airlock fitted with airtight doors. Personnel who enter this laboratory must wear a positive pressure suit ventilated by a life support system with HEPA filtered breathing air. The breathing air system must have redundant compressors, failure alarms and an emergency backup system.

   Rooms in the facility must be arranged to ensure sequential passage through the chemical shower, inner (dirty) change room, personal shower, and outer (clean) changing area upon exit.

   A chemical shower must be provided to decontaminate the surface of the positive pressure suit before the worker leaves the ABSL-4 laboratory. In the event of an emergency exit or failure of chemical shower, a method for decontaminating positive pressure suits, such as a gravity fed supply of chemical disinfectant, is needed.

   An automatically activated emergency power source must be provided at a minimum for the laboratory exhaust system, life support systems, alarms, lighting, entry and exit controls, BSCs, and door gaskets. Monitoring and control systems for air supply, exhaust, life support, alarms, entry and exit, and security systems should be on a UPS.

   A double-door autoclave, dunk tank, or fumigation chamber must be provided at the containment barrier for the passage of materials, supplies, or equipment.

2. Sinks inside the ABSL-4 laboratory must be placed near procedure areas, contain traps, and be connected to the wastewater decontamination system.

3. Walls, floors, and ceilings of the ABSL-4 laboratory must be constructed to form a sealed internal shell to facilitate fumigation and prohibit animal and insect intrusion. The internal surfaces of this shell must be resistant to liquids and chemicals used for cleaning and decontamination of the area. Floors must be monolithic, sealed and coved.

   All penetrations in the internal shell of the laboratory, suit storage room and the inner change room must be sealed.

   Drains, if present, in the laboratory floor must be connected directly to the liquid waste decontamination system. Sewer vents and other service lines must be protected by two HEPA filters in series and have protection against insect and animal intrusion.

   Services and plumbing that penetrate the laboratory walls, floors, or ceiling must be installed to ensure that no backflow from the laboratory occurs. These penetrations must be fitted with two (in series) backflow prevention devices. Consideration should be given to locating these devices outside of containment. Atmospheric venting systems must be provided with two HEPA filters in series and be sealed up to the second filter.

   Decontamination of the entire laboratory must be performed using a validated gaseous or vapor method when there have been significant changes in laboratory usage, before major renovations or maintenance shut downs, and in other situations, as determined by risk assessment.

4. Laboratory furniture must be of simple construction, capable of supporting anticipated loading and uses. Spaces between benches, cabinets, and equipment must be accessible for cleaning, decontamination and unencumbered movement of personnel. Chairs and other furniture should be covered with a non-porous material that can be easily decontaminated. Sharp edges and corners should be avoided.

5. Windows must be break-resistant and sealed.

6. BSCs and other primary containment barrier systems must be installed so that fluctuations of the room air supply and exhaust do not interfere with proper operations. BSCs should be located away from doors, heavily traveled laboratory areas, and other possible airflow disruptions.

7. Central vacuum systems are not recommended. If, however, there is a central vacuum system, it must not serve areas outside the ABSL-4 laboratory. Two in-line HEPA filters must be placed near each use point. Filters must be installed to permit in-place decontamination and replacement.

8. An eyewash station must be readily available in the laboratory area for use during maintenance and repair activities.

9. A dedicated non-recirculating ventilation system is provided. Only laboratories with the same HVAC requirements (i.e., other BSL-4 labs, ABSL-4, BSL-3-Ag labs) may share ventilation systems if gas tight dampers and HEPA filters isolate each individual laboratory system.

   The supply and exhaust components of the ventilation system must be designed to maintain the BSL-4/ABSL-4 laboratory at negative pressure to surrounding areas and provide correct differential pressure between adjacent areas within the laboratory.

   Redundant supply fans are recommended. Redundant exhaust fans are required. Supply and exhaust fans must be interlocked to prevent positive pressurization of the laboratory.

   The ventilation system must be monitored and alarmed to indicate malfunction or deviation from design parameters. A visual monitoring device must be installed near the clean change room so proper differential pressures within the laboratory may be verified.

   Supply air to the ABSL-4 laboratory, including the decontamination shower, must pass through a HEPA filter. All exhaust air from the BSL-4/ABSL-4 suit laboratory, decontamination shower and fumigation or decontamination chambers must pass through two HEPA filters, in series before discharge to the outside. The exhaust air discharge must be located away from occupied spaces and air intakes.

   All HEPA filters must be located as near as practicable to the areas where infectious materials and/or animals are housed or are manipulated in order to minimize the length of potentially contaminated ductwork. All HEPA filters must be tested and certified annually.

   The HEPA filter housings are designed to allow for *in situ* decontamination and validation of the filter prior to removal. The design of the HEPA filter housing must have gas-tight isolation dampers; decontamination ports; and ability to scan each filter assembly for leaks.

10. HEPA filtered exhaust air from a Class II BSC can be safely re-circulated back into the laboratory environment if the cabinet is tested and certified at least annually and operated according to the manufacturer's recommendations. Biological safety cabinets can also be connected to the laboratory exhaust system by either a thimble (canopy) connection or directly to the outside through an independent, direct (hard) connection. Provisions to assure proper safety cabinet performance and air system operation must be verified.

11. Pass through dunk tanks, fumigation chambers, or equivalent decontamination methods must be provided so that materials and equipment that cannot be decontaminated in the autoclave can be safely removed from the ABSL-4 laboratory. Access to the exit side of the pass-through shall be limited to those individuals authorized to be in the ABSL-4 laboratory.

12. Liquid effluents from chemical showers, sinks, floor drains, autoclave chambers, and other sources within the laboratory must be decontaminated by a proven method, preferably heat treatment, before being discharged to the sanitary sewer.

    Decontamination of all liquid wastes must be documented. The decontamination process for liquid wastes must be validated physically and biologically. Biological validation must be performed annually or more often as required by institutional policy.

    Effluents from personal body showers and toilets may be discharged to the sanitary sewer without treatment.

13. A double-door, pass through autoclave(s) must be provided for decontaminating materials passing out of the cabinet laboratory. Autoclaves that open outside of the laboratory must be sealed to the wall through which the autoclave passes. This bioseal must be durable and airtight. Positioning the bioseal so that the equipment can be accessed and maintained from outside the laboratory is strongly recommended. The autoclave doors must be interlocked so that only one can be opened at any time and be automatically controlled so that the outside door to the autoclave can only be opened after the decontamination cycle has been completed.

    The size of the autoclave should be sufficient to accommodate the intended usage, equipment size, and potential future increases in cage size. Autoclaves should facilitate isolation for routine servicing.

    Gas and liquid discharge from the autoclave chamber must be decontaminated. When feasible, autoclave decontamination processes should be designed so that over-pressurization cannot release unfiltered air or steam exposed to infectious material to the environment.

14. The ABSL-4 facility design parameters and operational procedures must be documented. The facility must be tested to verify that the design and operational parameters have been met prior to operation. Facilities must also be re-verified. Verification criteria should be modified as necessary by operational experience.

Consider placing ABSL-4 areas away from exterior walls of buildings to minimize the impact from the outside environmental and temperatures.

15. Appropriate communication systems must be provided between the laboratory and the outside (e.g., voice, fax, and computer). Provisions for emergency communication and access/egress should be considered.

References

1. Institute for Laboratory Animal Research. Guide for the care and use of laboratory animals. Washington, DC: National Academy Press; 1996.
2. Animal Welfare Act and Amendment, Title 9 CFR Subchapter A, Parts 1, 2, 3 (1976).
3. National Research Council; Institute for Laboratory Animal Research. Occupational health and safety in the care and use of research animals. Washington, DC: National Academy Press; 1997.
4. National Research Council; Institute for Laboratory Animal Research. Occupational health and safety in the care and use of nonhuman primates. Washington, DC: National Academy Press; 2003.
5. National Institutes of Health, Office of Laboratory Animal Welfare. Public Health Service policy on humane care and use of laboratory animals, Bethesda (MD); The National Institutes of Health (US); 2000.

Table 3. Summary of Recommended Animal Biosafety Levels for Activities in which Experimentally or Naturally Infected Vertebrate Animals are Used

BSL	Agents	Practices	Primary Barriers and Safety Equipment	Facilities (Secondary Barriers)
1	Not known to consistently cause diseases in healthy adults	Standard animal care and management practices, including appropriate medical surveillance programs	As required for normal care of each species ■ PPE: laboratory coats and gloves; eye, face protection, as needed	Standard animal facility: ■ No recirculation of exhaust air ■ Directional air flow recommended ■ Hand washing sink is available
2	■ Agents associated with human disease ■ Hazard: percutaneous injury, ingestion, mucous membrane exposure	ABSL-1 practice plus: ■ Limited access ■ Biohazard warning signs ■ "Sharps" precautions ■ Biosafety manual ■ Decontamination of all infectious wastes and animal cages prior to washing	ABSL-1 equipment plus primary barriers: ■ Containment equipment appropriate for animal special ■ PPE: Laboratory coats, gloves, face, eye and respiratory protection, as needed ■	ABSL-1 plus: ■ Autoclave available ■ Hand washing sink available ■ Mechanical cage washer recommended ■ Negative airflow into animal and procedure rooms recommended
3	Indigenous or exotic agents that may cause serious or potentially lethal disease through the inhalation route of exposure	ABSL-2 practice plus: ■ Controlled access ■ Decontamination of clothing before laundering ■ Cages decontaminated before bedding is removed ■ Disinfectant foot bath as needed	ABSL-2 equipment plus: ■ Containment equipment for housing animals and cage dumping activities ■ Class I, II or III BSCs available for manipulative procedures (inoculation, necropsy) that may create infectious aerosols ■ PPE: Appropriate respiratory protection	ABSL-2 facility plus: ■ Physical separation from access corridors ■ Self-closing, double-door access ■ Sealed penetrations ■ Sealed windows ■ Autoclave available in facility ■ Entry through ante-room or airlock ■ Negative airflow into animal and procedure rooms ■ Hand washing sink near exit of animal or procedure room
4	■ Dangerous/exotic agents which post high risk of aerosol transmitted laboratory infections that are frequently fatal, for which there are no vaccines or treatments ■ Agents with a close or identical antigenic relationship to an agent requiring BSL-4 until data are available to redesignate the level ■ Related agents with unknown risk of transmission	ABSL-3 practices plus: ■ Entrance through change room where personal clothing is removed and laboratory clothing is put on; shower on exiting ■ All wastes are decontaminated before removal from the facility	ABSL-3 equipment plus: ■ Maximum containment equipment (i.e., Class III BSC or partial containment equipment in combination with full body, air-supplied positive-pressure suit) used for all procedures and activities	ABSL-3 facility plus: ■ Separate building or isolated zone ■ Dedicated supply and exhaust, vacuum, and decontamination systems ■ Other requirements outlined in the text

Section VI—Principles of Laboratory Biosecurity

Since the publication of the 4th edition of BMBL in 1999, significant events have brought national and international scrutiny to the area of laboratory security. These events, including the anthrax attacks on U.S. citizens in October 2001 and the subsequent expansion of the United States Select Agent regulations in December 2003, have led scientists, laboratory managers, security specialists, biosafety professionals, and other scientific and institutional leaders to consider the need for developing, implementing and/or improving the security of biological agents and toxins within their facilities. Appendix F of BMBL 4th edition provided a brief outline of issues to consider in developing a security plan for biological agents and toxins capable of serious or fatal illness to humans or animals. In December 2002, Appendix F was updated and revised as a security and emergency response guidance for laboratories working with select agents.[1] Section VI replaces the previous appendices. The current Appendix F discusses Select Agent and Toxin regulations.

This section describes laboratory biosecurity planning for microbiological laboratories. As indicated below, laboratories with good biosafety programs already fulfill many of the basic requirements for security of biological materials. For laboratories not handling select agents, the access controls and training requirements specified for BSL-2 and BSL-3 in BMBL may provide sufficient security for the materials being studied. Security assessments and additional security measures should be considered when select agents, other agents of high public health and agriculture concern, or agents of high commercial value such as patented vaccine candidates, are introduced into the laboratory.

The recommendations presented in this section are advisory. Excluding the Select Agent regulations, there is no current federal requirement for the development of a biosecurity program. However, the application of these principles and the assessment process may enhance overall laboratory management. Laboratories that fall under the Select Agent regulations should consult Appendix F (42 CFR part 73; 7 CFR 331 and 9 CFR 121).[4,5,6]

The term "biosecurity" has multiple definitions. In the animal industry, the term biosecurity relates to the protection of an animal colony from microbial contamination. In some countries, the term biosecurity is used in place of the term biosafety. For the purposes of this section the term "biosecurity" will refer to the protection of microbial agents from loss, theft, diversion or intentional misuse. This is consistent with current WHO and American Biological Safety Association (ABSA) usage of this term.[2,3]

Security is not a new concept in biological research and medical laboratories. Several of the security measures discussed in this section are embedded in the biosafety levels that serve as the foundation for good laboratory practices throughout the biological laboratory community. Most biomedical and microbiological laboratories do not have select agents or toxins, yet maintain control over and account for research materials, protect relevant sensitive information, and work in facilities with access controls commensurate with the potential public health and economic impact of the biological agents in their collections. These measures are in place in most laboratories that apply good laboratory management practices and have appropriate biosafety programs.

Biosafety and Biosecurity

Biosafety and biosecurity are related, but not identical, concepts. Biosafety programs reduce or eliminate exposure of individuals and the environment to potentially hazardous biological agents. Biosafety is achieved by implementing various degrees of laboratory control and containment, through laboratory design and access restrictions, personnel expertise and training, use of containment equipment, and safe methods of managing infectious materials in a laboratory setting.

The objective of biosecurity is to prevent loss, theft or misuse of microorganisms, biological materials, and research-related information. This is accomplished by limiting access to facilities, research materials and information. While the objectives are different, biosafety and biosecurity measures are usually complementary.

Biosafety and biosecurity programs share common components. Both are based upon risk assessment and management methodology; personnel expertise and responsibility; control and accountability for research materials including microorganisms and culture stocks; access control elements; material transfer documentation; training; emergency planning; and program management.

Biosafety and biosecurity program risk assessments are performed to determine the appropriate levels of controls within each program. Biosafety looks at appropriate laboratory procedures and practices necessary to prevent exposures and occupationally-acquired infections, while biosecurity addresses procedures and practices to ensure that biological materials and relevant sensitive information remain secure.

Both programs assess personnel qualifications. The biosafety program ensures that staff are qualified to perform their jobs safely through training and documentation of technical expertise. Staff must exhibit the appropriate level of professional responsibility for management of research materials by adherence to appropriate materials management procedures. Biosafety practices require laboratory access to be limited when work is in progress. Biosecurity practices ensure that access to the laboratory facility and biological materials are limited and controlled as necessary. An inventory or material management process for control and tracking of biological stocks or other sensitive materials is also a component of both programs. For biosafety, the shipment of infectious biological materials must adhere to safe packaging, containment and appropriate transport procedures, while biosecurity ensures that transfers are controlled, tracked and documented commensurate with the potential risks. Both programs must engage laboratory personnel in the development of practices and procedures that fulfill the biosafety and biosecurity program objectives but that do not hinder research or clinical/diagnostic activities. The success of both of these programs hinges on a laboratory culture that understands and accepts the rationale for biosafety and biosecurity programs and the corresponding management oversight.

In some cases, biosecurity practices may conflict with biosafety practices, requiring personnel and management to devise policies that accommodate both sets of objectives. For example, signage may present a conflict between the two programs. Standard biosafety practice requires that signage be posted on laboratory doors to alert people to the hazards that may be present within the laboratory. The biohazard sign normally includes the name of the agent, specific hazards associated with the use or handling of the agent and contact information for the investigator. These practices may conflict with security objectives. Therefore, biosafety and biosecurity considerations must be balanced and proportional to the identified risks when developing institutional policies.

Designing a biosecurity program that does not jeopardize laboratory operations or interfere with the conduct of research requires a familiarity with microbiology and the materials that require protection. Protecting pathogens and other sensitive biological materials while preserving the free exchange of research materials and information may present significant institutional challenges. Therefore, a combination or tiered approach to protecting biological materials, commensurate with the identified risks, often provides the best resolution to conflicts that may arise. However, in the absence of legal requirements for a biosecurity program, the health and safety of laboratory personnel and the surrounding environment should take precedence over biosecurity concerns.

Risk Management Methodology

A risk management methodology can be used to identify the need for a biosecurity program. A risk management approach to laboratory biosecurity 1) establishes which, if any, agents require biosecurity measures to prevent loss, theft, diversion, or intentional misuse, and 2) ensures that the protective measures provided, and the costs associated with that protection, are proportional to the risk. The need for a biosecurity program should be based on the possible impact of the theft, loss, diversion, or intentional misuse of the materials, recognizing that different agents and toxins will pose different levels of risk. Resources are not infinite. Biosecurity policies and procedures should not seek to protect against every conceivable risk. The risks need to be identified, prioritized and resources allocated based on that prioritization. Not all institutions will rank the same agent at the same risk level. Risk management methodology takes into consideration available institutional resources and the risk tolerance of the institution.

Developing a Biosecurity Program

Management, researchers and laboratory supervisors must be committed to being responsible stewards of infectious agents and toxins. Development of a biosecurity program should be a collaborative process involving all stakeholders. The stakeholders include but are not limited to: senior management; scientific staff; human resource officials; information technology staff; and safety, security and engineering officials. The involvement of organizations and/or personnel responsible for a facility's overall security is critical because many potential biosecurity measures may already be in place as part of an existing safety or security program. This coordinated approach is critical in ensuring that the biosecurity program provides reasonable, timely and cost-effective solutions addressing the identified security risks without unduly affecting the scientific or business enterprise or provision of clinical and/or diagnostic services.

The need for a biosecurity program should reflect sound risk management practices based on a site-specific risk assessment. A biosecurity risk assessment should analyze the probability and consequences of loss, theft and potential misuse of pathogens and toxins.[7] Most importantly, the biosecurity risk assessment should be used as the basis for making risk management decisions.

Example Guidance: A Biosecurity Risk Assessment and Management Process

Different models exist regarding biosecurity risk assessment. Most models share common components such as asset identification, threat, vulnerability and mitigation. What follows is one example of how a biosecurity risk assessment may be conducted. In this example, the entire risk assessment and risk management process may be divided into five main steps, each of which can be further subdivided: 1) identify and prioritize biologicals and/or toxins; 2) identify and prioritize the adversary/threat to biologicals and/or toxins; 3) analyze the risk of specific security scenarios; 4) design and develop an overall risk management program; and 5) regularly evaluate the institution's risk posture and protection objectives. Example guidance for these five steps is provided below.

Step 1: Identify and Prioritize Biological Materials

- Identify the biological materials that exist at the institution, form of the material, location and quantities, including non-replicating materials (i.e., toxins).

- Evaluate the potential for misuse of these biologic materials.

- Evaluate the consequences of misuse of these biologic materials.

- Prioritize the biologic materials based on the consequences of misuse (i.e., risk of malicious use).

At this point, an institution may find that none of its biologic materials merit the development and implementation of a separate biosecurity program or the existing security at the facility is adequate. In this event, no additional steps would need to be completed.

Step 2: Identify and Prioritize the Threat to Biological Materials

- Identify the types of "Insiders" who may pose a threat to the biologic materials at the institution.

- Identify the types of "Outsiders" (if any) who may pose a threat to the biologic materials at the institution.

- Evaluate the motive, means, and opportunity of these various potential adversaries.

Step 3: Analyze the Risk of Specific Security Scenarios

- Develop a list of possible biosecurity scenarios, or undesired events that could occur at the institution (each scenario is a combination of an agent, an adversary, and an action). Consider:

  - access to the agent within your laboratory;
  - how the undesired event could occur;
  - protective measures in place to prevent occurrence;
  - how the existing protection measures could be breached (i.e., vulnerabilities).

- Evaluate the probability of each scenario materializing (i.e., the likelihood) and its associated consequences. Assumptions include:

  - although a wide range of threats are possible, certain threats are more probable than others;
  - all agents/assets are not equally attractive to an adversary; valid and credible threats, existing precautions, and the potential need for select enhanced precautions are considered.

- Prioritize or rank the scenarios by risk for review by management.

Step 4: Develop an Overall Risk Management Program

- Management commits to oversight, implementation, training and maintenance of the biosecurity program.

- Management develops a biosecurity risk statement, documenting which biosecurity scenarios represent an unacceptable risk and must be mitigated versus those risks appropriately handled through existing protection controls.

- Management develops a biosecurity plan to describe how the institution will mitigate those unacceptable risks including:

  - a written security plan, standard operating procedures, and incident response plans;
  - written protocols for employee training on potential hazards, the biosecurity program and incident response plans.

- Management ensures necessary resources to achieve the protection measures documented in the biosecurity plan.

Step 5: Re-evaluate the Institution's Risk Posture and Protection Objectives

- Management regularly reevaluates and makes necessary modifications to the:

  - biosecurity risk statement;
  - biosecurity risk assessment process;
  - the institution's biosecurity program/plan;
  - the institution's biosecurity systems.

- Management assures the daily implementation, training and annual re-evaluation of the security program.

Elements of a Biosecurity Program

Many facilities may determine that existing safety and security programs provide adequate mitigation for the security concerns identified through biosecurity risk assessment. This section offers examples and suggestions for components of a biosecurity program should the risk assessment reveal that further protections may be warranted. Program components should be site-specific and based upon organizational threat/vulnerability assessment and as determined appropriate by facility management. Elements discussed below should be implemented, as needed, based upon the risk assessment process. They should not be construed as "minimum requirements" or "minimum standards" for a biosecurity program.

Program Management

If a biosecurity plan is implemented, institutional management must support the biosecurity program. Appropriate authority must be delegated for implementation and the necessary resources provided to assure program goals are being met. An organizational structure for the biosecurity program that clearly defines the chain of command, roles, and responsibilities should be distributed to the staff. Program management should ensure that biosecurity plans are created, exercised, and revised as needed. The biosecurity program should be integrated into relevant institutional policies and plans.

Physical Security—Access Control and Monitoring

The physical security elements of a laboratory biosecurity program are intended to prevent the removal of assets for non-official purposes. An evaluation of the physical security measures should include a thorough review of the building and premises, the laboratories, and biological material storage areas. Many requirements for a biosecurity plan may already exist in a facility's overall security plan.

Access should be limited to authorized and designated employees based on the need to enter sensitive areas. Methods for limiting access could be as simple as locking doors or having a card key system in place. Evaluations of the levels of access should consider all facets of the laboratory's operations and programs (e.g., laboratory entrance requirements, freezer access). The need for entry by visitors, laboratory workers, management officials, students, cleaning/ maintenance staff, and emergency response personnel should be considered.

Personnel Management

Personnel management includes identifying the roles and responsibilities for employees who handle, use, store and transport dangerous pathogens and/or other important assets. The effectiveness of a biosecurity program against identified threats depends, first and foremost, on the integrity of those individuals who have access to pathogens, toxins, sensitive information and/or other assets. Employee screening policies and procedures are used to help evaluate these individuals. Policies should be developed for personnel and visitor identification, visitor management, access procedures, and reporting of security incidents.

Inventory and Accountability

Material accountability procedures should be established to track the inventory, storage, use, transfer and destruction of dangerous biological materials and assets when no longer needed. The objective is to know what agents exist at a facility, where they are located, and who is responsible for them. To achieve this, management should define: 1) the materials (or forms of materials) subject to accountability measures; 2) records to be maintained, update intervals and timelines for record maintenance; 3) operating procedures associated with inventory maintenance (e.g., how material is identified, where it can be used and stored); and 4) documentation and reporting requirements.

It is important to emphasize that microbiological agents are capable of replication and are often expanded to accommodate the nature of the work involving their use. Therefore, knowing the exact "working" quantity of organisms at any given time may be impractical. Depending on the risks associated with a pathogen or toxin, management can designate an individual who is accountable, knowledgeable about the materials in use, and responsible for security of the materials under his or her control.

Information Security

Policies should be established for handling sensitive information associated with the biosecurity program. For the purpose of these policies, "sensitive information" is that which is related to the security of pathogens and toxins, or other critical infrastructure information. Examples of sensitive information may include facility security plans, access control codes, agent inventories and storage locations. Discussion of information security in this section does not pertain to information which has been designated "classified" by the United States pursuant to Executive Order 12958, as amended, and is governed by United States law or to research-related information which is typically unregulated or unrestricted through the peer review and approval processes.

The objective of an information security program is to protect information from unauthorized release and ensure that the appropriate level of confidentiality is preserved. Facilities should develop policies that govern the identification, marking and handling of sensitive information. The information security program should be tailored to meet the needs of the business environment, support the mission of the organization, and mitigate the identified threats. It is critical that access to sensitive information be controlled. Policies for properly identifying and securing sensitive information including electronic files and removable electronic media (e.g., CDs, computer drives) should be developed.

Transport of Biological Agents

Material transport policies should include accountability measures for the movement of materials within an institution (e.g., between laboratories, during shipping and receiving activities) and outside of the facility (e.g., between institutions or locations). Transport policies should address the need for appropriate documentation and material accountability and control procedures for pathogens in transit between locations. Transport security measures should be instituted to ensure that appropriate authorizations have been received and that adequate communication between facilities has occurred before, during, and after transport of pathogens or other potentially hazardous biological materials. Personnel should be adequately trained and familiar with regulatory and institutional procedures for proper containment, packaging, labeling, documentation and transport of biological materials.

Accident, Injury and Incident Response Plans

Laboratory security policies should consider situations that may require emergency responders or public safety personnel to enter the facility in response to an accident, injury or other safety issue or security threat. The preservation of human life, the safety and health of laboratory employees and the surrounding community must take precedence in an emergency over biosecurity concerns. Facilities are encouraged to coordinate with medical, fire, police and other emergency officials when preparing emergency and security breach response plans. Standard Operation Procedures (SOPs) should be developed that minimize the potential exposure of responding personnel to potentially hazardous biological materials. Laboratory emergency response plans should be integrated with relevant facility-wide or site-specific security plans. These plans should also consider such adverse events as bomb threats, natural disasters and severe weather, power outages, and other facility emergencies that may introduce security threats.

Reporting and Communication

Communication is an important aspect of a biosecurity program. A "chain-of-notification" should be established in advance of an actual event. This communication chain should include laboratory and program officials, institution management, and any relevant regulatory or public authorities. The roles and responsibilities of all involved officials and programs should be clearly defined. Policies should address the reporting and investigation of potential security breaches (e.g., missing biological agents, unusual or threatening phone calls, unauthorized personnel in restricted areas).

Training and Practice Drills

Biosecurity training is essential for the successful implementation of a biosecurity program. Program management should establish training programs that inform and educate individuals regarding their responsibilities within the laboratory and the institution. Practice drills should address a variety of scenarios such as loss or theft of materials, emergency response to accidents and injuries, incident reporting and identification of and response to security breaches. These scenarios may be incorporated into existing emergency response drills such as fire drills or building evacuation drills associated with bomb threats. Incorporating biosecurity measures into existing procedures and response plans often provides efficient use of resources, saves time and can minimize confusion during emergencies.

Security Updates and Re-evaluations

The biosecurity risk assessment and program should be reviewed and updated routinely and following any biosecurity-related incident. Reevaluation is a necessary and on-going process in the dynamic environments of today's biomedical and research laboratories. Biosecurity program managers should develop and conduct biosecurity program audits and implement corrective actions as needed. Audit results and corrective actions should be documented. The appropriate program officials should maintain records.

Select Agents

If an entity possesses, uses or transfers select agents, it must comply with all requirements of the National Select Agent Program. See Appendix F for additional guidance on the CDC and USDA Select Agent Programs (42 CFR Part 73; 7 CFR 331 and 9 CFR 121).

References

1. Richmond, JY, Nesby-O'Dell, SL. Laboratory security and emergency response guidance for laboratories working with select agents. MMWR Recomm Rep. 2002;51:(RR-19):1-6.
2. Laboratory biosafety manual. 3rd ed. Geneva: World Health Organization; 2004.
3. American Biological Safety Association. ABSA biosecurity task force white paper: understanding biosecurity. Illinois: The Association; 2003.
4. Possession, use and transfer of select agents and toxins, 42 CFR Part 73 (2005).
5. Possession, use and transfer of biological agents and toxins, 7 CFR Part 331 (2005).
6. Possession, use and transfer of biological agents and toxins, 9 CFR Part 121 (2005).
7. Casadevall A, Pirofski L. The weapon potential of a microbe. Bethesda, The National Institutes of Health; 2005.

Section VII—Occupational Health and Immunoprophylaxis

The goal of medical support services in a biomedical research setting is to promote a safe and healthy workplace. This is accomplished by limiting opportunities for exposure, promptly detecting and treating exposures, and using information gained from work injuries to further enhance safety precautions. Occupational health and safety in biomedical research settings is a responsibility shared by healthcare providers, safety specialists, principal investigators, employers, and workplace personnel. Optimal worker protection depends on effective, ongoing collaboration among these groups. Supervisors, working with personnel representatives, should describe workers' proposed tasks and responsibilities. First line supervisors and safety professionals should identify the potential worksite health hazards. Principal investigators may serve as subject matter experts. The health provider should design medical support services in consultation with representatives from the institutional environmental health and safety program and the principal investigators. Workers should be fully informed of the available medical support services and encouraged to utilize them. Requisite occupational medical services are described below and expanded discussions of the principles of effective medical support services are available in authoritative texts.[1,2]

Services offered by the medical support team should be designed to be in compliance with United States Department of Labor (DOL), OSHA regulations, patient confidentiality laws, and the Americans with Disabilities Act of 1990.[3-8] Medical support services should be based upon detailed risk assessments and tailored to meet the organization's needs. Risk assessments should define potential hazards and exposures by job responsibility. They should be provided for all personnel regardless of employment status. Contracted workers, students, and visitors should be provided occupational medical care by their employer or sponsor equivalent to that provided by the host institution for exposures, injuries, or other emergencies experienced at the worksite.

Occupational medical services may be provided through a variety of arrangements (e.g., in-house or community based) as long as the service is readily available and allows timely, appropriate evaluation and treatment. The interaction between worker, healthcare provider and employer may be complex, such as a contract worker who uses his own medical provider or uses contract medical services. Thus, plans for providing medical support for workers should be completed before work actually begins. The medical provider must be knowledgeable about the nature of potential health risks in the work environment and have access to expert consultation.

Prevention is the most effective approach to managing biohazards. Prospective workers should be educated about the biohazards to which they may be occupationally exposed, the types of exposures that place their health at risk, the nature and significance of such risks, as well as the appropriate first aid and follow up for potential exposures. That information should be reinforced annually, at the time of any significant change in job responsibility, and following recognized and suspected exposures.[9-11]

Medical support services for biomedical research facilities should be evaluated annually. Joint annual review of occupational injury and illness reports by healthcare providers and environmental health and safety representatives can assist revision of exposure prevention strategies to minimize occupational health hazards that cannot be eliminated.

Occupational Health Support Service Elements

Preplacement Medical Evaluations

Workers who may be exposed to human pathogens should receive a preplacement medical evaluation. Healthcare providers should be cognizant of potential hazards encountered by the worker. A description of the requirements for the position and an understanding of the potential health hazards present in the work environment, provided by the worker's supervisor, should guide the evaluation. The healthcare provider should review the worker's previous and ongoing medical problems, current medications, allergies to medicines, animals, and other environmental proteins, and prior immunizations. With that information, the healthcare provider determines what medical services are indicated to permit the individual to safely assume the duties of the position. Occasionally, it may be useful to review pre-existing medical records to address specific concerns regarding an individual's medical fitness to perform the duties of a specific position. If pre-existing medical records are unavailable or are inadequate, the healthcare provider may need to perform a targeted medical exam. Comprehensive physical examinations are rarely indicated. During the visit, the healthcare provider should inform the worker of potential health hazards in the work area and review steps that should be taken in the event of an accidental exposure. This visit also establishes a link with the medical support services provider.

When occupational exposure to human pathogens is a risk, employers should consider collecting and storing a serum specimen prior to the initiation of work with the agent. It can be used to establish baseline sero-reactivity, should additional blood samples be collected for serological testing subsequent to a recognized or suspected exposure.

Occasionally, it is desirable to determine an individual's vulnerability to infection with specific agents prior to assigning work responsibilities. Some occupational exposures present substantially more hazard to identifiable sub-populations of workers. Immunodeficient workers or non-immune pregnant female workers may experience devastating consequences from exposures that pose a chance of risk to pregnant women with prior immunity and other immunocompetent workers (e.g., cytomegalovirus or toxoplasmosis). Serologic testing should be used to document baseline vulnerability to specific infections to which the worker might be exposed, and non-immune workers should be adequately informed about risks. In specific settings, serologic documentation that individual workers have pre-existing immunity to specific infections also may be required for the protection of research animals.[10]

Vaccines

Commercial vaccines should be made available to workers to provide protection against infectious agents to which they may be occupationally exposed.[12-16] The Advisory Committee on Immunization Practices (ACIP) provides expert advice to the Secretary of the DHHS, the Assistant Secretary for Health, and the CDC on the most effective means to prevent vaccine-preventable diseases and to increase the safe usage of vaccines and related biological products. The ACIP develops recommendations for the routine administration of vaccines to pediatric and adult populations, and schedules regarding the appropriate periodicity, dosage, and contraindications. The ACIP is the only entity in the federal government that makes such recommendations. The ACIP is available at the CDC Web site: *www.cdc.gov.*

If the potential consequences of infection are substantial and the protective benefit from immunization is proven, acceptance of such immunization may be a condition for employment. Current, applicable vaccine information statements must be provided whenever a vaccine is administered. Each worker's immunization history should be evaluated for completeness and currency at the time of employment and re-evaluated when the individual is assigned job responsibilities with a new biohazard.

When occupational exposure to highly pathogenic agents is possible and no commercial vaccine is available, it may be appropriate to immunize workers using vaccines or immune serum preparations that are investigational, or for which the specific indication constitutes an off-label use. Use of investigational products, or of licensed products for off-label indications must be accompanied by adequate informed consent outlining the limited availability of information on safety and efficacy. Use of investigational products should occur through Investigational New Drug (IND) protocols providing safety oversight by both the Food and Drug Administration (FDA) and appropriate Institutional Human Subjects Research Protection Committees.[17,18] Recommendation of investigational products, as well as commercial vaccines that are less efficacious, associated with high rates of local or systemic reactions, or that produce increasingly severe reactions with repeated use, should be considered carefully. Receipt of such vaccines is rarely justified as a job requirement.

Investigational vaccines for eastern equine encephalomyelitis (EEE) virus, Venezuelan equine encephalitis (VEE) virus, western equine encephalomyelitis (WEE) virus, and Rift Valley fever viruses (RVFV), may be available in limited quantities and administered on-site at the Special Immunization Program, United States Army Medical Research Institute of Infectious Diseases (USAMRIID).

Periodic Medical Evaluations

Routine, periodic medical evaluations generally are not recommended; however, limited periodic medical evaluations or medical clearances targeted to job requirements may occasionally be warranted (e.g., respirator usage).[3] In special circumstances, it may be appropriate to offer periodic laboratory testing to workers with substantial risk of exposure to infectious agents to detect pre-clinical or sub-clinical evidence for an occupationally acquired infection. Before asymptomatic workers without specific exposures are tested for seroreactivity, the benefit of such testing should be justified, plans for further investigation of indeterminate test results should be delineated, and clearly defined criteria for interpretation of results should be developed.

Medical Support for Occupational Illnesses and Injuries

Workers should be encouraged to seek medical evaluation for symptoms that they suspect may be related to infectious agents in their work area, without fear of reprisal. A high index of suspicion for potential occupational exposures should be maintained during any unexplained illness among workers or visitors to worksites containing biohazards. Modes of transmission, as well as the clinical presentation of infections acquired through occupational exposures, may differ markedly from naturally acquired infections. Fatal occupational infections have resulted from apparently trivial exposures. The healthcare provider should have a working understanding of the biohazards present in the workplace and remain alert for subtle evidence of infection and atypical presentations. A close working relationship with the research or clinical program in which the affected employee works is absolutely essential. In the event of injury, consultation between healthcare provider, employee, and the employee's supervisor is required for proper medical management and recordkeeping.

All occupational injuries, including exposures to human pathogens, should be reported to the medical support services provider. Strategies for responding to biohazard exposures should be formulated in advance. Proper post-exposure response is facilitated by exposure-specific protocols that define appropriate first aid, potential post-exposure prophylaxis options, recommended diagnostic tests, and sources of expert medical evaluation. These protocols should address how exposures that occur outside of regular work hours are handled and these protocols should be distributed to potential healthcare providers (e.g., local hospital emergency departments). In exceptional cases, the protocols should be reviewed with state and community public health departments. Emergency medical support training should be provided on a regular basis for both employees and healthcare providers.

The adequacy and timeliness of wound cleansing or other response after an exposure occurs may be the most critical determinant in preventing infection. First aid should be defined, widely promulgated, and immediately available to an injured worker. Barriers to subsequent medical evaluation and treatment should be identified and minimized to facilitate prompt, appropriate care. Laboratory SOPs should include a printed summary of the recommended medical response to specific exposures that can guide immediate response in the work place and that the injured worker can provide to the treating facility. The medical provider's description of the injury should include:

- The potential infectious agent.

- The mechanism and route of exposure (percutaneous, splash to mucous membranes or skin, aerosol, etc.).

- Time and place of the incident.

- Personal protective equipment used at the time of the injury.

- Prior first aid provided (e.g., nature and duration of cleaning and other aid, time that lapsed from exposure to treatment).

- Aspects of the worker's personal medical history relevant to risk of infection or complications of treatment.

First aid should be repeated if the initial adequacy is in question. Healthcare providers must use appropriate barrier precautions to avoid exposure to infectious agents and toxins.

In some instances, it may be possible to prevent or ameliorate illness through post-exposure prophylaxis. Protocols should be developed in advance that clearly identify the situations in which post-exposure prophylaxis are to be considered, the appropriate treatment, and the source of products and expert consultation. Accurate quantification of risk associated with all exposures is not possible, and the decision to administer post-exposure prophylaxis may have to be made quickly and in the absence of confirmatory laboratory testing. Post-exposure regimens may involve off-label use of licensed products (e.g., use of smallpox vaccine for workers exposed to monkeypox) in settings where there is insufficient experience to provide exact guidance on the safety or likely protective efficacy of the prophylactic regimen. Thus, protocols should exist that delineate the circumstances under which it would be appropriate to consider use of each product following exposure, as well as the limits of our understanding of the value of some post-exposure interventions. In these cases, consultations with subject matter experts are especially useful.

Estimating the significance of an exposure may be difficult, despite having established protocols. The clinician may need to make a "best-estimate" based upon knowledge of similar agents, exposure circumstances, and advice received from knowledgeable experts. Appropriate post-exposure prophylactic response is always pathogen and exposure dependent, and may be host-factor dependent and influenced by immediate post-exposure management. Before prophylactic treatment is undertaken, confirm the likelihood that an exposure occurred, that prophylaxis is indicated and is not contraindicated by past medical history. Conveying this information to the injured worker requires clear, honest communication. The clinical risk assessment and treatment decision process should be carefully explained, the worker's questions addressed with relevant, preprinted educational materials provided. Prompt treatment should be provided, with a mutually agreed plan to follow the individual's clinical course.

The applicable workers compensation claim form should be provided with appropriate explanations for its completion. The supervisor must receive a description of the accident or incident, confirm the circumstances of the injury or exposure and provide relevant advice. The report also should be distributed to all other relevant parties, such as the safety professional. Each incident should receive prompt reconsideration of the initial risk assessment and reevaluation of current strategies to reduce the possibility of future exposures.

Post-exposure serologic testing may be useful, but it is important to determine how information obtained from serologic testing will be interpreted. It is also essential to collect serum specimens at the appropriate interval for a given situation. Assessment of sero-reactivity in exposed workers is most helpful when the results of specimens collected over time can be compared. Ideally specimens collected prior to, at the time of and several weeks following exposure, should be tested simultaneously and results compared to assess changes in the pattern of sero-reactivity. Serum collected too early after exposure may fail to react even when infection has occurred, because antibodies have not yet been produced in detectable quantities. When immediate institution of post-exposure prophylaxis may delay seroconversion, or when the agent to which the worker was exposed results in seroconversion completed over months (e.g., retroviruses), testing of specimens collected late after exposure is particularly important.

Testing of a single serum specimen is generally discouraged and can result in misinterpretation of nonspecific sero-reactivity. Evidence of sero-conversion or a significant ($\geq$ 4 fold) increase in titer associated with a compatible clinical syndrome is highly suggestive of acute infection.

However, the significance of and appropriate response to sero-conversion in the absence of illness is not always clear. If sero-reactivity is evident in the earliest specimen, it is important to re-test that specimen in tandem with serum specimens archived prior to occupational exposure and/or collected serially over time to investigate whether a change in titer suggestive of new infection can be identified.

In some exposure situations, it may be appropriate to store serially collected serum samples, and to send them for testing as evidence of seroconversion only if symptoms develop that suggest an infection may have occurred (e.g., Monkey B virus exposures). Serum collected at the time of employment, and any other specimens not immediately tested should be stored frozen at a temperature of -20° C or lower in a freezer that does not experience freeze-thaw cycles. An inventory system should be established to ensure the accurate and timely retrieval of samples, while protecting patient privacy.

When investigational or other non-commercial assays are utilized, the importance of appropriate controls and the ability to compare serially collected specimens for quantification/characterization of reactivity is increased. The availability of aliquoted samples that allow additional testing may be essential to assist interpretation of ambiguous results. Caution should be taken to avoid placing more confidence in testing outcomes than can be justified by the nature of the assays.

Occupational Health in the BSL-4 Setting

Work with BSL-4 agents involves special challenges for occupational health. Infections of laboratory staff by such agents may be expected to result in serious or lethal disease for which limited treatment options exist. In addition, BSL-4 agents are frequently geographically exotic to the areas in which high containment labs are located but produce immediate public health concern if infections occur in laboratory staff. Potential (if unlikely) transmission from infected staff into the human or animal populations in the areas surrounding the laboratories may raise such concerns to higher levels. Thus, SOPs for BSL-4 settings require special attention to management of unexplained worker absence, including protocols for monitoring, medical evaluation, work-up, and follow-up of workers with unexplained nonspecific illness. Advance planning for the provision of medical care to workers potentially infected with BSL-4 agents is a fundamental component of an occupational health program for a BSL-4 facility.

References

1. Menckel E, Westerholm P, editors. Evaluation in occupational health practice. Oxford: Butterworth-Heinemann; 1999.
2. Levy B, Wegman DH, editors. Occupational health: recognizing and preventing work-related disease and injury, 4th ed. Philadelphia: Lippincott Williams & Wilkins; 2000.
3. Occupational Safety and Health Administration (www.osha.gov). Washington, DC: The Administration. Applicable OSHA Laws, Regulations and Interpretations; [about 2 screens]. Available at: www.osha.gov/comp-links.html.

4. American Medical Association (*www.ama-assn.org*). Chicago: The Association; 1995-2006 [updated 2005 Mar 7]. Patient Confidentiality; [about 6 screens] Available at: *www.ama-assn.org/ama/pub/category/4610.html*.
5. United States Department of Justice (*www.usdoj.gov*). Washington, DC: The Department; [updated 2006 July 28]. American Disabilities Act; [about seven screens] Available at: *http://www.usdoj.gov/crt/ada/adahom1.htm*.
6. Occupational Safety and Health Act of 1970, Pub. L. No. 91-596, 84 Stat. 1590 (December 29, 1970).
7. Non-mandatory Compliance Guidelines for Hazard Assessment and Personal Protective Equipment Selection, 29 CFR Part 1910 Subpart I App B (1994).
8. Multi-Employer Citation Policy, C.P.L. 02-00-124 (1999).
9. National Research Council (US), Committee on Occupational Safety and Health in Research Animal Facilities, Institute of Laboratory Animal Resources, Commission on Life Sciences. Occupational health and safety in the care and use of research animals. Washington, DC: The National Academies Press; 1997.
10. National Research Council (US), Committee on Occupational Health and Safety in the Care and Use of Nonhuman Primates, Institute for Laboratory Animal Research, Division on Earth and Life Studies. Occupational health and safety in the care and use of nonhuman primates. Washington, DC: National Academy Press; 2003.
11. Cohen JI, Davenport DS, Stewart JA, et al. Recommendations for prevention of and therapy for exposure to B virus (cercopithecine herpesvirus 1). Clin Infect Dis 2002;35:1191-203.
12. Atkinson WL, Pickering LK, Schwartz B, et al. General recommendations on immunization. Recommendations of the Advisory Committee on Immunization Practices (ACIP) and the American Academy of Family Physicians (AAFP). MMWR Recomm Rep. 2002;51:1-35.
13. Centers for Disease Control and Prevention. Update on adult immunization. Recommendations of the immunization practices advisory committee (ACIP). MMWR Recomm Rep. 1991;40:1-94.
14. Centers for Disease Control and Prevention. Immunization of health-care workers: recommendations of the Advisory Committee on Immunization Practices (ACIP) and the Hospital Infection Control Practices Advisory Committee (HICPAC). MMWR Morb Mortal Wkly Rep. 1997;46:1-42.
15. Centers for Disease Control and Prevention. Use of vaccines and immune globulins in persons with altered immunocompetence. Advisory Committee on Immunization Practices (ACIP). MMWR Morb Mortal Wkly Rep. 1993 Apr 9;42(RR-04):1-18.
16. Centers for Disease Control and Prevention. Update: vaccine side effects, adverse reactions, contraindications, and precautions. Recommendations of the Advisory Committee on Immunization Practices (ACIP). MMWR Recomm Rep. 1996 Sep 6; 45(RR-12):1-35. Erratum in: MMWR Morb Mortal Wkly Rep. 1997;46:227.

17. United States Department of Health and Human Services (*www.dhhs.gov*). Washington, DC: The Department; [updated 2006 Mar 6]. Office of Human Research Protections Guidance; [about two screens]. Available at: *http://www.hhs.gov/ohrp*.

18. United States Food and Drug Administration (*www.fda.gov*). Rockville, MD; The Administration; [updated 2005 June 13]. Information on Submitting an Investigational New Drug Application for a Biological Product [about three screens] Available at: *http://www.fda.gov/cber/ind/ind.htm*.

Section VIII—Agent Summary Statements

Section VIII-A: Bacterial Agents

Bacillus anthracis

Bacillus anthracis, a gram-positive, non-hemolytic, and non-motile bacillus, is the etiologic agent of anthrax, an acute bacterial disease of mammals, including humans. Like all members of the genus *Bacillus*, under adverse conditions *B. anthracis* has the ability to produce spores that allow the organism to persist for long periods until the return of more favorable conditions. Reports of suspected anthrax outbreaks date back to as early as 1250 BC. The study of anthrax and *B. anthracis* in the 1800s contributed greatly to our general understanding of infectious diseases. Much of Koch's postulates were derived from work on identifying the etiologic agent of anthrax. Louis Pasteur developed the first attenuated live vaccine for anthrax.

Most mammals are susceptible to anthrax; it mostly affects herbivores that ingest spores from contaminated soil and, to a lesser extent, carnivores that scavenge on the carcasses of diseased animals. Anthrax still occurs frequently in parts of central Asia and Africa. In the United States, it occurs sporadically in animals in parts of the West, Midwest and Southwest.

The infectious dose varies greatly from species to species and is route-dependent. The inhalation anthrax infectious dose (ID) for humans primarily has been extrapolated from inhalation challenges of nonhuman primates (NHP) or studies done in contaminated mills. Estimates vary greatly but the medium lethal dose (LD_{50}) is likely within the range of 2,500-55,000 spores.[1] It is believed that very few spores (10 or less) are required for cutaneous anthrax.[2]

Occupational Infections

Occupational infections are possible when in contact with contaminated animals, animal products or pure cultures of *B. anthracis*, and may include ranchers, veterinarians and laboratory workers. Numerous cases of laboratory-associated anthrax (primarily cutaneous) have been reported.[3,4] Recent cases include suspected cutaneous anthrax in a laboratory worker in Texas and a cutaneous case in a North Dakota male who disposed of five cows that died of anthrax.[5,6]

Natural Modes of Infection

The clinical forms of anthrax in humans that result from different routes of infection are: 1) cutaneous (via broken skin); 2) gastrointestinal (via ingestion); and 3) inhalation anthrax. Cutaneous anthrax is the most common and readily treatable form of the disease. Inhalation anthrax used to be known as "Woolsorter disease" due to its prevalence in textile mill workers handling wool and other contaminated animal products. While naturally occurring disease is no longer a significant public health problem in the United States, anthrax has become a bioterrorism concern. In 2001, 22 people were diagnosed with anthrax acquired from spores sent through the mail, including 11 cases of inhalation anthrax with five deaths and 11 cutaneous cases.[7]

Laboratory Safety and Containment Recommendations

B. anthracis may be present in blood, skin lesion exudates, cerebrospinal fluid, pleural fluid, sputum, and rarely, in urine and feces. The primary hazards to laboratory personnel are: direct and indirect contact of broken skin with cultures and contaminated laboratory surfaces, accidental parenteral inoculation and, rarely, exposure to infectious aerosols. Efforts should be made to avoid production of aerosols by working with infectious organisms in a BSC. In addition, all centrifugation should be done using aerosol-tight rotors that are opened within the BSC after each run.

BSL-2 practices, containment equipment, and facilities are recommended for activities using clinical materials and diagnostic quantities of infectious cultures. ABSL-2 practices, containment equipment and facilities are recommended for studies utilizing experimentally infected laboratory rodents. BSL-3 practices, containment equipment, and facilities are recommended for work involving production quantities or high concentrations of cultures, screening environmental samples (especially powders) from anthrax-contaminated locations, and for activities with a high potential for aerosol production. Workers who frequently centrifuge *B. anthracis* suspensions should use autoclavable aerosol-tight rotors. In addition, regular routine swabbing specimens for culture should be routinely obtained inside the rotor and rotor lid and, if contaminated, rotors should be autoclaved before re-use.

Special Issues

Vaccines A licensed vaccine for anthrax is available. Guidelines for its use in occupational settings are available from the ACIP.[8,9] Worker vaccination is recommended for activities that present an increased risk for repeated exposures to *B. anthracis* spores including: 1) work involving production quantities with a high potential for aerosol production; 2) handling environmental specimens, especially powders associated with anthrax investigations; 3) performing confirmatory testing for *B. anthracis*, with purified cultures; 4) making repeated entries into known *B. anthracis*-spore-contaminated areas after a terrorist attack; 5) work in other settings in which repeated exposure to aerosolized *B. anthracis* spores might occur. Vaccination is not recommended for workers involved in routine processing of clinical specimens or environmental swabs in general diagnostic laboratories.

Select Agent *B. anthracis* is a select agent requiring registration with CDC and/or USDA for possession, use, storage and/or transfer. See Appendix F for additional information.

Transfer of Agent Importation of this agent may require CDC and/or USDA importation permits. Domestic transport of this agent may require a permit from USDA/APHIS/VS. A Department of Commerce (DoC) permit may be required for the export of this agent to another country. See Appendix C for additional information.

Bordetella pertussis

Bordetella pertussis, an exclusively human respiratory pathogen of worldwide distribution, is the etiologic agent of whooping cough or pertussis. The organism is a fastidious, small gram-negative coccobacillus that requires highly specialized culture and transport media for cultivation in the laboratory. Its natural habitat is the human respiratory tract.

Occupational Infections

Occupational transmission of pertussis has been reported, primarily among healthcare workers.[10-16] Outbreaks, including secondary transmission, among workers have been documented in hospitals, long-term care institutions, and laboratories. Nosocomial transmissions have been reported in healthcare settings. Laboratory-acquired pertussis has been documented.[17,18]

Natural Modes of Infection

Pertussis is highly communicable, with person-to-person transmission occurring via aerosolized respiratory secretions containing the organism. The attack rate among susceptible hosts is affected by the frequency, proximity, and time of exposure to infected individuals. Although the number of reported pertussis cases declined by over 99% following the introduction of vaccination programs in the 1940s, the 3- to 4-year cycles of cases have continued into the post-vaccination era.[19-21]

Laboratory Safety and Containment Recommendations

The agent may be present in high levels in respiratory secretions, and may be found in other clinical material, such as blood and lung tissue in its infrequent manifestation of septicemia and pneumonia, respectively.[22,23] Because the natural mode of transmission is via the respiratory route, aerosol generation during the manipulation of cultures and contaminated clinical specimens generates the greatest potential hazard.

BSL-2 practices, containment equipment, and facilities are recommended for all activities involving the use or manipulation of known or potentially infectious clinical material and cultures. ABSL-2 practices and containment equipment should be employed for housing experimentally infected animals. Primary containment devices and equipment, including biological safety cabinets, safety centrifuge cups or safety centrifuges should be used for activities likely to generate potentially infectious aerosols. BSL-3 practices, containment equipment, and facilities are appropriate for production operations.

Special Issues

Vaccines Pertussis vaccines are available but are not currently approved or recommended for use in persons over six years of age. Because this recommendation may change in the near future, the reader is advised to review the current recommendations of the ACIP published in the Morbidity and Mortality Weekly Report (MMWR) and at the CDC Vaccines and Immunizations Web site for the latest recommendations for adolescents and adults.

Transfer of Agent Importation of this agent may require CDC and/or USDA importation permits. Domestic transport of this agent may require a permit from USDA/APHIS/VS. A DoC permit may be required for the export of this agent to another country. See Appendix C for additional information.

Brucella species

The genus *Brucella* consists of slow-growing, very small gram-negative coccobacilli whose natural hosts are mammals. Seven *Brucella* species have been described using epidemiologic and biological characteristics, although at the genetic level all brucellae are closely related. *B. melitensis* (natural host: sheep/goats), *B suis* (natural host: swine), *B. abortus* (natural host: cattle), *B. canis* (natural host: dogs), and *B. "maris"* (natural host: marine mammals) have caused illness in humans exposed to the organism including laboratory personnel.[24,25] Hypersensitivity to *Brucella* antigens is a potential but rare hazard to laboratory personnel. Occasional hypersensitivity reactions to *Brucella* antigens occur in workers exposed to experimentally and naturally infected animals or their tissues.

Occupational Infections

Brucellosis has been one of the most frequently reported laboratory infections in the past and cases continue to occur.[4,26-28] Airborne and mucocutaneous exposures can produce LAI. Accidental self-inoculation with vaccine strains is an occupational hazard for veterinarians.

Natural Modes of Infection

Brucellosis (Undulant fever, Malta fever, Mediterranean fever) is a zoonotic disease of worldwide occurrence. Mammals, particularly cattle, goats, swine, and sheep act as reservoirs for brucellae. Multiple routes of transmission have been identified, including direct contact with infected animal tissues or products, ingestion of contaminated milk, and airborne exposure in pens and stables.

Laboratory Safety and Containment Recommendations

Brucella infects the blood and a wide variety of body tissues, including cerebral spinal fluid, semen, pulmonary excretions, placenta, and occasionally urine. Most laboratory-associated cases occur in research facilities and involve exposures to *Brucella* organisms grown in large quantities or exposure to placental tissues containing *Brucella*. Cases have occurred in clinical laboratory settings from sniffing bacteriological cultures[29] or working on open bench tops.[30] Aerosols from, or direct skin contact with, cultures or with infectious clinical specimens from animals (e.g., blood, body fluids, tissues) are commonly implicated in human infections. Aerosols generated during laboratory procedures have caused multiple cases per exposure.[30,31] Mouth pipetting, accidental parenteral inoculations, and sprays into eyes, nose and mouth result in infection. The infectious dose of *Brucella* is 10-100 organisms by aerosol route and subcutaneous route in laboratory animals.[32,33]

BSL-2 practices, containment equipment, and facilities are recommended for routine clinical specimens of human or animal origin. Products of conception containing or believed to contain pathogenic *Brucella* should be handled with BSL-3 practices due to the high concentration of organisms per gram of tissue. BSL-3 and ABSL-3 practices, containment equipment, and facilities are recommended, for all manipulations of cultures of pathogenic *Brucella* spp. listed in this summary, and for experimental animal studies.

Special Issues

Vaccines Human *Brucella* vaccines have been developed and tested in other countries with limited success. A human vaccine is not available in the United States.[34]

Select Agent *Brucella abortus, Brucella melitensis,* and *Brucella suis* are select agents requiring registration with CDC and/or USDA for possession, use, storage and/or transfer. See Appendix F for additional information.

Transfer of Agent Importation of this agent may require CDC and/or USDA importation permits. Domestic transport of this agent may require a permit from USDA/APHIS/VS. A DoC permit may be required for the export of this agent to another country. See Appendix C for additional information.

Burkholderia mallei

Burkholderia mallei (formerly *Pseudomonas mallei*) is a non-motile gram-negative rod associated with glanders, a rare disease of equine species and humans. While endemic foci of infection exist in some areas of the world, glanders due to natural infection is extremely rare in he United States.

Occupational Infections

Glanders occurs almost exclusively among individuals who work with equine species and/or handle *B. mallei* cultures in the laboratory. *B. mallei* can be very infectious in the laboratory setting. The only reported case of human glanders in the United States over the past 50 years resulted from a laboratory exposure.[35] Modes of transmission may include inhalation and/or mucocutaneous exposure.

Natural Mode of Infection

Glanders is a highly communicable disease of horses, goats, and donkeys. Zoonotic transmission occurs to humans, but person-to-person transmission is rare. Clinical glanders no longer occurs in the Western Hemisphere or in most other areas of the world, although enzootic foci are thought to exist in Asia and the eastern Mediterranean.[36] Clinical infections in humans are characterized by tissue abscesses and tend to be very serious.

Laboratory Safety and Containment Recommendations

B. mallei can be very hazardous in a laboratory setting. In a pre-biosafety era report, one-half of the workers in a *B. mallei* research laboratory were infected within a year of working with the organism.[37] Laboratory-acquired infections have resulted from aerosol and cutaneous exposure.[37,38] Laboratory infections usually are caused by exposure to bacterial cultures rather than to clinical specimens. Workers should take precautions to avoid exposure to aerosols from bacterial cultures, and to tissues and purulent drainage from victims of this disease.

Primary isolations from patient fluids or tissues may be performed with BSL-2 practices, containment equipment, and facilities in a BSC. Procedures must be performed under BSL-3 containment whenever infectious aerosols or droplets are generated, such as during centrifugation or handling infected animals, or when large quantities of the agent are produced. Procedures conducted outside of a BSC (centrifugation, animal manipulation, etc.) that generate infectious aerosols require respiratory protection. Sealed cups should be used with all centrifuges and these should be opened only inside a BSC. Gloves should be worn when working with potentially infectious material or animals. Animal work with *B. mallei* should be done with ABSL-3 practices, containment equipment, and facilities.

Special Issues

Select Agent *B. mallei* is a select agent requiring registration with CDC and/or USDA for possession, use, storage and/or transfer. See Appendix F for additional information.

Transfer of Agent Importation of this agent may require CDC and/or USDA importation permits. Domestic transport of this agent may require a permit from

USDA/APHIS/VS. A DoC permit may be required for the export of this agent to another country. See Appendix C for additional information.

Burkholderia pseudomallei

Burkholderia pseudomallei (formerly *Pseudomonas pseudomallei*) is a motile gram-negative, oxidase-positive rod that is found in soil and water environments of equatorial regions, including Southeast Asia, Northern Australia, Central America and South America. This organism is the causative agent of melioidosis, an unusual bacterial disease characterized by abscesses in tissues and organs. Victims of the disease frequently exhibit recrudescence months or years after the initial infection.

Occupational Infections

Melioidosis is generally considered to be a disease associated with agriculture; however, *B. pseudomallei* can be hazardous for laboratory workers. There are two reports of melioidosis in laboratory workers who were infected by aerosols or via skin exposure.[39,40] Laboratory workers with diabetes are at increased risk of contracting melioidosis.[41]

Natural Modes of Infection

While primarily a disease found in Southeast Asia and Northern Australia, melioidosis can occasionally be found in the Americas.[42] Natural modes of transmission include the exposure of mucous membranes or damaged skin to soil or water containing the organism, the aspiration or ingestion of contaminated water, or the inhalation of dust from contaminated soil. In endemic areas, 5-20% of agricultural workers have antibody titers to *B. pseudomallei*, in the absence of overt disease.[43]

Laboratory Safety and Containment Recommendations

B. pseudomallei can cause a systemic disease in human patients. Infected tissues and purulent drainage from cutaneous or tissue abscesses can be sources of infection. Blood and sputum also are potential sources of infection.

Work with clinical specimens from patients suspected of having melioidosis and of *B. pseudomallei* cultures may be performed with BSL-2 practices, containment equipment, and facilities. Work should be done in a BSC. Gloves always should be worn when manipulating the microorganism. In cases where infectious aerosols or droplets could be produced, or where production quantities of the organism are generated, these procedures should be confined to BSL-3 facilities with all pertinent primary containment against escape of aerosols. Respiratory protection must be used if the microorganism is manipulated outside of a BSC, such as during centrifugation or handling infected animals. Sealed cups should be used in all centrifuges and these should be opened only in a BSC. Animal studies with this agent should be done at ABSL-3.

Special Issues

Select Agent *B. pseudomallei* is a select agent requiring registration with CDC and/or USDA for possession, use, storage and/or transfer. See Appendix F for additional information.

Transfer of Agent Importation of this agent may require CDC and/or USDA importation permits. Domestic transport of this agent may require a permit from USDA/APHIS/VS. A DoC permit may be required for the export of this agent to another country. See Appendix C for additional information.

Campylobacter (C. jejuni subsp. jejuni, C. coli, C. fetus subsp. fetus, C. upsaliensis)

Campylobacters are curved, S-shaped, or spiral rods associated with gastrointestinal infections (primarily *C. jejuni* subsp. *jejuni* and *C. coli*), bacteremia, and sepsis (primarily *C. fetus* subsp. *fetus* and *C. upsaliensis*). Organisms are isolated from stool specimens using selective media, reduced oxygen tension, and elevated incubation temperature (43°C).

Occupational Infections

These organisms rarely cause LAI, although laboratory-associated cases have been documented.[44-47] Experimentally infected animals also are a potential source of infection.[48]

Natural Modes of Infection

Numerous domestic and wild animals, including poultry, pets, farm animals, laboratory animals, and wild birds are known reservoirs and are a potential source of infection for laboratory and animal care personnel. While the infective dose is not firmly established, ingestion of as few as 500-800 organisms has caused symptomatic infection.[49-51] Natural transmission usually occurs from ingestion of organisms in contaminated food or water and from direct contact with infected pets, farm animals, or infants.[52]

Laboratory Safety and Containment Recommendations

Pathogenic *Campylobacter sp.* may occur in fecal specimens in large numbers. *C. fetus* subsp. *fetus* may also be present in blood, exudates from abscesses, tissues, and sputa. The primary laboratory hazards are ingestion and parenteral inoculation of *C. jejuni*. The significance of aerosol exposure is not known.

BSL-2 practices, containment equipment, and facilities are recommended for activities with cultures or potentially infectious clinical materials. ABSL-2 practices, containment equipment, and facilities are recommended for activities with naturally or experimentally infected animals.

Special Issues

Transfer of Agent Importation of this agent may require CDC and/or USDA importation permits. Domestic transport of this agent may require a permit from USDA/APHIS/VS. A DoC permit may be required for the export of this agent to another country. See Appendix C for additional information.

Chlamydia psittaci (Chlamydophila psittaci), C. trachomatis, C. pneumoniae (Chlamydophila pneumoniae)

Chlamydia psittaci, C. pneumoniae (sometimes called *Chlamydophila psittaci* and *Chlamydophila pneumoniae*) and *C. trachomatis* are the three species of *Chlamydia* known to infect humans. Chlamydiae are nonmotile, gram-negative bacterial pathogens with obligate intracellular life cycles. These three species of *Chlamydia* vary in host spectrum, pathogenicity, and in the clinical spectrum of disease. *C. psittaci* is a zoonotic agent that commonly infects psittacine birds and is highly pathogenic for humans. *C. trachomatis* is historically considered an exclusively human pathogen and is the most commonly reported bacterial infection in the United States. *C. pneumoniae* is considered the least pathogenic species, often resulting in subclinical or asymptomatic infections in both animals and humans.

Occupational Infections

Chlamydial infections caused by *C. psittaci* and *C. trachomatis* lymphogranuloma venereum (LGV) strains were at one time among the most commonly reported laboratory-associated bacterial infections.[26] In cases reported before 1955[4], the majority of infections were psittacosis, and these had the highest case fatality rate of laboratory-acquired infectious agents. The major sources of laboratory-associated psittacosis are contact with and exposure to infectious aerosols in the handling, care, or necropsy of naturally or experimentally infected birds. Infected mice and eggs also are important sources of *C. psittaci*. Most reports of laboratory-acquired infections with *C. trachomatis* attribute the infection to inhalation of large quantities of aerosolized organisms during purification or sonification procedures. Early reports commonly attributed infections to exposure to aerosols formed during nasal inoculation of mice or inoculation of egg yolk sacs and harvest of chlamydial elementary bodies. Infections are associated with fever, chills, malaise, and headache; a dry cough is also associated with *C. psittaci* infection. Some workers exposed to *C. trachomatis* have developed conditions including mediastinal and supraclavicular lymphadenitis, pneumonitis, conjunctivitis, and keratitis.[53] Seroconversion to chlamydial antigens is common and often striking although early antibiotic treatment may prevent an antibody response.

Laboratory-associated infections with *C. pneumoniae* have been reported.[54] Exposed workers were asymptomatic and infection was diagnosed by serology. The route of infection was attributed to inhalation of droplet aerosols created during procedures associated with culture and harvest of the agent from cell culture.

With all species of *Chlamydia*, mucosal tissues in the eyes, nose, and respiratory tract are most often affected by occupational exposures that can lead to infection.

Natural Modes of Infection

C. psittaci is the cause of psittacosis, a respiratory infection that can lead to severe pneumonia requiring intensive care support and possible death. Sequelae include endocarditis, hepatitis, and neurologic complications. Natural infections are acquired by inhaling dried secretions from infected birds. Psittacine birds commonly kept as pets (parrots, parakeets, cockatiels, etc.) and poultry are most frequently involved in transmission. *C. trachomatis* can cause a spectrum of clinical manifestations including genital tract infections, inclusion conjunctivitis, trachoma, pneumonia in infants, and LGV. The LGV strains cause more severe and systemic disease than do genital strains. *C. trachomatis* genital tract infections are sexually transmitted and ocular infections (trachoma) are transmitted by exposure to secretions from infected persons through contact or fomite transmission. *C. pneumoniae* is a common cause of respiratory infection; up to 50% of adults have serologic evidence of previous exposure. Infections with *C. pneumoniae* are transmitted by droplet aerosolization and are most often mild or asymptomatic, although there is a body of evidence associating this agent with chronic diseases such as atherosclerosis and asthma.

Laboratory Safety and Containment Recommendations

C. psittaci may be present in the tissues, feces, nasal secretions and blood of infected birds, and in blood, sputum, and tissues of infected humans. *C. trachomatis* may be present in genital, bubo, and conjunctival fluids of infected humans. Exposure to infectious aerosols and droplets, created during the handling of infected birds and tissues, are the primary hazards to laboratory personnel working with *C. psittaci*. The primary laboratory hazards of *C. trachomatis* and *C. pneumoniae* are accidental parenteral inoculation and direct and indirect exposure of mucous membranes of the eyes, nose, and mouth to genital, bubo, or conjunctival fluids, cell culture materials, and fluids from infected cell cultures or eggs. Infectious aerosols, including those that may be created as a result of centrifuge malfunctions, also pose a risk for infection.

BSL-2 practices, containment equipment, and facilities are recommended for personnel working with clinical specimens and cultures or other materials known or suspected to contain the ocular or genital serovars (A through K) of *C. trachomatis* or *C. pneumoniae*.

BSL-3 practices, containment equipment, and facilities are recommended for activities involving the necropsy of infected birds and the diagnostic examination of tissues or cultures known to contain or be potentially infected with *C. psittaci* strains of avian origin. Wetting the feathers of infected birds with a detergent-disinfectant prior to necropsy can appreciably reduce the risk of aerosols of infected feces and nasal secretions on the feathers and external surfaces of the bird. Activities involving non-avian strains of *C. psittaci* may be performed in a BSL-2 facility as long as BSL-3 practices are followed, including but not limited to the use of primary containment equipment such as BSCs. ABSL-3 practices, containment equipment, and facilities and respiratory protection are recommended for personnel working with naturally or experimentally infected caged birds.

BSL-3 practices and containment equipment are recommended for activities involving work with culture specimens or clinical isolates known to contain or be potentially infected with the LGV serovars (L$_1$ through L$_3$) of *C. trachomatis*. Laboratory work with the LGV serovars of *C. trachomatis* can be conducted in a BSL-2 facility as long as BSL-3 practices are followed when handling potentially infectious materials, including but not limited to use of primary containment equipment such as BSCs.

Gloves are recommended for the necropsy of birds and mice, the opening of inoculated eggs, and when there is the likelihood of direct skin contact with infected tissues, bubo fluids, and other clinical materials.

ABSL-2 practices, containment equipment, and facilities are recommended for activities with animals that have been experimentally infected with genital serovars of *C. trachomatis* or *C. pneumoniae*.

BSL-3 practices, containment equipment, and facilities are indicated for activities involving any of these species with high potential for droplet or aerosol production and for activities involving large quantities or concentrations of infectious materials.

Special Issues

Transfer of Agent Importation of this agent may require CDC and/or USDA importation permits. Domestic transport of this agent may require a permit from USDA/APHIS/VS. A DoC permit may be required for the export of this agent to another country. See Appendix C for additional information.

Neurotoxin-producing Clostridia species

Clostridium botulinum, and rare strains of *C. baratii* and *C. butyricum,* are anaerobic spore-forming species that cause botulism, a life-threatening food-borne illness. The pathogenicity of these organisms results from the production of botulinum toxin, one of the most highly potent neurotoxins currently recognized. Purified botulinum neurotoxin is a 150 kDa protein that acts selectively on peripheral cholinergic nerve endings to block neurotransmitter release.[55] The principal site of action is the neuromuscular junction, where blockade of transmission produces muscle weakness or paralysis. The toxin also acts on autonomic nerve endings where blockade of transmission can produce a variety of adverse effects. The toxin may also contain associated proteins that may increase its size to as high as 900 kDA.

Occupational Infections

There has been only one report of botulism associated with handling of the toxin in a laboratory setting.[56] However, concerns about potential use of the toxin as an agent of bioterrorism or biological warfare have led to increased handling of the substance by investigators studying mechanism of action and/or developing countermeasures to poisoning.[57]

Natural Modes of Infection

Botulinum toxin occurs in seven different serotypes (A to G), but almost all naturally-occurring human illness is due to serotypes A, B, E, and F.[58] Botulism occurs when botulinum toxin is released into circulation following ingestion of preformed toxin. However, animal studies have shown that botulism may occur through inhalation of preformed toxin. Use of appropriate personal protective equipment should prevent potential exposure through mucus membranes. Symptoms and even death are possible by accidental injection of botulinum toxin. Risk to toxin exposure is dependent on both route of exposure and toxin molecular weight size. Exposure to neurotoxin producing Clostridia species does not cause infection; however, in certain rare circumstances (Infant Botulism, Wound Botulism, and Adult colonization), the organism can colonize the intestinal tract and other sites and produce toxin. In Wound Botulism, exposure to toxin is caused by introduction of spores into puncture wounds and *in situ* production by the organism. Infants less than 1 year of age may be susceptible to intestinal colonization and develop the syndrome of Infant Botulism as a result of *in situ* production of toxin. Similarly to Infant Botulism, ingestion of spores by adults with a compromised gastrointestinal tract (GI), such as following GI surgery or long-term administration of antibiotics, may increase risk for intestinal infection and *in situ* production of toxin. See the *C. botulinum* Toxin Agent Summary Statement and Appendix I for additional information.

Laboratory Safety and Containment Recommendations

Neurotoxin producing *Clostridia* species or its toxin may be present in a variety of food products, clinical materials (serum, feces) and environmental samples (soil, surface water).[59] In addition, bacterial cultures may produce very high levels of toxin.[60] In healthy adults, it is typically the toxin and not the organism that causes disease. Risk of laboratory exposure is due to the presence of the toxin and not due to a potential of infection from the organisms that produce the toxin. Although spore-forming, there is no known risk to spore exposure except for the potential for the presence of residual toxin associated with pure spore preparations. Laboratory safety protocols should be developed with the focus on prevention of accidental exposure to the toxin produced by these *Clostridia* species.

BSL-2 practices, containment equipment, and facilities are recommended for activities that involve the organism or the toxin[61] including the handling of potentially contaminated food. Solutions of sodium hypochlorite (0.1%) or sodium hydroxide (0.1N) readily inactivate the toxin and are recommended for decontamination of work surfaces and for spills. Autoclaving of contaminated materials also is appropriate.

Additional primary containment and personnel precautions, such as those recommended for BSL-3, should be implemented for activities with a high potential for aerosol or droplet production, or for those requiring routine handling of larger quantities of the organism or of the toxin. ABSL-2 practices, containment equipment, and facilities are recommended for diagnostic studies and titration of toxin.

Special Issues

Vaccines A pentavalent (A, B, C, D and E) botulinum toxoid vaccine (PBT) is available through the CDC as an Investigational New Drug (IND). Vaccination is recommended for all personnel working in direct contact with cultures of neurotoxin producing Clostridia species or stock solutions of Botulinum neurotoxin. Due to a possible decline in the immunogenicity of available PBT stocks for some toxin serotypes, the immunization schedule for the PBT recently has been modified to require injections at 0, 2, 12, and 24 weeks, followed by a booster at 12 months and annual boosters thereafter. Since there is a possible decline in vaccine efficacy, the current vaccine contains toxoid for only 5 of the 7 toxin types, this vaccine should not be considered as the sole means of protection and should not replace other worker protection measures.

Post-Exposure Treatment An equine antitoxin product is available for treatment of patients with symptoms consistent with botulism. However, due to the risks inherent in equine products, treatment is not provided as a result of exposure unless botulism symptoms are present.

Select Agent Neurotoxin producing *Clostridia* species are select agents requiring registration with CDC and/or USDA for possession, use, storage and/or transfer. See Appendix F for additional information.

Transfer of Agent Importation of this agent may require CDC and/or USDA importation permits. Domestic transport of this agent may require a permit from USDA/APHIS/VS. A DoC permit may be required for the export of this agent to another country. See Appendix C for additional information.

Clostridium tetani and *Tetanus toxin*

Clostridium tetani is an anaerobic endospore-forming gram-positive rod found in the soil and an intestinal tract commensal. It produces a potent neurotoxin, tetanospasmin, which causes tetanus, an acute neurologic condition characterized by painful muscular contractions. Tetanospasmin is an exceedingly potent, high molecular weight protein toxin, consisting of a heavy chain (100kD) subunit that binds the toxin to receptors on neuronal cells and a light chain (50kD) subunit that blocks the release of inhibitory neural transmitter molecules within the central nervous system. The incidence of tetanus in the United States has declined steadily since the introduction of tetanus toxoid vaccines in the 1940's.[62]

Occupational Infections

Although the risk of infection to laboratory personnel is low, there have been five incidents of laboratory personnel exposure recorded.[4]

Natural Modes of Infection

Contamination of wounds by soil is the usual mechanism of transmission for tetanus. Of the 130 cases of tetanus reported to CDC from 1998 through 2000, acute injury (puncture, laceration, abrasion) was the most frequent predisposing condition. Elevated incidence rates also were observed for persons aged over 60 years, diabetics, and intravenous drug users.[63] When introduced into a suitable anaerobic or microaerophilic environment, *C. tetani* spores germinate and produce tetanospasmin. The incubation period ranges from 3 to 21 days. The observed symptoms are primarily associated with the presence of the toxin. Wound cultures are not generally useful for diagnosing tetanus.[64]

Laboratory Safety and Containment Recommendations

The organism may be found in soil, intestinal, or fecal samples. Accidental parenteral inoculation of the toxin is the primary hazard to laboratory personnel. Because it is uncertain if tetanus toxin can be absorbed through mucous membranes, the hazards associated with aerosols and droplets remain unclear.

BSL-2 practices, containment equipment, and facilities are recommended for activities involving the manipulation of cultures or toxin. ABSL-2 practices, containment equipment, and facilities are recommended for animal studies.

Special Issues

Vaccines The vaccination status of workers should be considered in a risk assessment for workers with this organism and/or toxin. While the risk of laboratory-associated tetanus is low, the administration of an adult diphtheria-tetanus toxoid at 10-year intervals further reduces the risk to laboratory and animal care personnel of toxin exposures and wound contamination, and is therefore highly recommended.[62] The reader is advised to consult the current recommendations of the ACIP. [65]

Transfer of Agent Importation of this agent may require CDC and/or USDA importation permits. Domestic transport of this agent may require a permit from USDA/APHIS/VS. A DoC permit may be required for the export of this agent to another country. See Appendix C for additional information.

Corynebacterium diphtheriae

Corynebacterium diphtheriae is a pleomorphic gram-positive rod that is isolated from the nasopharynx and skin of humans. The organism is easily grown in the laboratory on media containing 5% sheep blood. *C. diphtheriae* produces a potent exotoxin and is the causative agent of diphtheria, one of the most wide-spread bacterial diseases in the pre-vaccine era.

Occupational Infections

Laboratory-associated infections with *C. diphtheriae* have been documented, but laboratory animal-associated infections have not been reported.[4,66] Inhalation, accidental parenteral inoculation, and ingestion are the primary laboratory hazards.

Natural Modes of Infection

The agent may be present in exudates or secretions of the nose, throat (tonsil), pharynx, larynx, wounds, in blood, and on the skin. Travel to endemic areas or close contact with persons who have returned recently from such areas, increases risk.[67] Transmission usually occurs via direct contact with patients or carriers, and more rarely, with articles contaminated with secretions from infected people. Naturally occurring diphtheria is characterized by the development of grayish-white membranous lesions involving the tonsils, pharynx, larynx, or nasal mucosa. Systemic sequelae are associated with the production of diphtheria toxin. An effective vaccine has been developed for diphtheria and this disease has become a rarity in countries with vaccination programs.

BSL-2 practices, containment equipment, and facilities are recommended for all activities utilizing known or potentially infected clinical materials or cultures. ABSL-2 facilities are recommended for studies utilizing infected laboratory animals.

Special Issues

Vaccines A licensed vaccine is available. The reader is advised to consult the current recommendations of the CIP.[65] While the risk of laboratory-associated diphtheriai is low, the administration of an adult diphtheria-tentanus toxoid at 10-year intervals may further reduce the risk of illness to laboratory and animal care personnel.[62]

Transfer of Agent Importation of this agent may require CDC and/or USDA importation permits. Domestic transport of this agent may require a permit from USDA/APHIS/VS. A DoC permit may be required for the export of this agent to another country. See Appendix C for additional information.

Francisella tularensis

Francisella tularensis is a small gram-negative coccobacillus that is carried in numerous animal species, especially rabbits, and is the causal agent of tularemia (Rabbit fever, Deer fly fever, Ohara disease, or Francis disease) in humans. *F. tularensis* can be divided into three subspecies, *F. tularensis* (Type A), *F. holarctica* (Type B) and *F. novicida*, based on virulence testing, 16S sequence, biochemical reactions and epidemiologic features. Type A and Type B strains are highly infectious, requiring only 10-50 organisms to cause disease. Subspecies *F. novicida* is infrequently identified as the cause of human disease. Person-to-person transmission of tularemia has not been documented. The incubation period varies with the virulence of the strain, dose and route of introduction but ranges from 1-4 days with most cases exhibiting symptoms in 3-5 days.[68]

Occupational Infections

Tularemia has been a commonly reported laboratory-associated bacterial infection.[4] Most cases have occurred at facilities involved in tularemia research; however, cases have been reported in diagnostic laboratories as well. Occasional cases were linked to work with naturally or experimentally infected animals or their ectoparasites.

Natural Modes of Infection

Tick bites, handling or ingesting infectious animal tissues or fluids, ingestion of contaminated water or food and inhalation of infective aerosols are the primary transmission modes in nature. Occasionally, infections have occurred from bites or scratches by carnivores with contaminated mouthparts or claws.

Laboratory Safety and Containment Recommendations

The agent may be present in lesion exudates, respiratory secretions, cerebrospinal fluid (CSF), blood, urine, tissues from infected animals, fluids from infected animals, and fluids from infected arthropods. Direct contact of skin or mucous membranes with infectious materials, accidental parenteral inoculation, ingestion, and exposure to aerosols and infectious droplets has resulted in infection. Infection has been more commonly associated with cultures than with clinical materials and infected animals.[69]

BSL-2 practices, containment equipment, and facilities are recommended for activities involving clinical materials of human or animal origin suspected or known to contain *F. tularensis*. Laboratory personnel should be informed of the possibility of tularemia as a differential diagnosis when samples are submitted for diagnostic tests. BSL-3 and ABSL-3 practices, containment equipment, and facilities are recommended for all manipulations of suspect cultures, animal necropsies and for experimental animal studies. Preparatory work on cultures or contaminated materials for automated identification systems should be performed at BSL-3. Characterized strains of reduced virulence such as *F. tularensis* Type B (strain LVS) and *F. tularensis* subsp *novicida* (strain U112) can be manipulated in BSL-2. Manipulation of reduced virulence strains at high concentrations should be conducted using BSL-3 practices.

Special Issues

Select Agent *F. tularensis* is a select agent requiring registration with CDC and/or USDA for possession, use, storage and/or transfer. See Appendix F for additional information.

Transfer of Agent Importation of this agent may require CDC and/or USDA importation permits. Domestic transport of this agent may require a permit from USDA/APHIS/VS. A DoC permit may be required for the export of this agent to another country. See Appendix C for additional information.

Helicobacter species

Helicobacters are spiral or curved gram-negative rods isolated from gastrointestinal and hepatobiliary tracts of mammals and birds. There are currently 20 recognized species, including at least nine isolated from humans. Since its discovery in 1982, *Helicobacter pylori* has received increasing attention as an agent of gastritis.[70] The main habitat of *H. pylori* is the human gastric mucosa. Other *Helicobacter* spp. (*H. cinaedi*, *H. canadensis*, *H. canis*, *H. pullorum*, and *H. fennelliae*) may cause asymptomatic infection as well as proctitis, proctocolitis, enteritis and extraintestinal infections in humans.[71,72] *H. cinaedi* has been isolated from dogs, cats and Syrian hamsters.

Occupational Infections

Both experimental and accidental LAI with *H. pylori* have been reported.[73,74] Ingestion is the primary known laboratory hazard. The importance of aerosol exposures is unknown.

Natural Modes of Infection

Chronic gastritis and duodenal ulcers are associated with *H. pylori* infection. Epidemiologic associations have also been made with gastric adenocarcinoma. Human infection with *H. pylori* may be long in duration with few or no symptoms, or may present as an acute gastric illness. Transmission, while incompletely understood, is thought to be by the fecal-oral or oral-oral route.

Laboratory Safety and Containment Recommendations

H. pylori may be present in gastric and oral secretions and stool.[75] The enterohepatic helicobacters (e.g., *H. canadensis*, *H. canis*, *H, cinaedi*, *H. fennelliae*, *H. pullorum*, and *H. winghamensis*) may be isolated from stool specimens, rectal swabs, and blood cultures.[72] Protocols involving homogenization or vortexing of gastric specimens have been described for the isolation of *H. pylori*.[76] Containment of potential aerosols or droplets should be incorporated in these procedures.

BSL-2 practices, containment equipment, and facilities are recommended for activities with clinical materials and cultures known to contain or potentially contain the agents. ABSL-2 practices, containment equipment, and facilities are recommended for activities with experimentally or naturally infected animals.

Special Issues

Transfer of Agent Importation of this agent may require CDC and/or USDA importation permits. Domestic transport of this agent may require a permit from USDA/APHIS/VS. A DoC permit may be required for the export of this agent to another country. See Appendix C for additional information.

Legionella pneumophila and other *Legionella*-like Agents

Legionella are small, faintly staining gram-negative bacteria. They are obligately aerobic, slow-growing, nonfermentative organisms that have a unique requirement for L-cysteine and iron salts for *in vitro* growth. Legionellae are readily found in natural aquatic bodies and some species (*L. longbeachae*) have been recovered from soil.[77,78] They are able to colonize hot-water tanks at a temperature range from 40 to 50°C. There are currently 48 known *Legionella* species, 20 of which have been associated with human disease. *L. pneumophila* is the species most frequently encountered in human infections.[79-81]

Occupational Infections

Although laboratory-associated cases of legionellosis have not been reported in the literature, at least one case, due to presumed aerosol or droplet exposure during animal challenge studies with *L. pneumophila*, has been recorded.[82] Experimental infections have been produced in guinea pigs, mice, rats, embryonated chicken eggs, and human or animal cell lines.[83] A fatal case of pneumonia due to *L. pneumophila* was diagnosed in a calf, but only 1.7% (2/112) of the other cattle in the herd had serological evidence of exposure to *Legionella*.[84] The disease was linked to exposure to a hot water system colonized with *Legionella*. Animal-to-animal transmission has not been demonstrated.

Natural Modes of Infection

Legionella is commonly found in environmental sources, typically in man-made warm water systems. The mode of transmission from these reservoirs is aerosolization, aspiration or direct inoculation into the airway.[85] Direct person-to-person transmission does not occur. The spectrum of illness caused by *Legionella* species ranges from a mild, self-limited flu-like illness (Pontiac fever) to a disseminated and often fatal disease characterized by pneumonia and respiratory failure (Legionnaires disease). Although rare, *Legionella* has been implicated in cases of sinusitis, cellulitis, pericarditis, and endocarditis.[86] Legionellosis may be either community-acquired or nosocomial. Risk factors include smoking, chronic lung disease, and immunosuppression. Surgery, especially involving transplantation, has been implicated as a risk factor for nosocomial transmission.

Laboratory Safety and Containment Recommendations

The agent may be present in respiratory tract specimens (sputum, pleural fluid, bronchoscopy specimens, lung tissue), and in extrapulmonary sites. A potential hazard may exist for generation of aerosols containing high concentrations of the agent.

BSL-2 practices, containment equipment, and facilities are recommended for all activities involving the use or manipulation of potentially infectious materials, including minimizing the potential for dissemination of the organism from cultures of organisms known to cause disease. ABSL-2 practices, containment equipment and facilities are recommended for activities with experimentally-infected animals. Routine processing of environmental water samples for *Legionella* may be performed with standard BSL-2 practices. For activities likely to produce extensive aerosols and when large quantities of the pathogenic organisms are manipulated, BSL-2 with BSL-3 practices is recommended.

Special Issues

Transfer of Agent Importation of this agent may require CDC and/or USDA importation permits. Domestic transport of this agent may require a permit from USDA/APHIS/VS. A DoC permit may be required for the export of this agent to another country. See Appendix C for additional information.

Leptospira

The genus *Leptospira* is composed of spiral-shaped bacteria with hooked ends. Leptospires are ubiquitous in nature, either free-living in fresh water or associated with renal infection in animals. Historically, these organisms have been classified into pathogenic (*L. interrogans*) and saprophytic (*L. biflexa*) groups, but recent studies have identified more than 12 species based on genetic analysis. These organisms also have been characterized serologically, with more than 200 pathogenic and 60 saprophytic serovars identified as of 2003.[87] These organisms are the cause of leptospirosis, a zoonotic disease of worldwide distribution. Growth of leptospires in the laboratory requires specialized media and culture techniques, and cases of leptospirosis are usually diagnosed by serology.

Occupational Infections

Leptospirosis is a well-documented laboratory hazard. Approximately, 70 LAI and 10 deaths have been reported.[4,26] Direct and indirect contact with fluids and tissues of experimentally or naturally infected mammals during handling, care, or necropsy are potential sources of infection.[88-90] It is important to remember that rodents are natural carriers of leptospires. Animals with chronic renal infection shed large numbers of leptospires in the urine continuously or intermittently, for long periods of time. Rarely, infection may be transmitted by bites of infected animals.[88]

Natural Modes of Infection

Human leptospirosis typically results from direct contact with infected animals, contaminated animal products, or contaminated water sources. Common routes of infection include abrasions, cuts in the skin or via the conjunctiva. Higher rates of infection observed in agricultural workers and other occupations associated with animal contact.

Laboratory Safety and Containment Recommendations

The organism may be present in urine, blood, and tissues of infected animals and humans. Ingestion, accidental parenteral inoculation, and direct and indirect contact of skin or mucous membranes, particularly the conjunctiva, with cultures or infected tissues or body fluids are the primary laboratory hazards. The importance of aerosol exposure is not known.

BSL-2 practices, containment equipment, and facilities are recommended for all activities involving the use or manipulation of known or potentially infective tissues, body fluids, and cultures. The housing and manipulation of infected animals should be performed at ABSL-2. Gloves should be worn to handle and necropsy infected animals and to handle infectious materials and cultures in the laboratory.

Special Issues

Transfer of Agent Importation of this agent may require CDC and/or USDA importation permits. Domestic transport of this agent may require a permit from USDA/APHIS/VS. A DoC permit may be required for the export of this agent to another country. See Appendix C for additional information.

Listeria monocytogenes

Listeria monocytogenes is a gram-positive, non-spore-forming, aerobic bacillus; that is weakly beta-hemolytic on sheep blood agar and catalase-positive.[91] The organism has been isolated from soil, animal feed (silage) and a wide range of human foods and food processing environments. It may also be isolated from symptomatic/asymptomatic animals (particularly ruminants) and humans.[91,92] This organism is the causative agent of listeriosis, a food-borne disease of humans and animals.

Occupational Infections

Cutaneous listeriosis, characterized by pustular or papular lesions on the arms and hands, has been described in veterinarians and farmers.[93] Asymptomatic carriage has been reported in laboratorians.[94]

Natural Modes of Infection

Most human cases of listeriosis result from eating contaminated foods, notably soft cheeses, ready-to-eat meat products (hot dogs, luncheon meats), paté and smoked fish/seafood.[95] Listeriosis can present in healthy adults with symptoms of fever and gastroenteritis, pregnant women and their fetuses, newborns, and persons with impaired immune function are at greatest risk of developing severe infections including sepsis, meningitis, and fetal demise. In pregnant women, *Listeria monocytogenes* infections occur most often in the third trimester and may precipitate labor. Transplacental transmission of *L. monocytogenes* poses a grave risk to the fetus.[92]

Laboratory Safety and Containment Recommendations

Listeria monocytogenes may be found in feces, CSF, and blood, as well as numerous food and environmental samples.[91,92,96,97] Naturally or experimentally infected animals are a source of exposure to laboratory workers, animal care personnel and other animals. While ingestion is the most common route of exposure, *Listeria* can also cause eye and skin infections following direct contact with the organism.

BSL-2 practices, containment equipment, and facilities are recommended when working with clinical specimens and cultures known or suspected to contain the agent. Gloves and eye protection should be worn while handling infected or potentially infected materials. ABSL-2 practices, containment equipment and facilities are recommended for activities involving experimentally or naturally infected animals. Due to potential risks to the fetus, pregnant women should be advised of the risk of exposure to *L. monocytogenes*.

Special Issues

Transfer of Agent Importation of this agent may require CDC and/or USDA importation permits. Domestic transport of this agent may require a permit from USDA/APHIS/VS. A DoC permit may be required for the export of this agent to another country. See Appendix C for additional information.

Mycobacterium leprae

Mycobacterium leprae is the causative agent of leprosy (Hansen disease). The organism has not been cultivated in laboratory medium but can be maintained in a metabolically active state for some period. Organisms are recovered from infected tissue and can be propagated in laboratory animals, specifically armadillos and the footpads of mice. The infectious dose in humans is unknown. Although naturally occurring leprosy or leprosy-like diseases have been reported in armadillos[98] and in NHP,[99,100] humans are the only known important reservoir of this disease.

Occupational Infections

There are no cases reported as a result of working in a laboratory with biopsy or other clinical materials of human or animal origin. However, inadvertent human-to-human transmissions following an accidental needle stick by a surgeon and after use of a presumably contaminated tattoo needle were reported prior to 1950.[101,102]

Natural Modes of Infection

Leprosy is transmitted from person-to-person following prolonged exposure, presumably via contact with secretions from infected individuals.

Laboratory Safety and Containment Recommendations

The infectious agent may be present in tissues and exudates from lesions of infected humans and experimentally or naturally infected animals. Direct contact of the skin and mucous membranes with infectious materials and accidental parenteral inoculation are the primary laboratory hazards associated with handling infectious clinical materials. See Appendix B for appropriate tuberculocidal disinfectant.

BSL-2 practices, containment equipment, and facilities are recommended for all activities with known or potentially infectious materials from humans and animals. Extraordinary care should be taken to avoid accidental parenteral inoculation with contaminated sharp instruments. ABSL-2 practices, containment equipment, and facilities are recommended for animal studies utilizing rodents, armadillos, and NHP, because coughing with dissemination of infectious droplets does not occur in these species.

Special Issues

Transfer of Agent Importation of this agent may require CDC and/or USDA importation permits. Domestic transport of this agent may require a permit from USDA/APHIS/VS. A DoC permit may be required for the export of this agent to another country. See Appendix C for additional information.

Mycobacterium tuberculosis complex

The *Mycobacterium tuberculosis* complex includes *M. tuberculosis, M. bovis, M. africanum,* and *M. microti* that cause tuberculosis in humans, and more recently recognized *M. caprae* and *M. pinnipedii* that have been isolated from animals. *M. tuberculosis* grows slowly, requiring three weeks for formation of colonies on solid media. The organism has a thick, lipid-rich cell wall that renders bacilli resistant to harsh treatments including alkali and detergents and allows them to stain acid-fast.

Occupational Infections

M. tuberculosis and *M. bovis* infections are a proven hazard to laboratory personnel as well as others who may be exposed to infectious aerosols in the laboratory, autopsy rooms, and other healthcare facilities.[4,26,103-105] The incidence of tuberculosis in laboratory personnel working with *M. tuberculosis* has been reported to be three times higher than that of those not working with the agent.[106] Naturally or experimentally infected NHP are a proven source of human infection.[107] Experimentally infected guinea pigs or mice do not pose the same hazard because droplet nuclei are not produced by coughing in these species; however, litter from infected animal cages may become contaminated and serve as a source of infectious aerosols.

Natural Modes of Infection

M. tuberculosis is the etiologic agent of tuberculosis, a leading cause of morbidity and mortality worldwide. Persons infected with *M. tuberculosis* can develop active disease within months of infection or can remain latently infected and develop disease later in life. The primary focus of infection is the lungs, but most other organs can be involved. HIV infection is a serious risk factor for development of active disease. Infectious aerosols produced by coughing spread tuberculosis. *M. bovis* is primarily found in animals but also can produce tuberculosis in humans. It is spread to humans, primarily children, by consumption of non-pasteurized milk and milk products, by handling of infected carcasses, and by inhalation. Human-to-human transmission via aerosols also is possible.

Laboratory Safety and Containment Recommendations

Tubercle bacilli may be present in sputum, gastric lavage fluids, CSF, urine, and in a variety of tissues. Exposure to laboratory-generated aerosols is the most important hazard encountered. Tubercle bacilli may survive in heat-fixed smears[108] and may be aerosolized in the preparation of frozen sections and during manipulation of liquid cultures. Because of the low infective dose of *M. tuberculosis* (i.e., ID_{50} <10 bacilli), sputa and other clinical specimens from suspected or known cases of tuberculosis must be considered potentially infectious and handled with appropriate precautions. Accidental needle-sticks are also a recognized hazard.

BSL-2 practices and procedures, containment equipment, and facilities are required for non-aerosol-producing manipulations of clinical specimens such as preparation of acid-fast smears. All aerosol-generating activities must be conducted in a BSC. Use of a slide-warming tray, rather than a flame, is recommended for fixation of slides. Liquifaction and concentration of sputa for acid-fast staining may be conducted safely on the open bench by first treating the specimen in a BSC with an equal volume of 5% sodium hypochlorite solution (undiluted household bleach) and waiting 15 minutes before processing.[109,110]

BSL-3 practices, containment equipment, and facilities are required for laboratory activities in the propagation and manipulation of cultures of any of the subspecies of the *M. tuberculosis* complex and for animal studies using experimentally or naturally infected NHP. Animal studies using guinea pigs or mice can be conducted at ABSL-2.[111] BSL-3 practices should include the use of respiratory protection and the implementation of specific procedures and use of specialized equipment to prevent and contain aerosols. Disinfectants proven to be tuberculocidal should be used. See Appendix B for additional information.

Manipulation of small quantities of the attenuated vaccine strain *M. bovis* Bacillus Calmette-Guérin (BCG) can be performed at BSL-2 in laboratories that do not culture *M. tuberculosis* and do not have BSL-3 facilities. However, considerable care must be exercised to verify the identity of the strain and to ensure that cultures are not contaminated with virulent *M. tuberculosis* or other *M. bovis* strains. Selection of an appropriate tuberculocidal disinfectant is an important consideration for laboratories working with mycobacteria. See Appendix B for additional information.

Special Issues

Surveillance Annual or semi-annual skin testing with purified protein derivative (PPD) of previously skin-test-negative personnel can be used as a surveillance procedure.

Vaccines The attenuated live BCG, is available and used in other countries but is not used in the United States for immunization.

Transfer of Agent Importation of this agent may require CDC and/or USDA importation permits. Domestic transport of this agent may require a permit from USDA/APHIS/VS. A DoC permit may be required for the export of this agent to another country. See Appendix C for additional information.

Mycobacterium spp. other than M. tuberculosis complex and M. leprae

More than 100 species of mycobacteria are recognized. These include both slowly growing and rapidly growing species. In the past, mycobacterial isolates that were not identified as *M. tuberculosis* complex were often called atypical mycobacteria, but these are now more commonly referred to as nontuberculous mycobacteria or mycobacteria other than tuberculosis. Many of the species are common environmental organisms, and approximately 25 of them are associated with infections in humans. A number of additional species are associated with infections in immunocompromised persons, especially HIV-infected individuals. All of these species are considered opportunistic pathogens in humans and none are considered communicable. Mycobacteria are frequently isolated from clinical samples but may not be associated with disease. The most common types of infections and causes are:

1. pulmonary disease with a clinical presentation resembling tuberculosis caused by *M. kansasii, M. avium*, and *M. intracellulare;*

2. lymphadenitis associated with *M. avium* and *M. scrofulaceum;*

3. disseminated infections in immunocompromised individuals caused by *M. avium;*

4. skin ulcers and soft tissue wound infections including Buruli ulcer caused by *M. ulcerans,* swimming pool granuloma caused by *M. marinum* associated with exposure to organisms in fresh and salt water and fish tanks, and tissue infections resulting from trauma, surgical procedures, or injection of contaminated materials caused by *M. fortuitum, M. chelonei*, and *M. abscesens.*

Occupational Infections

Laboratory-acquired infections with *Mycobacterium* spp. other than
M. tuberculosis complex have not been reported.

Natural Modes of Infection

Person-to-person transmission has not been demonstrated. Presumably,
pulmonary infections are the result of inhalation of aerosolized bacilli, most likely
from the surface of contaminated water. Mycobacteria are widely distributed
in the environment and in animals. They are also common in potable water
supplies, perhaps as the result of the formation of biofilms. The source of
M. avium infections in immunocompromised persons has not been established.

Laboratory Safety and Containment Recommendations

Various species of mycobacteria may be present in sputa, exudates from
lesions, tissues, and in environmental samples. Direct contact of skin or mucous
membranes with infectious materials, ingestion, and accidental parenteral
inoculation are the primary laboratory hazards associated with clinical materials
and cultures. Aerosols created during the manipulation of broth cultures or
tissue homogenates of these organisms also pose a potential infection hazard.

   BSL-2 practices, containment equipment, and facilities are recommended
for activities with clinical materials and cultures of *Mycobacteria* spp. other than
M. tuberculosis complex. Clinical specimens may also contain *M. tuberculosis*
and care must be exercised to ensure the correct identification of cultures. Special
caution should be exercised in handling *M. ulcerans* to avoid skin exposure.
ABSL-2 practices, containment equipment, and facilities are recommended for
animal studies. Selection of an appropriate tuberculocidal disinfectant is an
important consideration for laboratories working with mycobacteria. See
Appendix B for additional information.

Special Issues

Transfer of Agent Importation of this agent may require CDC and/or USDA
importation permits. Domestic transport of this agent may require a permit from
USDA/APHIS/VS. A DoC permit may be required for the export of this agent to
another country. See Appendix C for additional information.

Neisseria gonorrhoeae

Neisseria gonorrhoeae is a gram-negative, oxidase-positive diplococcus associated
with gonorrhea, a sexually transmitted disease of humans. The organism may be
isolated from clinical specimens and cultivated in the laboratory using specialized
growth media.[112]

Occupational Infections

Laboratory-associated gonococcal infections have been reported in the United States and elsewhere.[113-116] These infections have presented as conjunctivitis, with either direct finger-to-eye contact or exposure to splashes of either liquid cultures or contaminated solutions proposed as the most likely means of transmission.

Natural Modes of Infection

Gonorrhea is a sexually transmitted disease of worldwide importance. The 2004 rate of reported infections for this disease in the United States was 112 per 100,000 population.[117] The natural mode of infection is through direct contact with exudates from mucous membranes of infected individuals. This usually occurs by sexual activity, although newborns may also become infected during birth.[112]

Laboratory Safety and Containment Recommendations

The agent may be present in conjunctival, urethral and cervical exudates, synovial fluid, urine, feces, and CSF. Accidental parenteral inoculation and direct or indirect contact of mucous membranes with infectious clinical materials are known primary laboratory hazards. Laboratory-acquired illness due to aerosol transmission has not been documented.

BSL-2 practices, containment equipment, and facilities are recommended for all activities involving the use or manipulation of clinical materials or cultures. Gloves should be worn when handling infected laboratory animals and when there is the likelihood of direct skin contact with infectious materials. Additional primary containment and personnel precautions such as those described for BSL-3 may be indicated when there is high risk of aerosol or droplet production, and for activities involving production quantities or high concentrations of infectious materials. Animal studies may be performed at ABSL-2.

Special Issues

Transfer of Agent Importation of this agent may require CDC and/or USDA importation permits. Domestic transport of this agent may require a permit from USDA/APHIS/VS. A DoC permit may be required for the export of this agent to another country. See Appendix C for additional information.

Neisseria meningitidis

Neisseria meningitidis is a gram-negative coccus responsible for serious acute meningitis and septicemia in humans. Virulence is associated with the expression of a polysaccharide capsule. Thirteen different capsular serotypes have been identified, with types A, B, C, Y, and W135 associated with the highest incidence of disease. The handling of invasive *N. meningitidis* isolates from blood or CSF represents an increased risk to microbiologists.[118,119]

Occupational Infections

Recent studies of LAI and exposures have indicated that manipulating suspensions of *N. meningitidis* outside a BSC is associated with a high risk for contracting meningococcal disease.[119] Investigations of potential laboratory-acquired cases of meningococcal diseases in the United States showed a many-fold higher attack rate for microbiologists compared to that of the United States general population age 30-59 years, and a case fatality rate of 50%, substantially higher than the 12-15% associated with disease among the general population. Almost all the microbiologists had manipulated sterile site isolates on an open laboratory bench.[120] While isolates obtained from respiratory sources are generally less pathogenic and consequently represent lower risk for microbiologists, rigorous protection from droplets or aerosols is mandated when microbiological procedures are performed on all *N. meningitidis* isolates, especially on those from sterile sites.

Natural Modes of Infection

The human upper respiratory tract is the natural reservoir for *N. meningitidis*. Invasion of organisms from the respiratory mucosa into the circulatory system causes infection that can range in severity from subclinical to fulminant fatal disease. Transmission is person-to-person and is usually mediated by direct contact with respiratory droplets from infected individuals.

Laboratory Safety and Containment Recommendations

N. meningitidis may be present in pharyngeal exudates, CSF, blood, and saliva. Parenteral inoculation, droplet exposure of mucous membranes, infectious aerosol and ingestion are the primary hazards to laboratory personnel. Based on the mechanism of natural infection and the risk associated with handling of isolates on an open laboratory bench, exposure to droplets or aerosols of *N. meningitidis* is the most likely risk for infection in the laboratory.

Specimens for *N. meningitidis* analysis and cultures of *N. meningitidis* not associated with invasive disease may be handled in BSL-2 facilities with rigorous application of BSL-2 standard practices, special practices, and safety equipment. All sterile-site isolates of *N. meningitidis* should be manipulated within a BSC. Isolates of unknown source should be treated as sterile-site isolates.

If a BSC is unavailable, manipulation of these isolates should be minimized, primarily focused on serogroup identification using phenolized saline solution while wearing a laboratory coat, gloves, and safety glasses or full-face splash shield. BSL-3 practices and procedures are indicated for activities with a high potential for droplet or aerosol production and for activities involving production quantities or high concentrations of infectious materials. Animal studies should be performed under ABSL-2 conditions.

Special Issues

Vaccines The quadrivalent meningococcal polysaccharide vaccine, which includes serogroups A, C, Y, and W-135, will decrease but not eliminate the risk of infection, because it is less than 100% effective and does not provide protection against serogroup B, which caused one-half of the laboratory-acquired cases in the United States in 2000.[118,120] Laboratorians who are exposed routinely to potential aerosols of *N. meningitidis* should consider vaccination.[118,121,122]

Transfer of Agent Importation of this agent may require CDC and/or USDA importation permits. Domestic transport of this agent may require a permit from USDA/APHIS/VS. A DoC permit may be required for the export of this agent to another country. See Appendix C for additional information.

Salmonella serotypes, other than S. Typhi

Salmonellae are gram-negative enteric bacteria associated with diarrheal illness in humans. They are motile oxidase-negative organisms that are easily cultivated on standard bacteriologic media, although enrichment and selective media may be required for isolation from clinical materials. Recent taxonomic studies have organized this genus into two species, *S. enterica* and *S. bongori*, containing more than 2500 antigenically distinct subtypes or serotypes.[123] *S. enterica* contains the vast majority of serotypes associated with human disease. *S. enterica* serotypes Typhimurium and Enteritidis (commonly designated *S. Typhimurium* and *S. Enteritidis*) are the serotypes most frequently encountered in the United States. This summary statement covers all pathogenic serotypes except *S. Typhi*.

Occupational Infections

Salmonellosis is a documented hazard to laboratory personnel.[4,26,124-125] Primary reservoir hosts include a broad spectrum of domestic and wild animals, including birds, mammals, and reptiles, all of which may serve as a source of infection to laboratory personnel. Case reports of laboratory-acquired infections indicate a presentation of symptoms (fever, severe diarrhea, abdominal cramping) similar to those of naturally-acquired infections, although one case also developed erythema nodosum and reactive arthritis.[126,127]

Natural Modes of Infection

Salmonellosis is a food borne disease of worldwide distribution. An estimated 5 million cases of salmonellosis occur annually in the United States. A wide range of domestic and feral animals (poultry, swine, rodents, cattle, iguanas, turtles, chicks, dogs, cats) may serve as reservoirs for this disease, as well as humans.[128] The most common mode of transmission is by ingestion of food from contaminated animals or contaminated during processing. The disease usually presents as an acute enterocolitis, with an incubation period ranging from 6 to 72 hours.

Laboratory Safety and Containment Recommendations

The agent may be present in feces, blood, urine, and in food, feed, and environmental materials. Ingestion or parenteral inoculation are the primary laboratory hazards. The importance of aerosol exposure is not known. Naturally or experimentally infected animals are a potential source of infection for laboratory and animal care personnel, and for other animals

Strict compliance with BSL-2 practices, containment equipment, and facilities are recommended for all activities utilizing known or potentially infectious clinical materials or cultures. This includes conducting procedures with aerosol or high splash potential in primary containment devices such as a BSCs or safety centrifuge cups. Personal protective equipment should be used in accordance with a risk assessment, including splash shields, face protection, gowns, and gloves. The importance of proper gloving techniques and frequent and thorough hand washing is emphasized. Care in manipulating faucet handles to prevent contamination of cleaned hands or the use of sinks equipped with remote water control devices, such as foot pedals, is highly recommended. Special attention to the timely and appropriate decontamination of work surfaces, including potentially contaminated equipment and laboratory fixtures, is strongly advised. ABSL-2 facilities and practices are recommended for activities with experimentally infected animals.

Special Issues

Transfer of Agent Importation of this agent may require CDC and/or USDA importation permits. Domestic transport of this agent may require a permit from USDA/APHIS/VS. A DoC permit may be required for the export of this agent to another country. See Appendix C for additional information.

Salmonella Typhi

Recent taxonomic studies have organized the genus *Salmonella* into two species, *S. enterica* and *S. bongori*, containing more than 2500 antigenically distinct subtypes or serotypes.[123] *S. enterica* contains the vast majority of serotypes associated with human disease. *S. enterica* serotype Typhi, commonly designated *S. Typhi*, is the causative agent of typhoid fever. *S. Typhi* is a motile gram-negative enteric bacterium that is easily cultivated on standard bacteriologic media, although enrichment and selective media may be required for isolation of this organism from clinical materials.

Occupational Infections

Typhoid fever is a demonstrated hazard to laboratory personnel.[4,129,130] Ingestion and less frequently, parenteral inoculation are the most significant modes of transmission in the laboratory. Secondary transmission to other individuals outside of the laboratory is also a concern.[131] Laboratory-acquired *S. Typhi* infections usually present with symptoms of septicemia, headache, abdominal pain, and high fever.[129]

Natural Modes of Infection

Typhoid fever is a serious, potentially lethal bloodstream infection of worldwide distribution. Humans are the sole reservoir and asymptomatic carriers may occur. The infectious dose is low (<103 organisms) and the incubation period may vary from one to six weeks, depending upon the dose of the organism. The natural mode of transmission is by ingestion of food or water contaminated by feces or urine of patients or asymptomatic carriers.[123]

Laboratory Safety and Containment Recommendations

The agent may be present in feces, blood, gallbladder (bile), and urine. Humans are the only known reservoir of infection. Ingestion and parenteral inoculation of the organism represent the primary laboratory hazards. The importance of aerosol exposure is not known.

Strict compliance with BSL-2 practices, containment equipment, and facilities are recommended for all activities utilizing known or potentially infectious clinical materials or cultures. This includes conducting procedures with aerosol or high splash potential in primary containment devices such as BSCs or safety centrifuge cups. Personal protective equipment should be used in accordance with a risk assessment, including splash shields, face protection, gowns, and gloves. The importance of proper gloving techniques and frequent and thorough hand washing is emphasized. Care in manipulating faucet handles to prevent contamination of cleaned hands or the use of sinks equipped with remote water control devices, such as foot pedals, is highly recommended. Special attention to the timely and appropriate decontamination of work surfaces, including potentially contaminated equipment and laboratory fixtures, is strongly advised. BSL-3 practices and equipment are recommended for activities likely to produce significant aerosols or for activities involving production quantities of organisms. ABSL-2 facilities, practices and equipment are recommended for activities with experimentally infected animals. ABSL-3 conditions may be considered for protocols involving aerosols.

Special Issues

Vaccines Vaccines for *S. Typhi* are available and should be considered for personnel regularly working with potentially infectious materials. The reader is advised to consult the current recommendations of the Advisory Committee on

Immunization Practices (ACIP) published in the CDC Morbidity and Mortality Weekly Report for recommendations for vaccination against *S. Typhi*.[132]

Transfer of Agent Importation of this agent may require CDC and/or USDA importation permits. Domestic transport of this agent may require a permit from USDA/APHIS/VS. A DoC permit may be required for the export of this agent to another country. See Appendix C for additional information.

Shiga toxin (Verocytotoxin)-producing Escherichia coli

Escherichi coli is one of five species in the gram-negative genus *Escherichia*. This organism is a common inhabitant of the bowel flora of healthy humans and other mammals and is one of the most intensively studied prokaryotes. An extensive serotyping system has been developed for *E. coli* based the O (somatic) and H (flagellar) antigens expressed by these organisms. Certain pathogenic clones of *E. coli* may cause urinary tract infections, bacteremia, meningitis, and diarrheal disease in humans, and these clones are associated with specific serotypes.

The diarrheagenic *E. coli* strains have been characterized into at least four basic pathogenicity groups: Shiga toxin (Verocytotoxin)-producing *E. coli* (a subset of which are referred to as enterohemorrhagic *E. coli*), enterotoxigenic *E. coli*, enteropathogenic *E. coli*, and enteroinvasive *E. coli*.[123] In addition to clinical significance, *E. coli* strains are commonly-used hosts for cloning experiments and other genetic manipulations in the laboratory. This summary statement provides recommendations for safe manipulation of Shiga toxin-producing *E. coli* strains. Procedures for safely handling laboratory derivatives of *E. coli* or other pathotypes of *E. coli* should be based upon a thorough risk assessment.

Occupational Infections

Shiga toxin-producing *E. coli* strains, including strains of serotype O157:H7, are a demonstrated hazard to laboratory personnel.[133-138] The infectious dose is estimated to be low—similar to that reported for *Shigella* spp., 10-100 organisms.[136] Domestic farm animals (particularly bovines) are significant reservoirs of the organisms; however, experimentally infected small animals are also sources of infection in the laboratory.[139] Verocytotoxin-producing *Escherichia coli* have also been in wild birds and rodents in close proximity to farms.[140]

Natural Modes of Infection

Cattle represent the most common natural reservoir of Shiga-toxin producing *E. coli*. Transmission usually occurs by ingestion of contaminated food, including raw milk, fruits, vegetables, and particularly ground beef. Human-to-human transmission has been observed in families, day care centers, and custodial institutions. Water-borne transmission has been reported from outbreaks associated with swimming in a crowded lake and drinking unchlorinated municipal water.[139] In a small proportion of patients (usually children) infected with these organisms, the disease progresses to hemolytic uremic syndrome or death.

Laboratory Safety and Containment Recommendations

Shiga toxin-producing *E. coli* are usually isolated from feces. However, a variety of food specimens contaminated with the organisms including uncooked ground beef, unpasteurized dairy products and contaminated produce may present laboratory hazards. This agent may be found in blood or urine specimens from infected humans or animals. Accidental ingestion is the primary laboratory hazard. The importance of aerosol exposure is not known.

Strict compliance with BSL-2 practices, containment equipment, and facilities are recommended for all activities utilizing known or potentially infectious clinical materials or cultures. Procedures with aerosol or high splash potential should be conducted with primary containment equipment or in devices such as a BSC or safety centrifuge cups. Personal protective equipment, such as splash shields, face protection, gowns, and gloves should be used in accordance with a risk assessment. The importance of proper gloving techniques and frequent and thorough hand washing is emphasized. Care in manipulating faucet handles to prevent contamination of cleaned hands or the use of sinks equipped with remote water control devices, such as foot pedals, is highly recommended. Special attention to the timely and appropriate decontamination of work surfaces, including potentially contaminated equipment and laboratory fixtures, is strongly advised. ABSL-2 practices and facilities are recommended for activities with experimentally or naturally infected animals.

Special Issues

Transfer of Agent Importation of this agent may require CDC and/or USDA importation permits. Domestic transport of this agent may require a permit from USDA/APHIS/VS. A DoC permit may be required for the export of this agent to another country. See Appendix C for additional information.

Shigella

The genus *Shigella* is composed of nonmotile gram-negative bacteria in the family Enterobacteriaceae. There are four subgroups that have been historically treated as separate species, even though more recent genetic analysis indicates that they are members of the same species. These include subgroup A (*Shigella dysenteriae*), subgroup B (*S. flexneri*), subgroup C (*S. boydii*), and subgroup D (*S. sonnei*). Members of the genus *Shigella* have been recognized since the late 19th century as causative agents of bacillary dysentery, or shigellosis.[123]

Occupational Infections

Shigellosis is one of the most frequently reported laboratory-acquired infections in the United States.[131,141] A survey of 397 laboratories in the United Kingdom revealed that in 1994-1995, four of nine reported laboratory-acquired infections were caused by *Shigella*.[142] Experimentally infected guinea pigs, other rodents, and NHP are proven sources of laboratory-acquired infection.[143,144]

Natural Modes of Infection

Humans and other large primates are the only natural reservoirs of *Shigella* bacteria. Most transmission is by fecal-oral route; infection also is caused by ingestion of contaminated food or water.[123] Infection with *Shigella dysenteriae* type 1 causes more severe, prolonged, and frequently fatal illness than does infection with other *Shigella*. Complications of shigellosis include hemolytic uremic syndrome, which is associated with *S. dysenteriae* 1 infection, and Reiter chronic arthritis syndrome, which is associated with *S. flexneri* infection.

Laboratory Safety and Containment Recommendations

The agent may be present in feces and, rarely, in the blood of infected humans or animals. Accidental ingestion and parenteral inoculation of the agent are the primary laboratory hazards. The 50% infectious dose (oral) of *Shigella* for humans is only a few hundred organisms.[143] The importance of aerosol exposure is not known.

Strict compliance with BSL-2 practices, containment equipment, and facilities are recommended for all activities utilizing known or potentially infectious clinical materials or cultures. Procedures with aerosol or high splash potential should be conducted with primary containment equipment such as a BSC or safety centrifuge cups. Personal protective equipment should be used in accordance with a risk assessment, including splash shields, face protection, gowns, and gloves. The importance of proper gloving techniques and frequent and thorough hand washing is emphasized. Care in manipulating faucet handles to prevent contamination of cleaned hands or the use of sinks equipped with remote water control devices, such as foot pedals, is highly recommended. Special attention to the timely and appropriate decontamination of work surfaces, including potentially contaminated equipment and laboratory fixtures, is strongly advised. ABSL-2 facilities and practices are recommended for activities with experimentally or naturally infected animals.

Special Issues

Vaccines Vaccines are currently not available for use in humans.

Transfer of Agent Importation of this agent may require CDC and/or USDA importation permits. Domestic transport of this agent may require a permit from

USDA/APHIS/VS. A DoC permit may be required for the export of this agent to another country. See Appendix C for additional information.

Treponema pallidum

Treponema pallidum is a species of extremely fastidious spirochetes that die readily upon desiccation or exposure to atmospheric levels of oxygen, and have not been cultured continuously *in vitro*.[145] *T. pallidum* cells have lipid-rich outer membranes and are highly susceptible to disinfection with common alcohols (i.e., 70% isopropanol). This species contains three subspecies including *T. pallidum* spp. *pallidum* (associated with venereal syphilis), *T. pallidum* spp. *endemicum* (associated with endemic syphilis), and *T. pallidum* spp. *pertenue* (associated with Yaws). These organisms are obligate human pathogens.

Occupational Infections

T. pallidum is a documented hazard to laboratory personnel. Pike lists 20 cases of LAI.[4] Syphilis has been transmitted to personnel working with a concentrated suspension of *T. pallidum* obtained from an experimental rabbit orchitis.[146] *T. pallidum* is present in the circulation during primary and secondary syphilis. The ID_{50} of *T. pallidum* needed to infect rabbits by subcutaneous injection has been reported to be as low as 23 organisms.[147] The concentration of *T. pallidum* in patients' blood during early syphilis, however, has not been determined. No cases of laboratory animal-associated infections are reported; however, rabbit-adapted *T. pallidum* (Nichols strain and possibly others) retains virulence for humans.

Natural Modes of Infection

Humans are the only known natural reservoir of *T. pallidum* and transmission occurs via direct sexual contact (venereal syphilis), direct skin contact (Yaws), or direct mucous contact (endemic syphilis). Venereal syphilis is a sexually transmitted disease that occurs in many areas of the world, whereas Yaws occurs in tropical areas of Africa, South America, the Caribbean, and Indonesia. Endemic syphilis is limited to arid areas of Africa and the Middle East.[145]

Laboratory Safety and Containment Recommendations

The agent may be present in materials collected from cutaneous and mucosal lesions and in blood. Accidental parenteral inoculation, contact with mucous membranes or broken skin with infectious clinical materials are the primary hazards to laboratory personnel.

BSL-2 practices, containment equipment, and facilities are recommended for all activities involving the use or manipulation of blood or other clinical samples from humans or infected rabbits. Gloves should be worn when there is a likelihood of direct skin contact with infective materials. Periodic serological monitoring should be considered in personnel regularly working with these materials. ABSL-2 practices, containment equipment, and facilities are recommended for work with infected animals.

Special Issues

Vaccines Vaccines are currently not available for use in humans.

Transfer of Agent Importation of this agent may require CDC and/or USDA importation permits. Domestic transport of this agent may require a permit from USDA/APHIS/VS. A DoC permit may be required for the export of this agent to another country. See Appendix C for additional information.

Vibrio enteritis species (V. cholerae, V. parahaemolyticus)

Vibrio species are straight or curved motile gram-negative rods. Growth of *Vibrio* species is stimulated by sodium and the natural habitats of these organisms are primarily aquatic environments. Although 12 different *Vibrio* species have been isolated from clinical specimens, *V. cholerae* and *V. parahaemolyticus* are the best-documented causes of human disease.[148] Vibrios may cause either diarrhea or extraintestinal infections.

Occupational Infections

Rare cases of bacterial enteritis due to LAI with either *V. cholerae* or *V. parahaemolyticus* have been reported from around the world.[4] Naturally and experimentally infected animals[149] and shellfish[150,151] are potential sources for such illnesses.

Natural Modes of Infection

The most common natural mode of infection is the ingestion of contaminated food or water. The human oral infecting dose of *V. cholerae* in healthy non-achlorhydric individuals is approximately 10^6-10^{11} colony forming units,[152] while that of *V. parahaemolyticus* ranges from 10^5-10^7 cells.[153] The importance of aerosol exposure is unknown although it has been implicated in at least one instance.[149] The risk of infection following oral exposure is increased in persons with abnormal gastrointestinal physiology including individuals on antacids, with achlorhydria, or with partial or complete gastrectomies.[154]

Laboratory Safety and Containment Recommendations

Pathogenic vibrios can be present in human fecal samples, or in the meats and the exterior surfaces of marine invertebrates such as shellfish. Other clinical specimens from which vibrios may be isolated include blood, arm or leg wounds, eye, ear, and gallbladder.[148] Accidental oral ingestion of *V. cholerae* or *V. parahaemolyticus* principally results from hands contaminated from the use of syringes or the handling of naturally contaminated marine samples without gloves.

BSL-2 practices, containment equipment, and facilities are recommended for activities with cultures or potentially infectious clinical materials. ABSL-2 practices, containment equipment, and facilities are recommended for activities with naturally or experimentally infected animals.

Special Issues

Vaccines The reader is advised to consult the current recommendations of the ACIP published in the MMWR for vaccination recommendations against *V. cholera*. There are currently no human vaccines against *V. parahaemolyticus*.

Transfer of Agent Importation of this agent may require CDC and/or USDA importation permits. Domestic transport of this agent may require a permit from USDA/APHIS/VS. A DoC permit may be required for the export of this agent to another country. See Appendix C for additional information.

Yersinia pestis

Yersinia pestis, the causative agent of plague, is a gram-negative, microaerophilic coccobacillus frequently characterized by a "safety pin" appearance on stained preparations from specimens. It is nonmotile and nonsporulating. There are three biotypes of *Y. pestis,* differentiated by their ability to ferment glycerol and reduce nitrate. All three biotypes are virulent. The incubation period for bubonic plague ranges from two to six days while the incubation period for pneumonic plague is one to six days. Pneumonic plague is transmissible person-to-person;[155] whereas bubonic plague is not. Laboratory animal studies have shown the lethal and infectious doses of *Y. pestis* to be quite low (less than 100 colony forming units).[156]

Occupational Infections

Y. pestis is a documented laboratory hazard. Prior to 1950, at least 10 laboratory-acquired cases were reported in the United States, four of which were fatal.[4,157] Veterinary staff and pet owners have become infected when handling domestic cats with oropharyngeal or pneumonic plague.

Natural Modes of Infection

Infective fleabites are the most common mode of transmission, but direct human contact with infected tissues or body fluids of animals and humans also may serve as sources of infection.

Primary pneumonic plague arises from the inhalation of infectious respiratory droplets or other airborne materials from infected animals or humans. This form of plague has a high case fatality rate if not treated and poses the risk of person-to-person transmission.

Laboratory Safety and Containment Recommendations

The agent has been isolated, in order of frequency of recovery, from bubo aspirate, blood, liver, spleen, sputum, lung, bone marrow, CSF, and infrequently from feces and urine, depending on the clinical form and stage of the disease. Primary hazards to laboratory personnel include direct contact with cultures and infectious materials from humans or animal hosts and inhalation of infectious aerosols or droplets generated during their manipulation. Laboratory and field personnel should be counseled on methods to avoid fleabites and accidental autoinoculation when handling potentially infected live or dead animals.

BSL-2 practices, containment equipment, and facilities are recommended for all activities involving the handling of potentially infectious clinical materials and cultures. In addition, because the infectious dose is so small, all work, including necropsies of potentially infected animals should be performed in a BSC. Special care should be taken to avoid generating aerosols or airborne droplets while handling infectious materials or when performing necropsies on naturally or experimentally infected animals. Gloves should be worn when handling potentially infectious materials including field or laboratory infected animals. BSL-3 is recommended for activities with high potential for droplet or aerosol production, and for activities involving large-scale production or high concentrations of infectious materials. Resistance of *Y. pestis* strains to antibiotics used in the treatment of plague should be considered in a thorough risk assessment and may require additional containment for personal protective equipment. For animal studies, a risk assessment that takes into account the animal species, infective strain, and proposed procedures should be performed in order to determine if ABSL-2 or ABSL-3 practices, containment equipment, and facilities should be employed. BSL-3 facilities and arthropod containment level 3 practices are recommended for all laboratory work involving infected arthropods.[157] See Appendix G for additional information on arthropod containment guidelines.

Special Issues

Select Agent *Yersinia pestis* is an HHS select agent requiring registration with CDC for the possession, use, storage and transfer. See Appendix F for further information.

Transfer of Agent Importation of this agent may require CDC and/or USDA importation permits. Domestic transport of this agent may require a permit from USDA/APHIS/VS. A DoC permit may be required for the export of this agent to another country. See Appendix C for additional information.

References

1. Inglesby TV, O'Toole T, Henderson DA, et al. Anthrax as a biological weapon. JAMA. 2002;287:2236-52.
2. Watson A, Keir D. Information on which to base assessments of risk from environments contaminated with anthrax spores. Epidemiol Infect. 1994;113:479-90.
3. Ellingson HV, Kadull PJ, Bookwalter HL, et al. Cutaneous anthrax: report of twenty-five cases. JAMA. 1946;131:1105-8.
4. Pike RM. Laboratory-associated infections: summary and analysis of 3,921 cases. Hlth Lab Sci 1976;13:105-14.
5. Centers for Disease Control and Prevention. Suspected cutaneous anthrax in a laboratory worker—Texas, 2002. MMWR Morb Mortal Wkly Rep. 2002;51:279-81.
6. Centers for Disease Control and Prevention. Human anthrax associated with an epizootic among livestock—North Dakota, 2000. MMWR Morb Mortal Wkly Rep. 2001;50:677-80.
7. Jernigan DB, Raghunathan PS, Bell BP, et al. Investigation of bioterrorism-related anthrax, United States, 2001: epidemiologic findings. Emerg Infect Dis. 2002;8:1019-28.
8. Centers for Disease Control and Prevention. Notice to readers: use of anthrax vaccine in response to terrorism: supplemental recommendations of the Advisory Committee on Immunization Practices (ACIP). MMWR Morb Mortal Wkly Rep. 2002;51:1024-6.
9. Centers for Disease Control and Prevention. Use of anthrax vaccine in the United States: recommendations of the Advisory Committee on Immunization Practices (ACIP). MMWR Morb Mortal Wkly Rep. 2000;49(RR15):1-20.
10. Addiss DG, Davis JP, Meade BD, et al. A pertussis outbreak in a Wisconsin nursing home. J Infect Dis. 1991;164:704-10.
11. Christie CD, Glover AM, Willke MJ, et al. Containment of pertussis in the regional pediatric hospital during the greater Cincinnati epidemic of 1993. Infect Control Hosp Epidemiol. 1995;16:556-63.
12. Kurt TL, Yeager AS, Guenette S, et al. Spread of pertussis by hospital staff. JAMA. 1972;221:264-7.
13. Linnemann CC Jr, Ramundo N, Perlstein PH, et al. Use of pertussis vaccine in an epidemic involving hospital staff. Lancet. 1975;2:540-3.
14. Shefer A, Dales L, Nelson M, et al. Use and safety of acellular pertussis vaccine among adult hospital staff during an outbreak of pertussis. J Infect Dis. 1995;171:1053-6.
15. Steketee RW, Wassilak SG, Adkins WN, et al. Evidence for a high attack rate and efficacy of erythromycin prophylaxis in a pertussis outbreak in a facility for the developmentally disabled. J Infect Dis. 1988;157:434-40.
16. Weber DJ, Rutala WA. Management of healthcare workers exposed to pertussis. Infect Control Hosp Epidemiol. 1994;15:411-5.

17. Beall B, Cassiday PK, Sanden GN. Analysis of *Bordetella pertussis* isolates from an epidemic by pulsed-field gel electrophoresis. J Clin Microbiol. 1995;33:3083-6.
18. Burstyn DG, Baraff LJ, Peppler MS, et al. Serological response to filamentous hemagglutinin and lymphocytosis-promoting toxin of *Bordetella pertussis*. Infect Immun. 1983;41:1150-6.
19. Farizo KM, Cochi SL, Zell ER, et al. Epidemiological features of pertussis in the United States, 1980-1989. Clin Infect Dis. 1992;14:708-19.
20. Guris D, Strebel PM, Bardenheier B, et al. Changing epidemiology of pertussis in the United States: increasing reported incidence among adolescents and adults, 1990-1996. Clin Infect Dis. 1999;28:1230-7.
21. Tanaka M, Vitek CR, Pascual FB, et al. Trends in pertussis among infants in the United States, 1980-1999. JAMA. 2003;290:2968-75.
22. Janda WM, Santos E, Stevens J, et al. Unexpected isolation of *Bordetella pertussis* from a blood culture. J Clin Microbiol. 1994;32:2851-3.
23. Centers for Disease Control and Prevention. Fatal case of unsuspected pertussis diagnosed from a blood culture—Minnesota, 2003. MMWR Morb Mortal Wkly Rep. 2004;53:131-2.
24. Morisset R, Spink WW. Epidemic canine brucellosis due to a new species, *Brucella canis*. Lancet. 1969;2:1000-2.
25. Spink WW. The nature of brucellosis. Minneapolis: The University of Minnesota Press; 1956.
26. Miller CD, Songer JR, Sullivan JF. A twenty-five year review of laboratory-acquired human infections at the National Animal Disease Center. Am Ind Hyg Assoc J. 1987;48:271-5.
27. Olle-Goig J, Canela-Soler JC. An outbreak of *Brucella melitensis* infection by airborne transmission among laboratory workers. Am J Publ Hlth 1987;77:335-8.
28. Memish ZA, Mah MW. Brucellosis in laboratory workers at a Saudi Arabian hospital. Am J Infect Control. 2001;29:48-52.
29. Grammont-Cupillard M, Berthet-Badetti L, Dellamonica P. Brucellosis from sniffing bacteriological cultures. Lancet. 1996;348:1733-4.
30. Huddleson IF, Munger M. A study of an epidemic of brucellosis due to *Brucella melitensis*. Am J Public Health. 1940;30:944-54.
31. Staszkiewicz J, Lewis CM, Coville J et al. Outbreak of *Brucella melitensis* among microbiology laboratory workers in a community hospital. J Clin Microbiol. 1991;29:278-90.
32. Pardon P, Marly J. Resistance of normal or immunized guinea pigs against a subcutaneous challenge of *Brucella abortus*. Ann Rech Vet. 1978;9:419-25.
33. Mense MG, Borschel RH, Wilhelmsen CL, et al. Pathologic changes associated with brucellosis experimentally induced by aerosol exposure in rhesus macaques (*Macaca mulatto*). Am J Vet Res. 2004;65:644-52.
34. Madkour MM. Brucellosis: overview. In: Madkour MM, editor. Madkour's Brucellosis. 2nd ed. Germany: Berlin Springer Verlag; 2001.

35. Centers for Disease Control and Prevention. Laboratory-acquired human glanders—Maryland, May 2000. MMWR Morb Mortal Wkly Rep. 2000;49:532-5.

36. Glanders. In: Chin J, Ascher M, editors. Control of communicable diseases. Washington, DC: American Public Health Association; 2000. p. 337-8.

37. Howe C, Miller WR. Human glanders: report of six cases. Ann Intern Med. 1947;26:93-115.

38. Srinivasan A, Kraus CN, DeShazer D, et al. Glanders in a military research microbiologist. N Engl J Med. 2001;345:256-8.

39. Green RN, Tuffnell PG. Laboratory acquired melioidosis. Am J Med. 1968;44:599-605.

40. Schlech WF 3rd, Turchik JB, Westlake RE, Jr., et al. Laboratory-acquired infection with *Pseudomonas pseudomallei* (melioidosis). N Engl J Med. 1981;305:1133-5.

41. Dance DA. Ecology of *Burkholderia pseudomallei* and the interactions between environmental *Burkholderia spp.* and human-animal hosts. Acta Trop. 2000;74:159-68.

42. Dorman SE, Gill VJ, Gallin JI, et al. *Burkholderia pseudomallei* infection in a Puerto Rican patient with chronic granulomatous disease: case report and review of occurrences in the Americas. Clin Infect Dis. 1998;26:889-94.

43. Melioidosis. In: Chin J, Ascher M, editors. Control of communicable diseases. Washington, DC: American Public Health Association. 2000. p. 335-7.

44. Masuda T, Isokawa T. Kansenshogaku Zasshi. [Biohazard in clinical laboratories in Japan]. J. Jpn Assoc Infect Dis. 1991;65:209-15.

45. Oates JD, Hodgin UG, Jr. Laboratory-acquired *Campylobacter* enteritis. South Med J. 1981;74:83.

46. Penner JL, Hennessy JN, Mills SD, Bradbury WC. Application of serotyping and chromosomal restriction endonuclease digest analysis in investigating a laboratory-acquired case of *Campylobacter jejuni* enteritis. J Clin Microbiol 1983;18:1427-8.

47. Prescott JF, Karmali MA. Attempts to transmit *Campylobacter* enteritis to dogs and cats. Can Med Assoc J. 1978;119:1001-2.

48. Young VB, Schauer DB, Fox JG. Animal models of *Campylobacter* infection. In: Nachamkin I, Blaser MJ, editors. *Campylobacter*. 2nd ed. Washington, DC: ASM Press; 2000.

49. Robinson DA. Infective dose of *Campylobacter jejuni* in milk. Brit Med J (Clin Res Ed). 1981;282:1584.

50. Black RE, Levine MM, Clements ML, et al. Experimental *Campylobacter jejuni* infections in humans. J Infect Dis. 1988;157:472-9.

51. Black RE, Perlman D, Clements ML, et al. Human volunteer studies with *Campylobacter jejuni*. In: Nachamkin I, Blaser MJ, Tompkins LS, editors. *Campylobacter jejuni*: Current status and future trends. Washington, DC: ASM Press. 1992. p. 207-15.

52. Nachamkin I, Blaser MJ, Tompkins LS. *Campylobacter jejuni*: current status and future trends. Washington, DC: ASM Press; 1992.

53. Bernstein DI, Hubbard T, Wenman W, et al. Mediastinal and supraclavicular lymphadenitis and pneumonitis due to *Chlamydia trachomatis* serovars L1 and L2. N Eng J Med. 1984;311:1543-6.

54. Hyman CL, Augenbraun MH, Roblin PM, et al. Asymptomatic respiratory tract infection with *Chlamydia pneumoniae* TWAR. J Clin Microbiol. 1991;29:2082-3.

55. Simpson LL. Identification of the major steps in botulinum toxin action. Ann Rev Pharmacol Toxicol. 2004;44:167-93.

56. Holzer VE. Botulismus durch inhalation. Med Klin. 1962;57:1735-8.

57. Arnon SS, Schecter R, Inglesby TV, et al. Botulinum toxin as a biological weapon. JAMA. 2001;285:1059-70.

58. Hatheway CL. Botulism: The current state of the disease. Curr Topics Microbiol Immunol. 1995:55-75.

59. Smith LDS, Sugiyama H. Botulism: the organism, its toxins, the disease. 2nd ed. Balows A, editor. Springfield (IL): Charles C Thomas; 1988.

60. Siegel LS, Metzger JF. Toxin production by clostridium botulinum type A under various fermentation conditions. Appl Environ Microbio; 1979;38:600-11.

61. Maksymowych AB, Simpson LL. A brief guide to the safe handling of biological toxins. In: Aktories K, editor. Bacterial toxins. London: Chapman and Hall; l997. p. 295-300.

62. Centers for Disease Control and Prevention. Diphtheria, tetanus, and pertussis: recommendations for vaccine use and other preventive measures: recommendations of the Immunization Practices Advisory Committee (ACIP). MMWR Morb Mortal Wkly Rep. 1991;40(RR10):1-28.

63. Centers for Disease Control and Prevention. Tetanus surveillance—United States, 1998-2000. MMWR Surveill Summ. 2003;52(SS-3):1-8.

64. Onderdonk A, Allen SD. Clostridium. In: Murray PR, Barron EJ, Pfaller M, et al, editors. Manual of clinical microbiology. 8th ed. Washington, DC: ASM Press. 2003. p. 574-86.

65. Centers for Disease Control and Prevention. Recommended childhood and adolescent immunization schedule—United States, January-June 2004. MMWR Morb Mortal Wkly Rep. 2004;53:Q1-4.

66. Geiss HK, Kiehl W, Thilo, W. A case report of laboratory-acquired diphtheria. Euro Surveill. 1997;2:67-8.

67. Centers for Disease Control and Prevention. Fatal respiratory diphtheria in a U.S. traveler to Haiti—Pennsylvania, 2003. MMWR Morb Mortal Wkly Rep. 2004; 52:1285-6

68. Dennis DT. Tularemia. In: Cohen J, Powderly WG, editors. Infectious diseases. Vol II. 2nd ed. Edinburgh: Mosby; 2004. p. 1649-53.

69. Burke DS. Immunization against tularemia: analysis of the effectiveness of live *Francisella tularensis* vaccine in prevention of laboratory-acquired tularemia. J Infect Dis. 1977;135:55-60.

70. Marshall BJ, Warren JR. Unidentified curved bacilli in the stomach of patients with gastritis and peptic ulceration. Lancet. 1984;1:1311-5.

71. Schauer DB. Enterohepatic *Helicobacter* species. In: Mobley HLT, Mendz GL, Hazell SL, editors. *Helicobacter pylori*: physiology and genetics. Washington, DC: ASM Press. 2001. p. 533-48.

72. Versalovic J, Fox JG. *Helicobacter*. In: Murray PR, Baron EJ, Jorgensen JH, et al, editors. Manual of clinical microbiology. 8th ed. Washington, DC: ASM Press. 2003. p. 915-28.

73. Marshall BJ, Armstrong JA, McGechie DB, et al. Attempt to fulfill Koch's postulates for pyloric *Campylobacter*. Med J Aust. 1985;142:436-9.

74. Matysiak-Budnik T, Briet F, Heyman M, et al. Laboratory-acquired *Helicobacter pylori* infection. Lancet. 1995;346:1489-90.

74. Mitchell HM. Epidemiology of infection. In: Mobley HLT, Mendz GL, Hazell SL, editors. *Helicobacter pylori*: physiology and genetics. Washington DC: ASM Press. 2001. p. 7-18, 90.

76. Perez-Perez GI. Accurate diagnosis of *Helicobacter pylori*. Culture, including transport. Gastroenterol Clin North Am. 2000;29:879-84.

77. Stout JE, Yu VL. Legionellosis. N Engl J Med. 1997;337:682-7.

78. Stout JE, Rihs JD, Yu VL. Legionella. In: Murray PR, Baron EJ, Jorgensen JH, et al, editors. Manual of clinical microbiology. 8th ed. Washington, DC: ASM Press. 2003. p. 809-23.

79. Benin AL, Benson RF, Beser RE. Trends in legionnaires' disease, 1980-1998: declining mortality and new patterns of diagnosis. Clin Infect Dis. 2002;35:1039-46.

80. Yu VL, Plouffe JF, Pastoris MC, et al. Distribution of *Legionella* species and serogroups isolated by culture in patients with sporadic community-acquired legionellosis: an international collaborative survey. J Infect Dis. 2002;186:127-8.

81. Fields BS, Benson RF, Besser RE. *Legionella* and legionnaires' disease: 25 years of investigation. Clin Microbiol Rev. 2002;15:506-26.

82. Centers for Disease Control and Prevention (CDC). 1976. Unpublished data. Center for Infectious Diseases. HEW, Public Health Service.

83. Friedman H, Yamamoto Y, Newton C, et al. Immunologic response and pathophysiology of *Legionella* infection. Semin Respir Infect. 1998;13:100-8.

84. Fabbi M, Pastoris MC, Scanziani E, et al. Epidemiological and environmental investigations of *Legionella pneumophila* infection in cattle and case report of fatal pneumonia in a calf. J Clin Microbiol. 1998;36:1942-7.

85. Muder RR, Yu VL, Woo A. Mode of transmission of *Legionella pneumophila*. A critical review. Arch Intern Med. 1986;146:1607-12.

86. Lowry PW, Tompkins LS. Nosocomial legionellosis: a review of pulmonary and extrapulmonary syndromes. Am J Infect Control. 1993;21:21-7.

87. Levett PN. *Leptospira* and *Leptonema*. In: Murray PR, Baron EJ, Jorgensen JH, et al, editors. Manual of clinical microbiology. 8th ed. Washington, DC: ASM Press. 2003. p. 929-36.

88. Barkin RM, Guckian JC, Glosser JW. Infection by *Leptospira ballum*: a laboratory-associated case. South Med J. 1974;67:155 passim.

89. Bolin CA, Koellner P. Human-to-human transmission of *Leptospira interrogans* by milk. J Infect Dis. 1988;158:246-7.
90. Stoenner HG, Maclean D. Leptospirosis (ballum) contracted from Swiss albino mice. AMA Arch Intern Med. 1958;101:606-10.
91. Schuchat A, Swaminathan B, Broome CV. Epidemiology of human listeriosis. Clin Microbiol Rev. 1991;4:169-83.
92. Armstrong D. *Listeria monocytogenes.* In: Mandell GL, Bennett JE, Dolin R, editors. Principles and practices of infectious diseases. 4th ed. NY: Churchill Livingstone. 1995. p. 1880-85.
93. McLauchlin J, Low JC. Primary cutaneous listeriosis in adults: an occupational disease of veterinarians and farmers. Vet Rec. 1994;135:615-7.
94. Ortel S. Listeriosis during pregnancy and excretion of *listeria* by laboratory workers. Zentralbl Bakteriol [Orig. A]. 1975;231:491-502.
95. Centers for Disease Control and Prevention. Update: foodborne listeriosis-United States, 1988-1990. MMWR Morb Mortal Wkly Rep. 1992;41:251, 257-8.
96. Ryser ET, Marth EH, editors. *Listeria*, Listeriosis, and food safety. 2nd ed. New York: Marcel Dekker, 1999.
97. Gellin BG, Broome CV. Listeriosis. JAMA. 1989;261:1313-20.
98. Walsh GP, Storrs EE, Burchfield HP, et al. Leprosy-like disease occurring naturally in armadillos. J Reticuloendothel Soc. 1975;18:347-51.
99. Donham KJ, Leininger JR. Spontaneous leprosy-like disease in a chimpanzee. J Infect Dis. 1977;136:132-6.
100. Meyers WM, Walsh GP, Brown HL, et al. Leprosy in a mangabey monkey-naturally acquired infection. Int J Lepr Other Mycobact Dis. 1985;53:1-14.
101. Marchoux PE. Un cas d'inoculation accidentelle du bacilli de Hanson en pays non lepreux. Int J Lepr. 1934;2:1-7.
102. Parritt RJ, Olsen RE. Two simultaneous cases of leprosy developing in tattoos. Am J Pathol. 1947;23:805-17.
103. Grist NR, Emslie JA. Infections in British clinical laboratories, 1982-3. J ClinPathol. 1985;38:721-5.
104. Muller HE. Laboratory-acquired mycobacterial infection. Lancet. 1988;2:331.
105. Pike RM, Sulkin SE, Schulze ML. Continuing importance of laboratory-acquired infections. Am J Public Health Nations Health. 1965;55:190-9.
106. Reid DD. Incidence of tuberculosis among workers in medical laboratories. Br Med J. 1957;(5035):10-4.
107. Kaufmann AF, Andersone DC. Tuberculosis control in nonhuman primates. In: Montali RJ, editor. Mycobacterial infections of zoo animals. Washington, DC: Smithsonian Institution Press. 1978. p. 227-34.
108. Allen BW. Survival of tubercle bacilli in heat-fixed sputum smears. J Clin Pathol. 1981;34:719-22.
109. Smithwick RW, Stratigos CB. Preparation of acid-fast microscopy smears for proficiency testing and quality control. J Clin Microbiol. 1978;8:110-1.
110. Oliver J, Reusser TR. Rapid method for the concentration of tubercle bacilli. Am Rev Tuberc. 1942;45:450-2.

111. Richmond JY, Knudsen RC, Good RC. Biosafety in the clinical mycobacteriology laboratory. Clin Lab Med. 1996;16:527-50.

112. Janda WM, Knapp JS. *Neisseria* and *Moraxella catarrhalis*. In: Murray PR, Baron EJ, Jorgensen JH, et al, editors. Manual of clinical microbiology. 8th ed. Washington, DC: ASM Press. 2003. p. 585-608.

113. Diena BB, Wallace R, Ashton FE, et al. Gonococcal conjunctivitis: accidental infection. Can Med Assoc J. 1976;115:609-12.

114. Malhotra R, Karim QN, Acheson JF. Hospital-acquired adult gonococcal conjunctivitis. J Infect. 1998;37:305.

115. Bruins SC, Tight RR. Laboratory acquired gonococcal conjunctivitis. JAMA. 1979;241:274.

116. Zajdowicz TR, Kerbs SB, Berg SW, et al. Laboratory-acquired gonococcal conjunctivitis: successful treatment with single-dose ceftriaxone. Sex Transm Dis. 1984;11:28-9.

117. Centers for Disease Control and Prevention [www.cdc.gov].Atlanta: The Centers; [updated 2006 April]. Gonorrhea—CDC Fact Sheet; [about five screens]. Available from: http://www.cdc.gov/std/Gonorrhea/STDFact-gonorrhea.htm.

118. Centers for Disease Control and Prevention. Prevention and control of meningococcal disease. Recommendations of the Advisory Committee on Immunization Practices (ACIP). MMWR Recomm Rep. 2000;49(RR-7):1-10.

119. Boutet R, Stuart JM, Kaczmarski EB, et al. Risk of laboratory-acquired meningococcal disease. J Hosp Infect. 2001;49:282-4.

120. Centers for Disease Control and Prevention. Laboratory-acquired meningococcal diseases—United States, 2000. MMWR Morb Mortal Wkly Rep. 2002;51:141-4.

121. Centers for Disease Control and Prevention. Immunization of health-care workers: recommendations of the Advisory Committee on Immunization Practices (ACIP) and the Hospital Infection Control Practices Advisory Committee (HICPAC). MMWR Recomm Rep. 1997;46(RR18):1-35.

122. Centers for Disease Control and Prevention. Laboratory-acquired meningococcemia—California and Massachusetts. MMWR Morb Mortal Wkly Rep. 1991;40:46-7,55.

123. Bopp CA, Brenner FW, Fields PI, et al. Escherichia, Shigella, and Salmonella. In: Murray PR, Baron EJ, Jorgenson JH, et al, editors. Manual of clinical microbiology. 8th ed. Washington, DC: ASM Press. 2003. p. 654-71.

124. Grist NR, Emslie JA. Infections in British clinical laboratories, 1984-5. J Clin Pathol. 1987;40:826-829. Nicklas W. Introduction of salmonellae into a centralized laboratory animal facility by infected day-old chicks. Lab Anim. 1987;21:161-3.

125. Steckelberg JM, Terrell CL, Edson RS. Laboratory-acquired Salmonella typhimurium enteritis: association with erythema nodosum and reactive arthritis. American J Med. 1988;85:705-7.

126. Baumberg S, Freeman R. Salmonella typhimurium strain LT-2 is still pathogenic for man. J Gen Microbiol. 1971;65:99-100.

127. Salmonellosis. In: Chin J, Ascher M, editors. Control of communicable diseases. Washington, DC: American Public Health Association. 2000. p. 440-4.

128. Blaser MJ, Hickman FW, Farmer JJ 3rd. Salmonella typhi: the laboratory as a reservoir of infection. J Infect Dis. 1980;142:934-8.

129. Grist NR, Emslie JA. Infections in British clinical laboratories, 1984-5. J Clin Pathol. 1987;40:826-829.

130. Sewell DL. Laboratory-associated infections and biosafety. Clin Microbiol Rev. 1995;8:389-405.

131. Centers for Disease Control and Prevention. Typhoid immunization recommendations of the Advisory Committee on Immunization Practices (ACIP). MMWR Morb Mortal Wkly Rep. 1994;43(RR14):1-7.

132. Laboratory acquired infection with Escherichia coli O157. Commun Dis Rep CDR Wkly 1994;4:29.

133. Escherichia coli O157 infection acquired in the laboratory. Commun Dis Rep CDR Wkly 1996;6:239.

134. Booth L, Rowe B. Possible occupational acquisition of Escherichia coli O157 infection. Lancet. 1993;342:1298-9.

135. Burnens AP, Zbinden R, Kaempf L, et al. A case of laboratory-acquired infection with Escherichia coli O157:H7. Zentralbl Bakteriol. 1993;279:512-7.

136. Rao GG, Saunders BP, Masterton RG. Laboratory-acquired verotoxin producing Escherichia coli (VTEC) infection. J Hosp Infect. 1996;33:228-30.

137. Spina N, Zansky S, Dumas N, et al. Four laboratory-associated cases of infection with Escherichia coli O157:H7. J Clin Microbiol. 2005;43:2938-9.

138. Chin J. Diarrhea caused by Escherichia coli. In: Chin J, ed. Control of Communicable Diseases. 17th ed. Washington, DC: American Public Health Association; 2000. p. 155-65.

139. Nielsen EM, Skov MN, Madsen JJ, et.al. Verocytotoxin-producing Escherichia coli in wild birds and rodents in close proximity to farms. Appl Envrion Microbiol. 2004;70:6944-7.

140. Mermel LA, Josephson SL, Dempsey J, et al. Outbreak of Shigella sonnei in a clinical microbiology laboratory. J Clin Microbiol. 1997;35:3163-5.

141. Walker D, Campbell D. A survey of infections in United Kingdom laboratories, 1994-1995. J Clin Pathol. 1999;52:415-8.

142. Parsot C, Sansonetti PJ. Invasion and the pathogenesis of Shigella infection. Curr Top Microbiol Immunol. 1996;209:25-42.

143. National Research Council. Zoonoses. In: Occupational health and safety in the care and use of research animals. Washington, DC: National Academy Press. 1997. p. 65-106.

144. Norris SJ, Pope V, Johnson RE, et al. Treponema and other human host-associated spirochetes. In: Murray PR, Barron EJ, Jorgensen JH, et al, editors. Manual of clinical microbiology. 8th ed. Washington, DC: ASM Press. 2003. p. 955-71.

145. Fitzgerald JJ, Johnson RC, Smith M. Accidental laboratory infection with Treponema pallidum, Nichols strain. J Am Vener Dis Assoc. 1976;3:76-8.

146. Magnuson HJ, Thomas EW, Olansky S, et al. Inoculation syphilis in human volunteers. Medicine. 1956;35:33-82.

147. Farmer JJ, Janda JM, Birkhead K. Vibrio. In: Murray PR, Baron EJ, Jorgensen JH, et al, editors. Manual of clinical microbiology. 8th ed. Washington DC: ASM Press. 2003. p. 706-18.

148. Sheehy TW, Sprinz H, Augerson WS, et al. Laboratory Vibrio cholerae infection in the United States. JAMA. 1966;197:321-6.

149. Lee KK, Liu PC, Huang CY. Vibrio parahaemolyticus infectious for both humans and edible mollusk abalone. Microbes Infect. 2003;5:481-5.

150. Morris JG Jr. Cholerae and other types of vibriosis: a story of human pandemics and oysters on the half shell. Clin Infect Dis. 2003;37:272-80.

151. Reidl J, Klose KE. Vibrio cholerae and cholera: out of the water and into the host. FEMS Microbiol Rev. 2002;26:125-39.

152. Daniels NA, Ray B, Easton A, et al. Emergence of a new Vibrio parahaemolyticus serotype in raw oysters. JAMA. 2000;284:1541-5. Erratum in: JAMA 2001;285:169.

153. American Public Health Association. Cholera and other vibrioses. In: Heymann HL, editor. Control of communicable diseases. 18th ed. Baltimore (MD): United Book Press, Inc. 2004. p. 103-15.

154. Dennis DT, Gage, KL. Plague. In: Cohen J, Powderly WG, editors. Infectious diseases. Vol II. 2nd ed. Edinburgh: Mosby; 2004. p. 1641-8.

155. Russell P, Eley SM, Bell DL, et al. 1996. Doxycycline or ciprofloxacin prophylaxis and therapy against Yersinia pestis infection in mice. J Antimicrobial Chemother. 1996;37:769-74.

156. Burmeister RW, Tigertt WD, Overholt EL. Laboratory-acquired pneumonic plague. Report of a case and review of previous cases. Ann Intern Med. 1962;56:789-800.

157. Higgs S. Arthropod containment guidelines. Vector-Borne and Zoonotic Diseases. 2003;3:57-98.

Section VIII-B: Fungal Agents

Blastomyces dermatitidis

Blastomyces dermatitidis is a dimorphic fungal pathogen existing in nature and in laboratory cultures at room temperature as a filamentous mold with asexual spores (conidia) that are the infectious particles; these convert to large budding yeasts under the appropriate culture conditions *in vitro* at 37°C and in the parasitic phase *in vivo* in warm-blooded animals. The sexual stage is an Ascomycete with infectious ascospores.

Occupational Infections

Three groups are at greatest risk of laboratory-acquired infection: microbiologists, veterinarians and pathologists.[1] Laboratory-associated local infections have been reported following accidental parenteral inoculation with infected tissues or cultures containing yeast forms of *B. dermatitidis*.[2-8] Pulmonary infections have occurred following the presumed inhalation of conidia from mold-form cultures; two persons developed pneumonia and one had an osteolytic lesion from which *B. dermatitidis* was cultured.[9,10] Presumably, pulmonary infections are associated only with sporulating mold forms.

Natural Modes of Infection

The fungus has been reported from multiple geographically separated countries, but is best known as a fungus endemic to North America and in association with plant material in the environment. Infections are not communicable, but require common exposure from a point source. Although presumed to dwell within the soil of endemic areas, *B. dermatitidis* is extremely difficult to isolate from soil. Outbreaks associated with the exposure of people to decaying wood have been reported.[11]

Laboratory Safety and Containment Recommendations

Yeast forms may be present in the tissues of infected animals and in clinical specimens. Parenteral (subcutaneous) inoculation of these materials may cause local skin infection and granulomas. Mold form cultures of *B. dermatitidis* containing infectious conidia, and processing of soil or other environmental samples, may pose a hazard of aerosol exposure.

BSL-2 and ABSL-2 practices, containment equipment, and facilities are recommended for activities with clinical materials, animal tissues, yeast-form cultures, and infected animals. BSL-3 practices, containment equipment, and facilities are required for handling sporulating mold-form cultures already identified as *B. dermatitidis* and soil or other environmental samples known or likely to contain infectious conidia.

Special Issues

Transfer of Agent Importation of this agent may require CDC and/or USDA importation permits. Domestic transport of this agent may require a permit from USDA/APHIS/VS. A DoC permit may be required for the export of this agent to another country. See Appendix C for additional information.

———————————

Coccidioides immitis and Coccidioides posadasii

Coccidioides spp. is endemic to lower sonoran deserts of the western hemisphere including northern Mexico, southern Arizona, central and southern California, and west Texas. The original species (*C. immitis*) has been divided into *C. immitis* and *C. posadasii*.[12] These species are dimorphic fungal pathogens existing in nature and in laboratory cultures at room temperature as filamentous molds with asexual spores (single-cell arthroconidia three to five microns in size) that are the infectious particles that convert to spherules under the appropriate culture conditions *in vitro* at 37°C and *in vivo* in warm-blooded animals.

Occupational Infections

Laboratory-associated coccidioidomycosis is a documented hazard of working with sporulating cultures of *Coccidioides* spp.[13-15] Occupational exposure has also been associated in endemic regions with archeology[16] and high dust exposure.[17] Attack rates for laboratory and occupational exposure are higher than for ambient exposure when large numbers of spores are inhaled. Smith reported that 28 of 31 (90%) laboratory-associated infections in his institution resulted in clinical disease, whereas more than half of infections acquired in nature were asymptomatic.[18] Risk of respiratory infection from exposure to infected tissue or aerosols of infected secretions is very low. Accidental percutaneous inoculation has typically resulted in local granuloma formation.[19]

Natural Modes of Infection

Single spores can produce ambient infections by the respiratory route. Peak exposures occur during arid seasons. *Coccidioides* spp. grow in infected tissue as larger multicellular spherules, up to 70 microns in diameter and pose little or no risk of infection from direct exposure.

The majority of ambient infections is subclinical and results in life-long protection from subsequent exposures. The incubation period is one to three weeks and manifests as a community-acquired pneumonia with immunologically mediated fatigue, skin rashes, and joint pain. One of the synonyms for coccidioidomycosis is desert rheumatism. A small proportion of infections is complicated by hematogenous dissemination from the lungs to other organs, most frequently skin, the skeleton, and the meninges. Disseminated infection is much more likely in persons with cellular immunodeficiencies (AIDS, organ transplant recipient, lymphoma).

Laboratory Safety and Containment Recommendations

Because of their size, the arthroconidia are conducive to ready dispersal in air and retention in the deep pulmonary spaces. The much larger size of the spherule considerably reduces the effectiveness of this form of the fungus as an airborne pathogen.

Spherules of the fungus may be present in clinical specimens and animal tissues, and infectious arthroconidia in mold cultures and soil or other samples from natural sites. Inhalation of arthroconidia from environmental samples or cultures of the mold form is a serious laboratory hazard. Personnel should be aware that infected animal or human clinical specimens or tissues stored or shipped in such a manner as to promote germination of arthroconidia pose a theoretical laboratory hazard.

BSL-2 practices, containment equipment, and facilities are recommended for handling and processing clinical specimens, identifying isolates, and processing animal tissues. ABSL-2 practices, containment equipment, and facilities are recommended for experimental animal studies when the route of challenge is parenteral.

BSL-3 practices, containment equipment, and facilities are recommended for propagating and manipulating sporulating cultures already identified as *Coccidioides* spp. and for processing soil or other environmental materials known to contain infectious arthroconidia. Experimental animal studies should be done at BSL-3 when challenge is via the intranasal or pulmonary route.

Special Issues

Select Agent Some *Coccidioides* spp. are select agents requiring registration with CDC and/or USDA for possession, use, storage and/or transfer. See Appendix F for additional information.

Transfer of Agent Importation of this agent may require CDC and/or USDA importation permits. Domestic transport of this agent may require a permit from USDA/APHIS/VS. A DoC permit may be required for the export of this agent to another country. See Appendix C for additional information.

Cryptococcus Neoformans

Cryptococcus neoformans is a monomorphic fungal pathogen existing in nature, in laboratory cultures at room temperature and *in vivo* as a budding yeast. The sexual stage is grouped with the Basidiomycetes and is characterized by sparse hyphal formation with basidiospores. Both basidiospores and asexual yeasts are infectious.

Occupational Infections

Accidental inoculation of a heavy inoculum of *C. neoformans* into the hands of laboratory workers has occurred during injection or necropsy of laboratory animals.[20,21] Either a local granuloma or no lesion was reported, suggesting low pathogenicity by this route. Respiratory infections as a consequence of laboratory exposure have not been recorded.

Natural Modes of Infection

The fungus is distributed worldwide in the environment and is associated with pigeon feces. Infections are not transmissible from person-to-person, but require common exposure via the respiratory route to a point source.

Laboratory Safety and Containment Recommendations

Accidental parenteral inoculation of cultures or other infectious materials represents a potential hazard to laboratory personnel, particularly to those who may be immunocompromised. Bites by experimentally infected mice and manipulations of infectious environmental materials (e.g., pigeon feces) may also represent a potential hazard to laboratory personnel. *C. neoformans* has been isolated from bedding of cages housing mice with pulmonary infection indicating the potential for contamination of cages and animal facilities by infected animals.[22] Reports of cutaneous cryptococcal infection following minor skin injuries suggests that localized infection may complicate skin injuries incurred in laboratories that handle *C. neoformans*.[23]

BSL-2 and ABSL-2 practices, containment equipment, and facilities are recommended for activities with known or potentially infectious clinical, environmental, or culture materials and with experimentally infected animals. This agent and any samples that may contain this agent should also be handled in a Class II BSC.

Transfer of Agent Importation of this agent may require CDC and/or USDA importation permits. Domestic transport of this agent may require a permit from USDA/APHIS/VS. A DoC permit may be required for the export of this agent to another country. See Appendix C for additional information.

Histoplasma capsulatum

Histoplasma capsulatum is a dimorphic fungal pathogen existing in nature and in laboratory cultures at room temperature as a filamentous mold with asexual spores (conidia); these are the infectious particles that convert to small budding yeasts under the appropriate culture conditions *in vitro* at 37°C and in the parasitic phase *in vivo*. The sexual stage is an Ascomycete with infectious ascospores.

Occupational Infections

Laboratory-associated histoplasmosis is a documented hazard in facilities conducting diagnostic or investigative work.[24-27] Pulmonary infections have resulted from handling mold form cultures.[28,29] Local infection has resulted from skin puncture during autopsy of an infected human,[30] from accidental needle inoculation of a viable culture,[31] and from spray from a needle into the eye.[32] Collecting and processing soil samples from endemic areas has caused pulmonary infections in laboratory workers.[33] Conidia are resistant to drying and may remain viable for long periods of time. The small size of the infective conidia (less than 5 microns) is conducive to airborne dispersal and intrapulmonary retention. Work with experimental animals suggests that hyphal fragments are capable of serving as viable inocula.[24]

Natural Modes of Infection

The fungus is distributed worldwide in the environment and is associated with starling and bat feces. It has been isolated from soil, often in river valleys, between latitudes 45°N and 45°S. Histoplasmosis is naturally acquired by the inhalation of infectious particles, usually microconidia.[24] Infections are not transmissible from person-to-person, but require common exposure to a point source.

Laboratory Safety and Containment Recommendations

The infective stage of this dimorphic fungus (conidia) is present in sporulating mold form cultures and in soil from endemic areas. The yeast form in tissues or fluids from infected animals may produce local infection following parenteral inoculation or splash onto mucous membranes.

BSL-2 and ABSL-2 practices, containment equipment, and facilities are recommended for handling and processing clinical specimens, identifying isolates, animal tissues and mold cultures, identifying cultures in routine diagnostic laboratories, and for inoculating experimental animals, regardless of route. Any culture identifying dimorphic fungi should be handled in a Class II BSC.

BSL-3 practices, containment equipment, and facilities are recommended for propagating sporulating cultures of *H. capsulatum* in the mold form, as well as processing soil or other environmental materials known or likely to contain infectious conidia.

Special Issues

Transfer of Agent Importation of this agent may require CDC and/or USDA importation permits. Domestic transport of this agent may require a permit from USDA/APHIS/VS. A DoC permit may be required for the export of this agent to another country. See Appendix C for additional information.

Sporothrix schenckii

Sporothrix schenckii is a dimorphic fungal pathogen existing in nature and in laboratory cultures at room temperature as a filamentous mold with asexual spores (conidia); these are the infectious particles that convert to small budding yeasts in the parasitic phase *in vivo*. The sexual stage is unknown.

Occupational Infections

Most cases of sporotrichosis are reported sporadically following accidental inoculation with contaminated material. Large outbreaks have been documented in persons occupationally or recreationally exposed to soil or plant material containing the fungus. However, *S. schenckii* has caused a substantial number of local skin or eye infections in laboratory personnel.[34] Most occupational cases have been associated with accidents and have involved splashing culture material into the eye,[35,36] scratching,[37] or injecting[38] infected material into the skin or being bitten by an experimentally infected animal.[39,40] Skin infections in the absence of trauma have resulted also from handling cultures[41-43] or necropsy of animals[44] without any apparent trauma.

Natural Modes of Infection

The fungus is distributed worldwide in the environment and is associated with sphagnum moss and gardening, often involving sphagnum moss and traumatic implantation. Infections are not transmissible from person-to-person, but require common exposure to a point source. Rare respiratory and zoonotic infections occur. It is thought that naturally occurring lung disease results from inhalation.

Laboratory Safety and Containment Recommendations

Although localized skin and eye infections have occurred in an occupational setting, no pulmonary infections have been reported as a result from laboratory exposure. It should be noted that serious disseminated infections have been reported in immunocompromised persons.[45]

BSL-2 and ABSL-2 practices, containment equipment, and facilities are recommended for laboratory handling of suspected clinical specimens, soil and vegetation, and experimental animal activities with *S. schenckii*. Gloves should be worn during manipulation of *S. schenckii* and when handling experimentally infected animals. Any culture identifying dimorphic fungi should be handled in a Class II BSC.

Special Issues

Transfer of Agent Importation of this agent may require CDC and/or USDA importation permits. Domestic transport of this agent may require a permit from USDA/APHIS/VS. A DoC permit may be required for the export of this agent to another country. See Appendix C for additional information.

Dermatophytes (Epidermophyton, Microsporum, and Trichophyton)

The dermatophytes are biologically related species of the genera, *Epidermophyton, Microsporum,* and *Trichophyton* that exist as monomorphic pathogens in nature, in laboratory cultures at room temperature and *in vivo* as filamentous molds. The sexual stages, when known, are Ascomycetes with infectious ascospores. These fungi are distributed worldwide, with particular species being endemic in particular regions. The species are grouped by natural environment habitat as being primarily associated with humans (anthrophilic), other animals (zoophilic), or soil (geophilic).

Occupational Infections

Although skin, hair, and nail infections by these molds are among the most prevalent of human infections, the processing of clinical material has not been associated with laboratory infections. Infections have been acquired through contacts with naturally or experimentally infected laboratory animals (mice, rabbits, guinea pigs, etc.) and, occasionally, with handling cultures.[26,29,45,46]

Systemic dermatophytosis is a rare condition. Superficial chronic infections occur frequently among immunocompromised individuals as well as elderly and diabetic persons. Susceptible individuals should use extra caution.[47-50]

Natural Modes of Infection

Infections can be transmissible from person-to-person, or acquired from common exposure to a point source. The dermatophytes cause infection (dermatophytosis) by invading the keratinized tissues of living animals and are among the most common infectious agents of humans. This fungal group encompasses members of three genera: *Epidermophyton, Microsporum,* and *Trichophyton.* The severity of infection depends on the infective species or strain, the anatomic site and other host factors. One of the most severe dermatophytoses is favus, a disfiguring disease of the scalp caused by *Trychophyton schoenleinii.*

Laboratory Safety and Containment Recommendations

Dermatophytes pose a moderate potential hazard to individuals with normal immune status. In the clinical laboratory setting, the inappropriate handling of cultures is the most common source of infection for laboratory personnel. The most common laboratory procedure for detection of the infective dermatophyte is the direct microscopic examination of contaminated skin, hair, and nails, followed by its isolation and identification on appropriated culture media. Direct contact with contaminated skin, hair, and nails of humans could be another source of infection.[48,49] In research laboratories, dermatophytosis can be acquired by contact with contaminated soil (source of infection: geophilic species) or animal hosts (source of infection: zoophilic species).

BSL-2 and ABSL-2 practices, containment equipment, and facilities are recommended for handling cultures and soil samples. Any culture identifying dimorphic fungi should be handled in a Class II BSC.

Special Issues

Transfer of Agent Importation of this agent may require CDC and/or USDA importation permits. Domestic transport of this agent may require a permit from USDA/APHIS/VS. A DoC permit may be required for the export of this agent to another country. See Appendix C for additional information.

_____

Miscellaneous Molds

Several molds have caused serious infection in immunocompetent hosts following presumed inhalation or accidental subcutaneous inoculation from environmental sources. These agents include the dimorphic mold, *Penicillium marneffei*, and the dematiaceous (brown-pigmented) molds, *Bipolaris* species, *Cladophialophora bantiana*, *Exophiala (Wangiella) dermatitidis*, *Exserohilum* species, *Fonsecaea pedrosoi*, *Ochroconis gallopava (Dactylaria gallopava)*, *Ramichloridium mackenziei (Ramichloridium obovoideum)*, *Rhinocladiella atrovirens*, and *Scedosporium prolificans*.[51]

Occupational Infections

Even though no laboratory-acquired infections appear to have been reported with most of these agents, the gravity of naturally-acquired illness is sufficient to merit special precautions in the laboratory. *Penicillium marneffei* has caused a localized infection in a laboratory worker.[52] It also caused a case of laboratory-acquired disseminated infection following presumed inhalation when an undiagnosed HIV-positive individual visited a laboratory where students were handling cultures on the open bench.[53]

Natural Modes of Infection

The natural mode of infection varies by specific species; most are poorly characterized.

Laboratory Safety and Containment Recommendations

Inhalation of conidia from sporulating mold cultures or accidental injection into the skin during infection of experimental animals are potential risks to laboratory personnel.

   BSL-2 practices, containment equipment, and facilities are recommended for propagating and manipulating cultures known to contain these agents. Any culture identifying dimorphic fungi should be handled in a Class II BSC.

Special Issues

Transfer of Agent Importation of this agent may require CDC and/or USDA importation permits. Domestic transport of this agent may require a permit from USDA/APHIS/VS. A DoC permit may be required for the export of this agent to another country. See Appendix C for additional information.

References

1. DiSalvo AF. The epidemiology of blastomycosis. In: Al-Doory Y, Di Salvo AF, editors. Blastomycosis. New York: Plenum Medical Book Company; 1992. p. 75-104.
2. Evans N. A clinical report of a case of blastomycosis of the skin from accidental inoculation. JAMA. 1903;40:1172-5.
3. Harrekkm ER. The known and the unknown of the occupational mycoses. In: Occupational diseases acquired from animals. Continued education series no. 124. Ann Arbor: University of Michigan School of Public Health; 1964. p. 176-8.
4. Larsh HW, Schwarz J. Accidental inoculation blastomycosis. Cutis. 1977;19:334-6.
5. Larson DM, Eckman MR, Alber RL, et al. Primary cutaneous (inoculation) blastomycosis: an occupational hazard to pathologists. Amer J Clin Pathol. 1983;79:253-5.
6. Wilson JW, Cawley EP, Weidman FD, et al. Primary cutaneous North American blastomycosis. AMA Arch Dermatol. 1955;71:39-45.
7. Graham WR, Callaway JL. Primary inoculation blastomycosis in a veterinarian. J Am Acad Dermatol.1982;7:785-6.
8. Schwarz J, Kauffman CA. Occupational hazards from deep mycoses. Arch Dermatol. 1977;113:1270-5.

9.   Baum GL, Lerner PI. Primary pulmonary blastomycosis: a laboratory acquired infection. Ann Intern Med. 1970;73:263-5.

10. Denton JF, Di Salvo AF, Hirsch ML. Laboratory-acquired North American blastomycosis. JAMA. 1967;199:935-6.

11. Sugar AM, Lyman CA. A practical guide to medically important fungi and the diseases they cause. Philadelphia: Lippincott-Raven; 1997.

12   Fisher MC, Koenig GL, White TJ, et al. Molecular and phenotypic description of *Coccidioides posadasii* sp nov, previously recognized as the non-California population of *Coccidioides immitis*. Mycologia. 2002;94:73-84.

13. Pappagianis D. Coccidioidomycosis (San Joaquin or Valley Fever). In: DiSalvo A, editor. Occupational mycoses. Philadelphia: Lea and Febiger; 1983. p. 13-28.

14. Nabarro JDN. Primary pulmonary coccidioidomycosis: case of laboratory infection in England. Lancet. 1948;1:982-4.

15. Smith CE. The hazard of acquiring mycotic infections in the laboratory. Presented at 78th Annual Meeting American Public Health Association; 1950; St. Louis, MO;1950.

16. Werner SB, Pappagianis D, Heindl I, et al. An epidemic of coccidioidomycosis among archeology students in northern California. N Engl J Med. 1972;286:507-12.

17. Crum N, Lamb C, Utz G, et al. Coccidioidomycosis outbreak among United States Navy SEALs training in a *Coccidioides immitis*-endemic area-Coalinga, California. J Infect Dis. 2002;186:865-8.

18. Wilson JW, Smith CE, Plunkett OA. Primary cutaneous coccidioidomycosis: the criteria for diagnosis and a report of a case. Calif Med. 1953;79:233-9.

19. Tomlinson CC, Bancroft P. Granuloma *Coccidioides*: report of a case responding favorably to antimony and potassium tartrate. JAMA. 1928; 91:947-51.

20. Halde C. Percutaneous *Cryptococcus neoformans* inoculation without infection. Arch Dermatol. 1964; 89:545.

21. Casadevall A, Mukherjee J, Yuan R, et al. Management of injuries caused by *Cryptococcus neoformans*-contaminated needles. Clin Infect Dis. 1994;19:951-3.

22. Nosanchuk JD, Mednick A, Shi L, et al. Experimental murine cryptococcal infection results in contamination of bedding with *Cryptococcus neoformans*. Contemp Top Lab Anim Sci. 2003;42;9-412.

23. Bohne T, Sander A, Pfister-Wartha A, et al. Primary cutaneous cryptococcosis following trauma of the right forearm. Mycoses. 1996;39:457-9.

24. Furcolow ML. Airborne histoplasmosis. Bacteriol Rev. 1961;25:301-9.

25. Pike RM. Past and present hazards of working with infectious agents. Arch Path Lab Med. 1978;102:333-6.

26. Pike RM. Laboratory-associated infections: Summary and analysis of 3,921 cases. Hlth Lab Sci. 1976;13:105-14.

27. Schwarz J, Kauffman CA. Occupational hazards from deep mycoses. Arch Dermatol. 1977;113:1270-5.

28. Murray JF, Howard D. Laboratory-acquired histoplasmosis. Am Rev Respir Dis. 1964;89:631-40.
29. Sewell DL. Laboratory-associated infections and biosafety. Clin Micro Rev. 1995;8:389-405.
30. Tosh FE, Balhuizen J, Yates JL, et al. Primary cutaneous histoplasmosis: report of a case. Arch Intern Med. 1964;114:118-0.
31. Tesh RB, Schneidau JD. Primary cutaneous histoplasmosis. N Engl J Med. 1966;275:597-9.
32. Spicknall CG, Ryan RW, Cain A. Laboratory-acquired histoplasmosis. N Eng J Med. 1956;254:210-4.
33. Vanselow NA, Davey WN, Bocobo FC. Acute pulmonary histoplasmosis in laboratory workers: report of two cases. J Lab Clin Med. 1962; 59:236-43.
34. Ishizaki H, Ikeda M, Kurata Y. Lymphocutaneous sporotrichosis caused by accidental inoculation. J Dermatol. 1979;6:321-3.
35. Fava A. Un cas de sporotrichose conjonctivale et palpebrale primitives. Ann Ocul (Paris). 1909;141:338-43.
36. Wilder WH, McCullough CP. Sporotrichosis of the eye. JAMA. 1914;62:1156-60.
37. Carougeau, M. Premier cas Africain de sporotrichose de deBeurmann: transmission de la sporotrichose du mulet a l'homme. Bull Mem Soc Med Hop (Paris). 1909;28:507-10.
38. Thompson DW, Kaplan W. Laboratory-acquired sporotrichosis. Sabouraudia. 1977;15:167-70.
39. Jeanselme E, Chevallier P. Chancres sporotrichosiques des doigts produits par la morsure d'un rat inocule de sporotrichose. Bull Mem Soc Med Hop (Paris). 1910;30:176-78.
40. Jeanselme E, Chevallier P. Transmission de la sporotrichose a l'homme par les morsures d'un rat blanc inocule avec une nouvelle variete de Sporotrichum: Lymphangite gommeuse ascendante. Bull Mem Soc Med Hop (Paris). 1911;31:287-301.
41. Meyer KF. The relationship of animal to human sporotrichosis: studies on American sporotrichosis III. JAMA. 1915;65:579-85.
42. Norden A. Sporotrichosis: clinical and laboratory features and a serologic study in experimental animals and humans. Acta Pathol Microbiol Scand. 1951;89:3-119.
43. Cooper CR, Dixon DM, Salkin IF. Laboratory-acquired sporotrichosis. J Med Vet Mycol. 1992;30:169-71.
44. Fielitz H. Ueber eine Laboratoriumsinfektion mit dem Sporotrichum de Beurmanni. Centralbl Bakteriol Parasitenk Abt I Orig. 1910;55:361-70.
45. Sugar AM, Lyman CA. A practical guide to medically important fungi and the diseases they cause. Philadelphia: Lippincott-Raven; 1997. p. 86-7.
46. Voss A, Nulens E. Prevention and control of laboratory-acquired infections. In: Murray PR, Baron EJ, Jorgensen JH et al, editors. Manual of Clinical Microbiology. 8th edition. Washington, DC: ASM Press; 2003. p. 109-20.
47. Kamalam A, Thambiah AS. *Trichophyton* simii infection due to laboratory accident. Dermatologica. 1979; 159:180-1.

48. Scher RK, Baran R. Onychomycosis in clinical practice: factors contributing to recurrence. Br J Dermatol. 2003;149 Suppl 65:5-9.
49. Gupta AK. Onychomycosis in the elderly. Drugs Aging. 2000;16:397-407.
50. Romano C, Massai L, Asta F, et al. Prevalence of dermatophytic skin and nail infections in diabetic patients. Mycoses. 2001;44:83-6.
51. Brandt ME, Warnock DW. Epidemiology, clinical manifestations, and therapy of infections caused by dematiaceous fungi. J Chemother. 2003;1536-47.
52. Segretain G. *Penicillium marneffii*, n. sp., agent of a mycosis of the reticuloendothelial system. Mycopathologia. 1959;11:327-53. [French]
53. Hilmarsdottir I, Coutell A, Elbaz J, et al. A French case of laboratory-acquired disseminated *Penicillium marneffei* infection in a patient with AIDS. Clin Infect Dis. 1994;19:357-58.

Section VIII-C: Parasitic Agents

General Issues

Additional details about occupationally-acquired cases of parasitic infections, as well as recommendations for post exposure management, are provided elsewhere.[1-3] Effective antimicrobial treatment is available for most parasitic infections.[4] Immunocompromised persons should receive individualized counseling (specific to host and parasite factors) from their personal healthcare provider and their employer about the potential risks associated with working with live organisms.

BSL-2 and ABSL-2 practices,[5] containment equipment, and facilities are recommended for activities with infective stages of the parasites discussed in this chapter.

Microsporidia, historically considered parasites, are now recognized by most experts to be fungi; however, microsporidia are maintained in the parasitic agent section is this edition. These organisms are discussed here because a laboratory-acquired case of infection has been reported,[6] and most persons currently still look for microsporidia associated with discussion of parasitic agents.

Importation of parasitic agents may require CDC and/or USDA importation permits. Domestic transport of this agent may require a permit from USDA/APHIS/VS.

Blood and Tissue Protozoal Parasites

Blood and tissue protozoal parasites that pose greatest occupational risk include *Babesia, Leishmania, Plasmodium, Toxoplasma*, and *Trypanosoma*. Other tissue protozoa of potential concern include free-living ameba (*Acanthamoeba, Balamuthia mandrillaris, Naegleria fowleri*) and some species of microsporidia including *Encephalitozoon cuniculi* that commonly cause extraintestinal infection.

Leishmania spp. cause human leishmaniasis; *Plasmodium* spp. cause human malaria, or some, such as *P. cynomolgi* cause nonhuman primate malaria; *Toxoplasma gondii* causes toxoplasmosis; *Trypanosoma cruzi* causes American trypanosomiasis or Chagas disease; and *Trypanosoma brucei gambiense* and *T. b. rhodesiense* cause African trypanosomiasis or (African) sleeping sickness. With the exception of *Leishmania* and *Toxoplasma*, these agents are classically thought of as bloodborne and have stages that circulate in the blood. Although not always recognized, both *Leishmania* and *Toxoplasma* may have stages that circulate in the blood. Some, such as *Plasmodium* and *Trypanosoma cruzi*, also have tissue stages. *Leishmania* spp. are well recognized to have skin and deep tissue stages and *Toxoplasma gondii* forms tissue cysts, including in the central nervous system.

Occupational Infections

Laboratory-acquired infections with *Leishmania* spp., *Plasmodium* spp., *Toxoplasma gondii*, and *Trypanosoma* spp. have been reported; the majority of these involved needle-stick or other cutaneous exposure to infectious stages of the organisms through abraded skin, including microabrasions.[1,2]

Laboratory-acquired infections may be asymptomatic. If clinically manifest, they may exhibit features similar to those seen in naturally acquired infections, although bypassing natural modes of infection could result in atypical signs and symptoms. Cutaneous leishmaniasis could manifest as various types of skin lesions (e.g., nodules, ulcers, plaques), while visceral leishmaniasis may result in fever, hepatosplenomegaly, and pancytopenia. However, only one of the laboratorians known to have become infected with *L. (L.) donovani*, an organism typically associated with visceral leishmaniasis, developed clinical manifestations of visceral involvement (e.g., fever, splenomegaly, leukopenia).[1] The other laboratorians developed skin lesions. Laboratory-acquired malaria infections may result in fever and chills, fatigue, and hemolytic anemia. Laboratorians can become infected with *T. gondii* through accidental ingestion of sporulated oocysts, but also may become infected through skin or mucous membrane contact with either tachyzoites or bradyzoites in human or animal tissue or culture. Symptoms in laboratory-acquired *T. gondii* infections may be restricted to flu-like conditions with enlarged lymph nodes, although rash may be present. *Trypanosoma cruzi* infection could manifest initially as swelling and redness at the inoculation site, fever, rash, and adenopathy. Myocarditis and electrocardiographic changes may develop. Infection with *T. b. rhodesiense* and *T. b. gambiense* also may cause initial swelling and redness at the inoculation site, followed by fever, rash, adenopathy, headache, fatigue and neurologic signs.

Blood and tissue protozoal infections associated with exposure to laboratory animals are not common. Potential direct sources of infection for laboratory personnel include accidental needle-stick while inoculating or bleeding animals, contact with lesion material from cutaneous leishmaniasis, and contact with blood of experimentally or naturally infected animals. In the case of rodents experimentally inoculated with *Toxoplasma gondii* via the intraperitoneal route, contact with peritoneal fluid could result in exposure to infectious organisms. Mosquito-transmitted malaria infections can occur under laboratory conditions as nearly half of the occupationally acquired malaria infections were reported to be vector borne, and contact with body fluids (including feces) of reduviids (triatomines) experimentally or naturally infected with *T. cruzi* poses a risk to laboratory personnel.

Babesia microti and other *Babesia* spp. can cause human babesiosis or piroplasmosis. Under natural conditions, *Babesia* is transmitted by the bite of an infected tick, or by blood transfusion; in the United States, hard ticks (*Ixodes*) are the principal vectors. Although no laboratory infections with *Babesia* have been reported, they could easily result from accidental needle-stick or other cutaneous exposure of abraded skin to blood containing parasites. Persons who are asplenic, immunocompromised, or elderly have increased risk for severe illness if infected.

Natural Modes of Infection

Leishmaniasis is endemic in parts of the tropics, subtropics, and southern Europe, while malaria is widely distributed throughout the tropics. However, the prevalence of these diseases varies widely among endemic areas; the diseases can be very focal in nature. The four species of malaria that infect humans have no animal reservoir hosts. Some *Leishmania* spp. may have a number of important mammalian reservoir hosts, including rodents and dogs. Only cats and other felines can serve as definitive hosts for *Toxoplasma gondii*, which is distributed worldwide. Birds and mammals, including sheep, pigs, rodents, cattle, deer, and humans can be infected from ingestion of tissue cysts or fecal oocysts and subsequently develop tissue cysts throughout the body. Chagas disease occurs from Mexico southward throughout most of Central and South America, with the exception of the southern-most tip of South America. It has been characterized in some accounts as a zoonotic infection, yet the role of animals in maintaining human infection is unclear. A variety of domestic and wild animals are found naturally infected with *T. cruzi*, but human infection undoubtedly serves as the major source of infection for other humans. African trypanosomiasis is endemic in sub-Saharan Africa but is extremely focal in its distribution. Generally, *T. b. gambiense* occurs in West and Central Africa while *T. b. rhodesiense* occurs in East and Southeast Africa. *T. b. rhodesiense* is a zoonotic infection with cattle or, in a more limited role, game animals serving as reservoir hosts, whereas humans are the only epidemiologically important hosts for *T. b. gambiense*.

Leishmania, Plasmodium, and both American and African trypanosomes are all transmitted in nature by blood-sucking insects. Sandflies in the genera *Phlebotomus* and *Lutzomyia* transmit *Leishmania*; mosquitoes in the genus *Anopheles* transmit *Plasmodium*; reduviid (triatomine) bugs such as *Triatoma, Rhodnius*, and *Panstrongylus* transmit *T. cruzi* (in the feces rather than the saliva of the bug), and tsetse flies in the genus *Glossina* transmit African trypanosomes.

Laboratory Safety and Containment Recommendations

Infective stages may be present in blood, CSF, bone marrow, or other biopsy tissue, lesion exudates, and infected arthropods. Depending on the parasite, the primary laboratory hazards are skin penetration through wounds or microabrasions, accidental parenteral inoculation, and transmission by arthropod vectors. Aerosol or droplet exposure of organisms to the mucous membranes of the eyes, nose, or mouth are potential hazards when working with cultures of

Leishmania, Toxoplasma gondii, or *T. cruzi,* or with tissue homogenates or blood containing hemoflagellates. Immuno-compromised persons should avoid working with live organisms.

Because of the potential for grave consequences of toxoplasmosis in the developing fetus, women who are or might become pregnant and who are at risk for infection with *T. gondii* should receive counseling from their personal physician and employer regarding appropriate means of mitigating the risk (including alternate work assignments, additional PPE, etc.). Working with infectious oocysts poses the greatest risk of acquiring infection; needle-sticks with material containing tachyzoites or bradyzoites also pose a significant risk. Infection with tachyzoites or bradyzoites through mucous membranes or skin abrasions is also possible. Kittens and cats that might be naturally infected with *Toxoplasma* pose some risk to personnel.[5] Good hygiene and use of personal protection measures would reduce the risk.

One laboratory infection with microsporidia has been reported, associated with conjunctival exposure to spores leading to the development of keratoconjunctivitis. Infection could also result from ingestion of spores in feces, urine, sputum, CSF, or culture. No laboratory-acquired infections have been reported with *Acanthamoeba* spp., *Balamuthia mandrillaris* or *Naegleria fowleri*; however, the possibility of becoming infected by inhalation, by accidental needle-sticks, or through exposure to mucous membranes or microabrasions of the skin should be considered.

BSL-2 and ABSL-2 practices, containment equipment, and facilities are recommended for activities with infective stages of the parasites listed.[5] Infected arthropods should be maintained in facilities that reasonably preclude the exposure of personnel or the escape of insects. (See Appendix E.) Personal protection (e.g., lab coat, gloves, face shield), in conjunction with containment in a BSC, is indicated when working with cultures, tissue homogenates, or blood containing organisms.

Special Issues

Treatment Highly effective medical treatment for most protozoal infections exists.[4] An importation or domestic transfer permit for this agent can be obtained from USDA/APHIS/VS.

Transfer of Agent Importation of this agent may require CDC and/or USDA importation permits. Domestic transport of this agent may require a permit from USDA/APHIS/VS. A DoC permit may be required for the export of this agent to another country. See Appendix C for additional information.

Intestinal Protozoal Parasites

Intestinal protozoal parasites that pose greatest occupational risk include *Cryptosporidium, Isospora, Entamoeba histolytica*, and *Giardia*. Other intestinal pathogens of concern are some species of microsporidia, specifically *Septata intestinalis* and *Enterocytozoon bieneusi*. *Cryptosporidium parvum, C. hominis*, and *Isospora belli* cause intestinal coccidiosis, most often referred to as cryptosporidiosis and isosporiasis, respectively. *Entamoeba histolytica* can cause both intestinal and extraintestinal infection (e.g., liver abscess) called amebiasis, and *Giardia intestinalis* causes giardiasis.

Occupational Infections

Laboratory-acquired infections with *Cryptosporidium* spp., *E. histolytica*, *G. intestinalis*, and *I. belli* have been reported.[1-3] The mode of exposure in laboratory-acquired infections in this group of agents mimics the natural infection routes for the most part, and consequently, clinical symptoms are typically very similar to those seen in naturally acquired infections. For *Cryptosporidium*, *E. histolytica*, *G. intestinalis*, and *I. belli*, the common clinical manifestations are symptoms of gastroenteritis (e.g., diarrhea, abdominal pain and cramping, loss of appetite). Infection with *E. histolytica* may result in bloody stools.

Laboratory animal-associated infections with this group of organisms have been reported and provide a direct source of infection for laboratory personnel who are exposed to feces of experimentally or naturally infected animals.[3] Handling *Cryptosporidium* oocysts requires special care, as laboratory-acquired infections have occurred commonly in personnel working with this agent, especially if calves are used as the source of oocysts. Other experimentally infected animals pose potential risks as well. Circumstantial evidence suggests that airborne transmission of oocysts of this small organism (i.e., 4-6 μm diameter) may occur. Rigid adherence to protocol should reduce the occurrence of laboratory-acquired infection in laboratory and animal care personnel.

Natural Modes of Infection

All of these intestinal protozoa have a cosmopolitan distribution, and in some settings, including developed countries, the prevalence of infection can be high. The natural mode of infection for this group of organisms is typically ingestion of an environmentally hardy oocyst (for the coccidia) or cyst (for *E. histolytica* and *G. intestinalis*). The ID_{50}, best established for *Cryptosporidium*, has been shown for some strains to be 5-10 oocysts.[7] This suggests that even a single oocyst might pose a risk for infection in an exposed laboratorian. The infectious dose for other parasites in this group is not as well established, but is probably in the same range. Further, because these protozoa multiply in the host, ingestion of even small inocula can cause infection and illness. The role for animal reservoir hosts is diverse in this group of organisms. In the case of *C. hominis*, principally humans are infected, whereas for *C. parvum*, humans, cattle, and other mammals can be infected and serve as reservoir hosts for human infection. In the case of *E. histolytica*, humans serve as the only significant source of infection, and there is no convincing evidence that any animal serves as reservoir host for *I. belli*. The extent to which *Giardia* spp. parasitizing animals can infect humans is only now becoming better understood, but most human infection seems to be acquired from human-to-human transmission. The organisms in this group do not require more than one host to complete their life cycle because they infect, develop, and result in shedding of infectious stages all in a single host. Ingestion of contaminated drinking or recreational water has also been a common source of cryptosporidiosis and giardiasis.

Laboratory Safety and Containment Recommendations

Infective stages may be present in the feces or other body fluids and tissues. Depending on the parasite, ingestion is the primary laboratory hazard. Immunocompromised persons should avoid working with live organisms. Laboratorians who work only with killed or inactivated parasite materials, or parasite fractions, are not at significant risk.

Similarly, no accidental laboratory infection with *Sarcocystis* has been reported, although care should be exercised when working with infected meat products to avoid accidental ingestion. It is not known if laboratorians could be accidentally infected through parenteral inoculation of *Sarcocystis*; nevertheless caution should be exercised when working with cultures, homogenates, etc.

BSL-2 and ABSL-2 practices, containment equipment, and facilities are recommended for activities with infective stages of the parasites listed.[5] Primary containment (e.g., BSC) or personal protection (e.g., face shield) is especially important when working with *Cryptosporidium*. Oocysts are infectious when shed (i.e., are already sporulated and do not require further development time outside the host), often are present in stool in high numbers, and are environmentally hardy.

Commercially available iodine-containing disinfectants are effective against *E. histolytica* and *G. intestinalis*, when used as directed, as are high concentrations of chlorine (1 cup of full-strength commercial bleach [~5% chlorine] per gallon of water [1:16, vol/vol]).[1,2]

If a laboratory spill contains *Cryptosporidium* oocysts, the following approach is recommended.[2] A conventional laboratory detergent/cleaner should be used to remove contaminating matter from surfaces (e.g., of bench tops and equipment). After organic material has been removed, 3% hydrogen peroxide (i.e., undiluted, commercial hydrogen peroxide, identified on the bottle as 3% or "10 vol" hydrogen peroxide) can be used to disinfect surfaces; dispensing bottles that contain undiluted hydrogen peroxide should be readily available in laboratories in which surfaces could become contaminated.

Affected surfaces should be flooded (i.e., completely covered) with hydrogen peroxide. If a large volume of liquid contaminates surfaces, to avoid diluting the hydrogen peroxide, absorb the bulk of the spill with disposable paper towels. Dispense hydrogen peroxide repeatedly, as needed, to keep affected surfaces covered (i.e., wet/moist) for ~30 minutes. Absorb residual hydrogen peroxide with disposable paper towels and allow surfaces to dry thoroughly (10 to 30 minutes) before use. All paper towel litter and other disposable materials should be autoclaved or similarly disinfected before disposal. Reusable laboratory items can be disinfected and washed in a laboratory dishwasher by using the "sanitize" cycle and a detergent containing chlorine. Alternatively, immerse contaminated items for ~1 hour in a water bath preheated to 50° C; thereafter, wash them in a detergent/disinfectant solution.

Special Issues

Treatment Highly effective medical treatment exists for most protozoal infections; treatment with nitazoxanide for *Cryptosporidium* is now available, but efficacy has not been proven.[4]

Transfer of Agent Importation of these agents may require CDC and/or USDA importation permits. Domestic transport of these agents may require a permit from USDA/APHIS/VS. A DoC permit may be required for the export of this agent to another country. See Appendix C for additional information.

Trematode Parasites

Trematode parasites that pose greatest occupational risk are the *Schistosoma* spp., although others including *Fasciola* are of concern. *Schistosoma mansoni* causes intestinal schistosomiasis or bilharziasis, also known as Manson's blood fluke, in which the adult flukes reside in the venules of the bowel and rectum. *Fasciola hepatica*, the sheep liver fluke, causes fascioliasis, where the adult flukes live in the common and hepatic bile ducts of the human or animal host.

Occupational Infections

Laboratory-acquired infections with *S. mansoni* and *F. hepatica* have been reported, but accidental infections with other *Schistosoma* spp. could also occur.[1,2] By nature of the infection, none have been directly associated with laboratory animals, with the exception of infected mollusk intermediate hosts.

Laboratory-acquired infections with *F. hepatica* may be asymptomatic, but could have clinical manifestations such as right upper quadrant pain, biliary colic, obstructive jaundice, elevated transaminase levels, and other pathology associated with hepatic damage resulting from migration of the fluke through the liver en route to the bile duct. Most laboratory exposures to schistosomes would result in predictably low worm burdens with minimal disease potential. However, clinical manifestations of infection with *S. mansoni* could include dermatitis, fever, cough, hepatosplenomegaly, and adenopathy.

Natural Modes of Infection

Fasciola hepatica has a cosmopolitan distribution and is most common in sheep-raising areas, although other natural hosts include goats, cattle, hogs, deer, and rodents. Snails in the family Lymnaeidae, primarily species of *Lymnaea*, are intermediate hosts for *F. hepatica*, and release cercariae that encyst on vegetation. Persons become infected with *F. hepatica* by eating raw or poorly cooked vegetation, especially green leafy plants such as watercress, on which metacercariae have encysted.

Schistosoma mansoni is widely distributed in Africa, South America, and the Caribbean; the prevalence of infection has been rapidly changing in some areas. Infection occurs when persons are exposed to free-swimming cercariae in contaminated bodies of water; cercariae can penetrate intact skin. The natural snail hosts capable of supporting development of *S. mansoni* are various species of *Biomphalaria*.

Laboratory Safety and Containment Recommendations

Infective stages of *F. hepatica* (metacercariae) and *S. mansoni* (cercariae) may be found, respectively, encysted on aquatic plants or in the water in laboratory aquaria used to maintain snail intermediate hosts. Ingestion of fluke metacercariae and skin penetration by schistosome cercariae are the primary laboratory hazards. Dissection or crushing of schistosome-infected snails may also result in exposure of skin or mucous membrane to cercariae-containing droplets. Additionally, metacercariae may be inadvertently transferred from hand to mouth by fingers or gloves, following contact with contaminated aquatic vegetation or aquaria.

All reported cases of laboratory-acquired schistosomiasis have been caused by *S. mansoni*, which probably reflects the fact that many more laboratories work with *S. mansoni* than with other *Schistosoma* spp. However, accidental infection with *S. haematobium*, *S. japonicum*, and *S. mekongi* could easily occur in the same manner as described for *S. mansoni*.

Exposure to cercariae of non-human species of schistosomes (e.g., avian species) may cause mild to severe dermatitis (swimmer's itch).

BSL-2 and ABSL-2 practices, containment equipment and facilities are recommended for laboratory work with infective stages of the parasites listed.[5] Gloves should be worn when there may be direct contact with water containing cercariae or vegetation with encysted metacercariae from naturally or experimentally infected snail intermediate hosts. Long-sleeved laboratory coats or other protective garb should be worn when working in the immediate area of aquaria or other water sources that may contain schistosome cercariae. Water from laboratory aquaria containing snails and cercariae should be decontaminated (e.g., ethanol, hypochlorite, iodine, or heat) before discharged to sanitary sewers.

Special Issues

Treatment Highly effective medical treatment for most trematode infections exists.[4]

Transfer of Agent Importation of these agents may require CDC and/or USDA importation permits. Domestic transport of these agents may require a permit from USDA/APHIS/VS. A DoC permit may be required for the export of this agent to another country. See Appendix C for additional information.

Cestode Parasites

Cestode parasites of potential risk for laboratorians include *Echinococcus* spp., *Hymenolepis nana,* and *Taenia solium.* Echinococcosis is an infection caused by cestodes in the genus *Echinococcus*; *E. granulosus* causes cystic echinococcosis, *E. multilocularis* causes alveolar echinococcosis, and *E. vogeli* and *E. oligarthrus* cause polycystic echinococcosis. Humans serve as intermediate hosts and harbor the metacestode or larval stage, which produces a hydatid cyst. *Hymenolepis nana*, the dwarf tapeworm, is cosmopolitan in distribution and produces hymenolepiasis, or intestinal infection with the adult tapeworm. *Taenia solium*, the pork tapeworm, causes both taeniasis (infection of the intestinal tract with the adult worm), and cysticercosis (infection of subcutaneous, intermuscular, and central nervous system with the metacestode stage or cysticercus).

Occupational Infections

No laboratory-acquired infections have been reported with any cestode parasite.

Natural Modes of Infection

The infectious stage of *Echinococcus, Hymenolepis*, and *Taenia* is the oncosphere contained within the egg. *Hymenolepis nana* is a one-host parasite and does not require an intermediate host; it is directly transmissible by ingestion of feces of infected humans or rodents. The life cycles of *Echinococcus* and *Taenia* require two hosts. Canids, including dogs, wolves, foxes, coyotes, and jackals, are the definitive hosts for *E. granulosus*, and various herbivores such as sheep, cattle, deer, and horses are the intermediate hosts. Foxes and coyotes are the principal definitive hosts for *E. multilocularis*, although dogs and cats also can become infected and rodents serve as the intermediate hosts. Bush dogs and pacas serve as the definitive and intermediate hosts, respectively, for *E. vogeli*. Dogs also may be infected. *Echinococcus oligarthrus* uses wild felines, including cougar, jaguarondi, jaguar, ocelot, and pampas cat, as definitive hosts and various rodents such as agoutis, pacas, spiny rats, and rabbits serve as intermediate hosts. People become infected when eggs shed by the definitive host are accidentally ingested. For *T. solium*, people can serve both as definitive host (harbor the adult tapeworm), and as accidental intermediate host (harbor the larval stages cysticerci). Pigs are the usual intermediate host, becoming infected as they scavenge human feces containing eggs.

Laboratory Safety and Containment Recommendations

Infective eggs of *Echinococcus* spp. may be present in the feces of carnivore definitive hosts.[3] *Echinococcus granulosus* poses the greatest risk because it is the most common and widely distributed species, and because dogs are the primary definitive hosts. For *T. solium*, infective eggs in the feces of humans serve as the source of infection. Accidental ingestion of infective eggs from these sources is the primary laboratory hazard. Ingestion of cysticerci of *T. solium (Cysticercus cellulosae)* leads to human infection with the adult tapeworm. For those cestodes listed, the ingestion of a single infective egg from the feces of the definitive host could potentially result in serious disease. Ingestion of the eggs of *H. nana* in the feces of definitive hosts (humans or rodents) could result in intestinal infection.

Although no laboratory-acquired infections with either *Echinococcus* spp. or *T. solium* have been reported, the consequences of such infections could be serious. Laboratory-acquired infections with cestodes could result in various clinical manifestations, depending upon the type of cestode. Human infection with *Echinococcus* spp. could range from asymptomatic to severe. The severity and nature of the signs and symptoms depends upon the location of the cysts, their size, and condition (alive versus dead). Clinical manifestations of a liver cyst could include hepatosplenomegaly, right epigastric pain, and nausea, while a lung cyst may cause chest pain, dyspnea, and hemoptysis. For *T. solium*, ingestion of eggs from human feces can result in cysticercosis, with cysts located in subcutaneous and intermuscular tissues, where they may be asymptomatic. Cysts in the central nervous system may cause seizures and other neurologic symptoms. Ingestion of tissue cysts of *T. solium* can lead to development of adult worms in the intestine of humans. Immunocompromised persons working with these cestodes must take special care as the asexual multiplication of the larval stages of these parasites makes them especially dangerous to such persons.

BSL-2 and ABSL-2 practices, containment equipment, and facilities are recommended for work with infective stages of these parasites.[5] Special attention should be given to personal hygiene (e.g., hand washing) and laboratory practices that would reduce the risk of accidental ingestion of infective eggs. Gloves are recommended when there may be direct contact with feces or with surfaces contaminated with fresh feces of carnivores infected with *Echinococcus* spp., humans infected with *T. solium*, or humans or rodents infected with *H. nana*.

Special Issues

Treatment Highly effective medical treatment for most cestode infections exists.[4]

Transfer of Agent Importation of these agents may require CDC and/or USDA importation permits. Domestic transport of these agents may require a permit from USDA/APHIS/VS. A DoC permit may be required for the export of this agent to another country. See Appendix C for additional information.

Nematode Parasites

Nematode parasites that pose greatest occupational risk include the ascarids, especially *Ascaris* and *Baylisascaris*; hookworms, both human and animal; *Strongyloides*, both human and animal; *Enterobius*; and the human filariae, primarily *Wuchereria* and *Brugia*. *Ancylostoma braziliense* and *A. caninum* cause hookworm infection in cats and dogs, respectively. *Ascaris lumbricoides* causes ascariasis and is known as the large intestinal roundworm of humans. *Enterobius vermicularis*, known as the human pinworm or seatworm, causes enterobiasis or oxyuriasis. *Strongyloides*, the threadworm, causes strongyloidiasis. *Ancylostoma*, *Ascaris*, and *Strongyloides* reside as adults in the small intestine of their natural hosts, whereas *E. vermicularis* colonizes the cecum and appendix.

Occupational Infections

Laboratory-associated infections with *Ancylostoma* spp., *A. lumbricoides*, *E. vermicularis*, and *Strongyloides* spp. have been reported.[1-3] Laboratory infections with hookworms and *Strongyloides* presumptively acquired from infected animals have been reported. Allergic reactions to various antigenic components of human and animal ascarids (e.g., aerosolized antigens) may pose risk to sensitized persons.

Laboratory-acquired infections with these nematodes can be asymptomatic, or can present with a range of clinical manifestations dependent upon the species and their location in host. Infection with hookworm of animal origin can result in cutaneous larva migrans or creeping eruption of the skin. Infection with *A. lumbricoides* may produce cough, fever, and pneumonitis as larvae migrate through the lung, followed by abdominal cramps and diarrhea or constipation from adult worms in the intestine. Infection with *E. vermicularis* usually causes perianal pruritis, with intense itching. Infection with animal *Strongyloides* spp. may induce cutaneous larva migrans.

Natural Modes of Infection

Ancylostoma infection in dogs and cats is endemic worldwide. Human infection occurs through penetration of the skin. Cutaneous larva migrans or creeping eruption occurs when infective larvae of animal hookworms, typically dog and cat hookworms, penetrate the skin and begin wandering. *Ancylostoma* larvae can also cause infection if ingested. These larvae do not typically reach the intestinal tract, although *A. caninum* has on rare occasions developed into non-gravid adult worms in the human gut.

Ascaris lumbricoides infection is endemic in tropical and subtropical regions of the world. Infection occurs following accidental ingestion of infective eggs. Unembryonated eggs passed in the stool require two to three weeks to become infectious, and *Ascaris* eggs are very hardy in the environment.

Enterobius vermicularis occurs worldwide, although infection tends to be more common in school-age children than adults, and in temperate than tropical regions. Pinworm infection is acquired by ingestion of infective eggs, most often on contaminated fingers following scratching of the perianal skin. Eggs passed by female worms are not immediately infective, but only require several hours' incubation to become fully infectious. Infection with this worm is relatively short (60 days on average), and reinfection is required to maintain an infection.

Strongyloides infection in animals is endemic worldwide. People become infected with animal *Strongyloides* when infective, filariform larvae penetrate the skin, and can develop cutaneous creeping eruption (larva currens).

Laboratory Safety and Containment Recommendations

Eggs and larvae of most nematodes are not infective in freshly passed feces; development to the infective stages may require from one day to several weeks. Ingestion of the infective eggs or skin penetration by infective larvae are the primary hazards to laboratory staff and animal care personnel. Development of hypersensitivity is common in laboratory personnel with frequent exposure to aerosolized antigens of ascarids.

Ascarid eggs are sticky, and special care should be taken to ensure thorough cleaning of contaminated surfaces and equipment. Caution should be used even when working with formalin-fixed stool samples because ascarid eggs can remain viable and continue to develop to the infective stage in formalin.[8]

Working with infective eggs of other ascarids, such as *Toxocara* and *Baylisascaris*, poses significant risk because of the potential for visceral migration of larvae, including invasion of the eyes and central nervous system. *Strongyloides stercoralis* is of particular concern to immuno-suppressed persons because potentially life-threatening systemic hyperinfection can occur. Lugol's iodine kills infective larvae and should be sprayed onto skin or laboratory surfaces that are contaminated accidentally. The larvae of *Trichinella* in fresh or digested tissue could cause infection if accidentally ingested. Arthropods infected with filarial parasites pose a potential hazard to laboratory personnel.

BSL-2 and ABSL-2 practices, containment equipment, and facilities are recommended for activities with infective stages of the nematodes listed here.[5] Exposure to aerosolized sensitizing antigens of ascarids should be avoided. Primary containment (e.g., BSC) is recommended for work that may result in aerosolization of sensitization from occurring.

Special Issues

Treatment Highly effective medical treatment for most nematode infections exists.[4]

Transfer of Agent Importation of these agents may require CDC and/or USDA importation permits. Domestic transport of these agents may require a permit from USDA/APHIS/VS. A DoC permit may be required for the export of this agent to another country. See Appendix C for additional information.

References

1. Herwaldt BL. Laboratory-acquired parasitic infections from accidental exposures. Clin Microbiol Rev. 2001:14:659-88.
2. Herwaldt BL. Protozoa and helminths. In Fleming DO and Hunt DL (eds). 4th ed. Biological Safety: Principles and Practices. Washington, DC, ASM Press, 2006.
3. Pike RM. Laboratory-associated infections: summary and analysis of 3921 cases. Health Lab Sci 1976;13:105-14.
4. The Medical Letter [*www.medicalletter.org*] New York: The Medical Letter; [cited 2007]. Drugs for Parasitic Infections; [about 24 screens].
5. Hankenson FC, Johnston NA, Weigler BJ, et al. Zoonoses of occupational health importance in contemporary laboratory animal research. Comp Med. 2003;53:579-601.
6. van Gool T, Biderre C, Delbac F, Wentink-Bonnema E, Peek R, Vivares CP. Serodiagnostic studies in an immunocompetent individual infected with *Encephalitozoon cuniculi*. J Infect Dis 2004;189:2243-9.
7. Messner MJ, Chappel CL, Okhuysen PC. Risk assessment for *Cryptosporidium*: a hierarchical Bayesian analysis of human dose response data. Water Res 2001;35:3934-40.
8. Ash LR, Orihel TC. Parasites, a guide to laboratory procedures and identification. Chicago: ASCP Press; 1991.

Section VIII-D: Rickettsial Agents

Coxiella burnetii

Coxiella burnetii is the etiologic agent of Q fever. *C. burnetii* is a bacterial obligate intracellular pathogen that undergoes its developmental cycle within an acidic vacuolar compartment exhibiting many characteristics of a phagolysosome. The developmental cycle consists of a large (approximately 1 μm in length) cell variant that is believed to be the more metabolically active, replicative cell type and a smaller, more structurally stable cell variant that is highly infectious and quite resistant to drying and environmental conditions.[1-4] The organism undergoes a virulent (Phase I) to avirulent (Phase II) transition upon serial laboratory passage in eggs or tissue culture.

The infectious dose of virulent Phase I organisms in laboratory animals has been calculated to be as small as a single organism.[5] The estimated human infectious dose for Q fever by inhalation is approximately 10 organisms.[6] Typically, the disease manifests with flu-like symptoms including fever, headache, and myalgia but can also cause pneumonia and hepatomegaly. Infections range from sub-clinical to severe although primary infections respond readily to antibiotic treatment. Although rare, *C. burnetii* is known to cause chronic infections such as endocarditis or granulomatous hepatitis.[7]

Occupational Infections

Q fever is the second most commonly reported LAI in Pike's compilation. Outbreaks involving 15 or more persons were recorded in several institutions.[8,9] Infectious aerosols are the most likely route of laboratory-acquired infections. Experimentally infected animals also may serve as potential sources of infection or laboratory and animal care personnel. Exposure to naturally infected, often fasymptomatic sheep and their birth products is a documented hazard to personnel.[10,11]

Natural Modes of Infection

Q fever (Q for query) occurs worldwide. Broad ranges of domestic and wild mammals are natural hosts for Q fever and sources of human infection. Parturient animals and their birth products are common sources of infection. The placenta of infected sheep may contain as many as 109 organisms per gram of tissue[12] and milk may contain 105 organisms per gram. The resistance of the organism to drying and its low infectious dose can lead to dispersal from contaminated sites.

Laboratory Safety and Containment Recommendations

The necessity of using embryonated eggs or cell culture techniques for the propagation of *C. burnetii* leads to extensive purification procedures. Exposure to infectious aerosols and parenteral inoculation cause most infections in laboratory and animal care personnel.[8,9] The agent may be present in infected arthropods and in the blood, urine, feces, milk, and tissues of infected animals or human hosts. Exposure to naturally infected, often asymptomatic, sheep and their birth products is a documented hazard to personnel.[10,11] Recommended precautions for facilities using sheep as experimental animals are described elsewhere.[10,13]

BSL-2 practices and facilities are recommended for nonpropagative laboratory procedures, including serological examinations and staining of impression smears. BSL-3 practices and facilities are recommended for activities involving the inoculation, incubation, and harvesting of embryonated eggs or cell cultures, the necropsy of infected animals and the manipulation of infected tissues. Experimentally infected animals should be maintained under ABSL-3 because infected rodents may shed the organisms in urine or feces.[8] A specific plaque-purified clonal isolate of an avirulent (Phase II) strain (Nine Mile) may be safely handled under BSL-2 conditions.[14]

Special Issues

Vaccines An investigational Phase I, Q fever vaccine (IND) is available on a limited basis from the Special Immunizations Program (301-619-4653) of the USAMRIID, Fort Detrick, Maryland, for at-risk personnel under a cooperative agreement with the individual's requesting institution. The use of this vaccine should be restricted to those who are at high risk of exposure and who have no demonstrated sensitivity to Q fever antigen. The vaccine can be reactogenic in those with prior immunity, thus requires skin testing before administration. The vaccine is only administered at USAMRIID and requires enrollment in their Q fever IND Immunization Program. For at-risk laboratory workers to participate in this program, fees are applicable. Individuals with valvular heart disease should not work with *C. burnetii*. (See Section VII.)

Select Agent *C. burnetii* is a select agent requiring registration with CDC and/or USDA for possession, use, storage and/or transfer. See Appendix F for additional information.

Transfer of Agent Importation of this agent may require CDC and/or USDA importation permits. Domestic transport of this agent may require a permit from USDA/APHIS/VS. A DoC permit may be required for the export of this agent to another country. See Appendix C for additional information.

Rickettsia prowazekii; Rickettsia typhi (R. mooseri); Orientia (Rickettsia) tsutsugamushi and Spotted Fever Group agents of human disease; Rickettsia rickettsii, Rickettsia conorii, Rickettsia akari, Rickettsia australis, Rickettsia siberica, and Rickettsia japonicum

Rickettsia prowazekii, Rickettsia typhi (R. mooseri), Orientia (Rickettsia) tsutsugamushi and the Spotted Fever Group agents of human disease (*Rickettsia rickettsii, Rickettsia conorii, Rickettsia akari, Rickettsia australis, Rickettsia siberica,* and

Rickettsia japonicum) are the etiologic agents of epidemic typhus, endemic (murine) typhus), scrub typhus, Rocky Mountain spotted fever, Mediterranean spotted fever, rickettsialpox, Queensland tick typhus, and North Asian spotted fever, respectively.

Rickettsia spp. are bacterial obligate intracellular pathogens that are transmitted by arthropod vectors and replicate within the cytoplasm of eukaryotic host cells. Two groups are recognized within the genus, the typhus group and the spotted fever group. The more distantly related scrub typhus group is now considered a distinct genus, *Orientia.* Rickettsiae are primarily associated with arthropod vectors in which they may exist as endosymbionts that infect mammals, including humans, through the bite of infected ticks, lice, or fleas.[15]

Occupational Infections

Pike reported 57 cases of laboratory-associated typhus (type not specified), 56 cases of epidemic typhus with three deaths, and 68 cases of murine typhus.[8] Three cases of murine typhus have been reported from a research facility.[16] Two were associated with handling of infectious materials on the open bench; the third case resulted from an accidental parenteral inoculation. These three cases represented an attack rate of 20% in personnel working with infectious materials. Rocky Mountain spotted fever is a documented hazard to laboratory personnel. Pike reported 63 laboratory-associated cases, 11 of which were fatal.[8] Oster reported nine cases occurring over a six-year period in one laboratory. All were believed to have been acquired because of exposure to infectious aerosols.[17]

Natural Modes of Infection

The epidemiology of rickettsial infections reflects the prevalence of rickettsiae in the vector population and the interactions of arthropod vectors with humans. Epidemic typhus is unusual among rickettsiae in that humans are considered the primary host. Transmission is by the human body louse; thus, outbreaks are now associated with breakdowns of social conditions. Endemic typhus is maintained in rodents and transmitted to humans by fleas. The various spotted fever group rickettsiae are limited geographically, probably by the distribution of the arthropod vector, although specific spotted fever group rickettsiae are found on all continents.[15]

Laboratory Safety and Containment Recommendations

The necessity of using embryonated eggs or cell culture techniques for the propagation of *Rickettsia* spp. incorporates extensive purification procedures. Accidental parenteral inoculation and exposure to infectious aerosols are the most likely sources of LAI.[18] Aerosol transmission of *R. rickettsii* has been experimentally documented in nonhuman primates.[19] Five cases of rickettsialpox recorded by Pike were associated with exposure to bites of infected mites.[8] Naturally and experimentally infected mammals, their ectoparasites, and their infected tissues are potential sources of human infection. The organisms are relatively unstable under ambient environmental conditions.

BSL-2 practices, containment equipment, and facilities are recommended for nonpropagative laboratory procedures, including serological and fluorescent antibody procedures, and for the staining of impression smears. BSL-3 practices, containment equipment, and facilities are recommended for all other manipulations of known or potentially infectious materials, including necropsy of experimentally infected animals and trituration of their tissues, and inoculation, incubation, and harvesting of embryonated eggs or cell cultures. ABSL-2 practices, containment equipment, and facilities are recommended for the holding of experimentally infected mammals other than arthropods. BSL-3 practices, containment equipment, and facilities are recommended for animal studies with arthropods naturally or experimentally infected with rickettsial agents of human disease. (See Appendix E.)

Several species, including *R. montana, R. rhipicephali, R. belli*, and *R. canada*, are not known to cause human disease and may be handled under BSL-2 conditions. New species are being described frequently and should be evaluated for appropriate containment on a case-by-case basis. Because of the proven value of antibiotic therapy in the early stages of ricketsial infection, it is essential that laboratories have an effective system for reporting febrile illnesses in laboratory personnel, medical evaluation of potential cases and, when indicated, institution of appropriate antibiotic therapy.

Special Issues

Medical Response Under natural circumstances, the severity of disease caused by rickettsial agents varies considerably. In the laboratory, very large inocula are possible, which might produce unusual and perhaps very serious responses. Surveillance of personnel for laboratory-associated infections with rickettsial agents can dramatically reduce the risk of serious consequences of disease. Experience indicates that infections adequately treated with specific anti-rickettsial chemotherapy on the first day of disease do not generally present serious problems. However, delay in instituting appropriate chemotherapy may result in debilitating or severe acute disease ranging from increased periods of convalescence in typhus and scrub typhus to death in *R. rickettsii* infections. The key to reducing the severity of disease from laboratory-associated infections is a reliable medical response which includes: 1) round-the-clock availability of an experienced medical officer; 2) indoctrination of all personnel on the potential hazards of working with rickettsial agents and advantages of early therapy; 3) a reporting system for all recognized overt exposures and accidents; 4) the reporting of all febrile illnesses, especially those associated with headache, malaise, and prostration when no other certain cause exists; and 5) an open and non-punitive atmosphere that encourages reporting of any febrile illness.

Select Agent *R. prowazekii* and *R. rickettsii* are select agents requiring registration with CDC and/or USDA for possession, use, storage and/or transfer. See Appendix F for additional information.

Transfer of Agent Importation of this agent may require CDC and/or USDA importation permits. Domestic transport of this agent may require a permit from USDA/APHIS/VS.

References

1. Babudieri B. Q fever: a zoonosis. Adv Vet Sci. 1959;5:81-182.
2. Ignatovich VF. The course of inactivation of *Rickettsia burnetii* in fluid media. Zh Mikrobiol Epidemiol Immunobiol. 1959;30:134-41.
3. Sawyer LA, Fishbein DB, McDade JE. Q fever: current concepts. Rev Infect Dis. 1987;9:935-46.
4. Heinzen RA, Hackstadt T, Samuel JE. Developmental biology of *Coxiella burnetii*. Trends Microbiol. 1999;7:149-54.
5. Ormsbee R, Peacock M, Gerloff R, et al. Limits of rickettsial infectivity. Infect Immun. 1978;19:239-45.
6. Wedum AG, Barkley WE, Hellman A. Handling of infectious agents. J Am Vet Med Assoc. 1972;161:1557-67.
7. Maurin M, Raoult D. Q fever. Clin Microbiol Rev. 1999;12:518-53.
8. Pike RM. Laboratory-associated infections: Summary and analysis of 3,921 cases. Health Lab Sci. 1976;13:105-14.
9. Johnson JE, Kadull PJ. Laboratory-acquired Q fever. A report of fifty cases. Am J Med. 1966;41:391-403.
10. Spinelli JS, Ascher MS, Brooks DL, et al. Q fever crisis in San Francisco: controlling a sheep zoonosis in a lab animal facility. Lab Anim. 1981;10:24-7.
11. Meiklejohn G, Reimer LG, Graves PS, et al. Cryptic epidemic of Q fever in a medical school. J Infect Dis. 1981;144:107-13.
12. Welsh HH, Lennette EH, Abinanti FR, et al. Q fever in California IV. Occurrence of *Coxiella burnetii* in the placenta of naturally infected sheep. Public Health Rep. 1951;66:1473-7.
13. Bernard KW, Helmick CG, Kaplan JE, et al. Q fever control measures: Recommendations for research facilities using sheep. Infect Control. 1982;3:461-5.
14. Hackstadt T. Biosafety concerns and *Coxiella burnetii*. Trends Microbiol. 1996;4:341-2.
15. Hackstadt T. The biology of rickettsiae. Infect Agents Dis. 1996;5:127-43.
16. Centers for Disease Control and Prevention. Laboratory-acquired endemic typhus. MMWR Morb Mortal Wkly Rep. 1978;27:215-6.
17. Oster CN, Burke DS, Kenyon RH, et al. Laboratory-acquired Rocky Mountain spotted fever. The hazard of aerosol transmission. N Engl J Med. 1977;297:859-63.
18. Hattwick MA, O'Brien RJ, Hanson BF. Rocky Mountain spotted fever: epidemiology of an increasing problem. Ann Intern Med. 1976;84:732-9.
19. Saslaw S, Carlisle HN. Aerosol infection of monkeys with *Rickettsia rickettsii*. Bacteriol Rev. 1966;30:636-45.

Section VIII-E: Viral Agents

Hantaviruses

Hantaviruses are negative sense RNA viruses belonging to the genus *Hantavirus* within the family *Bunyaviridae*. The natural hosts of hantaviruses are rodent species and they occur worldwide. Hantavirus pulmonary syndrome (HPS) is a severe disease caused by hantaviruses such as Sin Nombre virus or Andes virus whose hosts are rodents in the subfamily *Sigmodontinae*. This subfamily only occurs in the New World, so HPS is not seen outside North and South America. Hantaviruses in Europe and Asia frequently cause kidney disease, called nephropathica epidemica in Europe, and hemorrhagic fever with renal syndrome (HFRS) in Asia.

Occupational Infections

Documented laboratory-acquired infections have occurred in individuals working with hantaviruses.[1-4] Extreme caution must be used in performing any laboratory operation that may create aerosols (centrifugation, vortex-mixing, etc.). Operations involving rats, voles, and other laboratory rodents, should be conducted with special caution because of the extreme hazard of aerosol infection, especially from infected rodent urine.

Natural Modes of Infection

HPS is a severe, often fatal disease that is caused by Sin Nombre and Andes or related viruses.[5,6] Most cases of human illness have resulted from exposures to naturally infected wild rodents or to their excreta. Person-to-person transmission does not occur, with the exception of a few rare instances documented for Andes virus.[7] Arthropod vectors are not known to transmit hantaviruses.

Laboratory Safety and Containment Recommendations

Laboratory transmission of hantaviruses from rodents to humans via the aerosol route is well documented.[4-7] Exposures to rodent excreta, especially aerosolized infectious urine, fresh necropsy material, and animal bedding are presumed to be associated with risk. Other potential routes of laboratory infection include ingestion, contact of infectious materials with mucous membranes or broken skin and, in particular, animal bites. Viral RNA has been detected in necropsy specimens and in patient blood and plasma obtained early in the course of HPS;[8,9] however, the infectivity of blood or tissues is unknown.

BSL-2 practices, containment equipment, and facilities are recommended for laboratory handling of sera from persons potentially infected with hantaviruses. The use of a certified BSC is recommended for all handling of human body fluids when potential exists for splatter or aerosol.

Potentially infected tissue samples should be handled in BSL-2 facilities following BSL-3 practices and procedures. Cell-culture virus propagation and purification should be carried out in a BSL-3 facility using BSL-3 practices, containment equipment and procedures.

Experimentally infected rodent species known not to excrete the virus can be housed in ABSL-2 facilities using ABSL-2 practices and procedures. Primary physical containment devices including BSCs should be used whenever procedures with potential for generating aerosols are conducted. Serum or tissue samples from potentially infected rodents should be handled at BSL-2 using BSL-3 practices, containment equipment and procedures. All work involving inoculation of virus-containing samples into rodent species permissive for chronic infection should be conducted at ABSL-4.

Special Issues

Transfer of Agent Importation of this agent may require CDC and/or USDA importation permits. Domestic transport of this agent may require a permit from USDA/APHIS/VS.

Hendra Virus (formerly known as Equine Morbillivirus) and Nipah Virus

Hendra virus and *Nipah* virus are members of a newly recognized genus called *Henipavirus,* within the family *Paramyxoviridae.* Outbreaks of a previously unrecognized paramyxovirus, at first called equine morbillivirus, later named Hendra virus, occurred in horses in Australia in 1994 and 1995. During 1998-1999, an outbreak of illness caused by a similar but distinct virus, now known as Nipah virus, occurred in Malaysia and Singapore. Human illness, characterized by fever, severe headache, myalgia and signs of encephalitis occurred in individuals in close contact with pigs (i.e., pig farmers and abattoir workers).[10-14] A few patients developed a respiratory disease. Approximately 40% of patients with encephalitis died. Recently, cases of Nipah virus infection were described in Bangladesh, apparently the result of close contact with infected fruit bats without an intermediate (e.g., pig) host.

Occupational Infections

No laboratory-acquired infections are known to have occurred because of Hendra or Nipah virus exposure; however, three people in close contact with ill horses developed encephalitis or respiratory disease and two died.[15-20]

Natural Modes of Infection

The natural reservoir hosts for the Hendra and Nipah viruses appear to be fruit bats of the genus *Pteropus.*[21-23] Studies suggest that a locally occurring member of the genus, *Pteropus giganteus*, is the reservoir for the virus in Bangladesh.[24] Individuals who had regular contact with bats had no evidence of infection (antibody) in one study in Australia.[25]

Laboratory Safety and Containment Recommendations

The exact mode of transmission of these viruses has not been established. Most clinical cases to date have been associated with close contact with horses, their blood or body fluids (Australia) or pigs (Malaysia/Singapore) but presumed direct transmission from *Pteropus* bats has been recorded in Bangladesh. Hendra and Nipah viruses have been isolated from tissues of infected animals. In the outbreaks in Malaysia and Singapore, viral antigen was found in central nervous system, kidney and lung tissues of fatal human cases[26] and virus was present in secretions of patients, albeit at low levels.[27] Active surveillance for infection of healthcare workers in Malaysia has not detected evidence of occupationally acquired infections in this setting.[28]

Because of the unknown risks to laboratory workers and the potential impact on indigenous livestock should the virus escape a diagnostic or research laboratory, health officials and laboratory managers should evaluate the need to work with the virus and the containment capability of the facility before undertaking any work with Hendra, Nipah or suspected related viruses. BSL-4 is required for all work with these viruses. Once a diagnosis of Nipah or Hendra virus is suspected, all diagnostic specimens also must be handled at BSL-4. ABSL-4 is required for any work with infected animals.

Special Issues

Select Agent Hendra and Nipah virus are select agents requiring registration with CDC and/or USDA for possession, use, storage and/or transfer. See Appendix F for additional information.

Transfer of Agent Importation of this agent may require CDC and/or USDA importation permits. Domestic transport of this agent may require a permit from USDA/APHIS/VS. A DoC permit may be required for the export of this agent to another country. See Appendix C for additional information.

Hepatitis A Virus, Hepatitis E Virus

Hepatitis A virus is a positive single-stranded RNA virus, the type species of the Hepatovirus genus in the family Picornaviridae. Hepatitis E virus is a positive single-stranded RNA virus, the type species of the genus Hepevirus, a floating genus not assigned to any family.

Occupational Infections

Laboratory-associated infections with hepatitis A or E viruses do not appear to be an important occupational risk among laboratory personnel. However, hepatitis A is a documented hazard in animal handlers and others working with naturally or experimentally infected chimpanzees and other nonhuman primates.[29] Workers handling other recently captured, susceptible primates (owl monkeys, marmosets) also may be at risk for hepatitis A infection. Hepatitis E virus appears to be less of a risk to personnel than hepatitis A virus, except during pregnancy, when infection can result in severe or fatal disease.

Natural Modes of Infection

Most infections with hepatitis A are foodborne and occasionally water-borne. The virus is present in feces during the prodromal phase of the disease and usually disappears once jaundice occurs. Hepatitis E virus causes acute enterically-transmitted cases of hepatitis, mostly waterborne. In Asia, epidemics involving thousands of cases have occurred.

Laboratory Safety and Containment Recommendations

The agents may be present in feces and blood of infected humans and nonhuman primates. Feces, stool suspensions, and other contaminated materials are the primary hazards to laboratory personnel. Care should be taken to avoid puncture wounds when handling contaminated blood from humans or nonhuman primates. There is no evidence that aerosol exposure results in infection.

BSL-2 practices, containment equipment, and facilities are recommended for the manipulation of hepatitis A and E virus, infected feces, blood or other tissues. ABSL-2 practices and facilities are recommended for activities using naturally or experimentally-infected nonhuman primates or other animal models that may shed the virus.

Special Issues

Vaccines A licensed inactivated vaccine against hepatitis A is available. Vaccines against hepatitis E are not currently available.

Transfer of Agent Importation of this agent may require CDC and/or USDA importation permits. Domestic transport of this agent may require a permit from USDA/APHIS/VS. A DoC permit may be required for the export of this agent to another country. See Appendix C for additional information.

Hepatitis B Virus, Hepatitis C Virus (formerly known as nonA nonB Virus), Hepatitis D Virus

Hepatitis B virus (HBV) is the type species of the *Orthohepadnavirus* genus in the family *Hepadnaviridae*. Hepatitis C virus (HCV) is the type species of the *Hepacivirus* genus in the family *Flaviviridae*. Hepatitis D virus (HDV) is the only member of the genus *Deltavirus*.

These viruses are naturally acquired from a carrier during blood transfusion, vaccination, tattooing, or body piercing with inadequately sterilized instruments. Non-parenteral routes, such as domestic contact and unprotected (heterosexual and homosexual) intercourse, are also major modes of transmission.

Individuals who are infected with the HBV are at risk of infection with HDV, a defective RNA virus that requires the presence of HBV virus for replication. Infection with HDV usually exacerbates the symptoms caused by HBV infection.

Occupational Infections

Hepatitis B has been one of the most frequently occurring laboratory-associated infections, and laboratory workers are recognized as a high-risk group for acquiring such infections.[30]

Hepatitis C virus infection can occur in the laboratory situation as well.[31] The prevalence of antibody to hepatitis C (anti-HCV) is slightly higher in medical care workers than in the general population. Epidemiologic evidence indicates that HCV is spread predominantly by the parenteral route.[32]

Laboratory Safety and Containment Recommendations

HBV may be present in blood and blood products of human origin, in urine, semen, CSF and saliva. Parenteral inoculation, droplet exposure of mucous membranes, and contact exposure of broken skin are the primary laboratory hazards.[33] The virus may be stable in dried blood or blood components for several days. Attenuated or avirulent strains have not been identified.

HCV has been detected primarily in blood and serum, less frequently in saliva and rarely or not at all in urine or semen. It appears to be relatively unstable to storage at room temperature and repeated freezing and thawing.

BSL-2 practices, containment equipment, and facilities are recommended for all activities utilizing known or potentially infectious body fluids and tissues. Additional primary containment and personnel precautions, such as those described for BSL-3, may be indicated for activities with potential for droplet or aerosol production and for activities involving production quantities or concentrations of infectious materials. ABSL-2 practices, containment equipment and facilities are recommended for activities utilizing naturally or experimentally infected chimpanzees or other NHP. Gloves should be worn when working with infected animals and when there is the likelihood of skin contact with infectious materials. In addition to these recommended precautions, persons working with HBV, HCV, or other bloodborne pathogens should consult the OSHA Bloodborne Pathogen Standard.[34] Questions related to interpretation of this Standard should be directed to federal, regional or state OSHA offices.

Special Issues

Vaccines Licensed recombinant vaccines against hepatitis B are available and are highly recommended for and offered to laboratory personnel.[35] Vaccines against hepatitis C and D are not yet available for use in humans, but vaccination against HBV will also prevent HDV infection.

Transfer of Agent Importation of this agent may require CDC and/or USDA importation permits. Domestic transport of this agent may require a permit from USDA/APHIS/VS.

Herpesvirus Simiae (Cerocopithecine Herpesvirus I, Herpes B Virus)

B virus is a member of the *alphaherpesvirus* genus (simplexvirus) in the family *Herpesviridae.* It occurs naturally in macaque monkeys, of which there are nine distinct species. Macaques may have primary, recurrent, or latent infections often with no apparent symptoms or lesions. B virus is the only member of the family of simplex herpesviruses that can cause zoonotic infections. Human infections have been identified in at least 50 instances, with approximately 80% mortality when untreated. There remains an approximate 20% mortality in the absence of timely treatment with antiviral agents.[36] There have been no reported cases where prompt first aid with wound or exposure site cleansing was performed, and no cases where cleaning and post exposure prophylaxis were done. Cases prior to 1970 were not treated with antiviral agents because none were available. Morbidity and mortality associated with zoonotic infection results from invasion of the central nervous system, resulting in ascending paralysis ultimately with loss of ability to sustain respiration in the absence of mechanical ventilation. From 1987-2004, five additional fatal infections bring the number of lethal infections to 29 since the discovery of B virus in 1933.

Occupational Infections

B virus is a hazard in facilities where macaque monkeys are present. Mucosal secretions (saliva, genital secretions, and conjunctival secretions) are the primary body fluids associated with risk of B virus transmission. However, it is possible for other materials to become contaminated. For instance, a research assistant at the Yerkes Primate Center who died following mucosal splash without injury in 1997 was splashed with something in the eye while transporting a caged macaque. In part on this basis, the eye splash was considered low risk. However, feces, urine or other fluids may be contaminated with virus shed from mucosal fluids. Zoonoses have been reported following virus transmission through a bite, scratch, or splash accident. Cases of B virus have also been reported after exposure to monkey cell cultures and to central nervous system tissue. There is often no apparent evidence of B virus infection in the animals or their cells and tissues, making it imperative that all suspect exposures be treated according to recommended standards.[36] The risks associated with this hazard are, however, readily reduced by practicing barrier precautions and by rapid and thorough cleansing immediately following a possible site contamination. Precautions should be followed when work requires the use of any macaque species, even antibody negative animals. In most documented cases of B virus zoonosis, virus was not recovered from potential sources except in four cases, making speculations that some macaque species may be safer than others unfounded. The loss of five lives in the past two decades underscores that B virus infections have a low probability of occurrence, but when they do occur it is with high consequences.

Specific, regular training in risk assessments for B virus hazards including understanding the modes of exposure and transmission should be provided to individuals encountering B virus hazards. This training should include proper use of personal protective equipment, which is essential to prevention. Immediate and thorough cleansing following bites, scratches, splashes, or contact with potential fomites in high-risk areas appears to be helpful in prevention of B virus infections.[37] First aid and emergency medical assistance procedures are most effective when institutions set the standard to be practiced by all individuals encountering B virus hazards.

Natural Modes of Infection

B virus occurs as a natural infection of Asiatic macaque monkeys, and some 10% of newly caught rhesus monkeys have antibodies against the virus, which is frequently present in kidney cell cultures of this animal.

Reservoir species include *Macaca mulatta, M. fascicularis, M.fusata, M. arctoides, M. cyclopsis* and *M. radiata.* In these species the virus causes vesicular lesions on the tongue and lips, and sometimes of the skin. B virus is not present in blood or serum in infected macaques. Transmission of B virus appears to increase when macaques reach sexual maturity.

Laboratory Safety and Containment Recommendations

The National Academies Press has recently published ILAR's guidelines for working with nonhuman primates.[38] Additional resources are provided in the references following this agent summary statement. Asymptomatic B virus shedding accounts for most transmission among monkeys and human workers, but those working in the laboratory with potentially infected cells or tissues from macaques are also at risk. Exposure of mucous membranes or through skin breaks provides this agent access to a new host, whether the virus is being shed from a macaque or human, or present in or on contaminated cells, tissues, or surfaces.[36] B virus is not generally found in serum or blood, but these products obtained through venipuncture should be handled carefully because contamination of needles via skin can occur. When working with macaques directly, virus can be transmitted through bites, scratches, or splashes only when the animal is shedding virus from mucosal sites. Fomites, or contaminated surfaces (e.g., cages, surgical equipment, tables), should always be considered sources of B virus unless verified as decontaminated or sterilized. Zoonotically infected humans should be cautioned about autoinoculation of other susceptible sites when shedding virus during acute infection.

BSL-2 practices and facilities are suitable for all activities involving the use or manipulation of tissues, cells, blood, or serum from macaques with appropriate personal protective equipment. BSL-3 practices are recommended for handling materials from which B virus is being cultured using appropriate personal protective equipment, and BSL-4 facilities are recommended for propagation of virus obtained from diagnostic samples or stocks. Experimental infections of macaques as well as small animal models with B virus are recommended to be restricted to BSL-4 containment.

All macaques regardless of their origin should be considered potentially infected. Animals with no detectable antibody are not necessarily B virus-free. Macaques should be handled with strict barrier precaution protocols and injuries should be tended immediately according to the recommendations of the B Virus Working Group led by NIH and CDC.[36]

Barrier precautions and appropriate first aid are the keys to prevention of severe morbidity and mortality often associated with B virus zoonoses. These prevention tools were not implemented in each of the five B virus fatalities during the past two decades. Guidelines are available for safely working with macaques and should be consulted.[36,39] The correct use of gloves, masks, and protective coats, gowns, aprons, or overalls is recommended for all personnel while working with non-human primates, especially macaques and other Old World species, including for all persons entering animal rooms where non-human primates are housed. To minimize the potential for mucous membrane exposure, some form of barrier is required to prevent droplet splashes to eyes, mouth, and nasal passages. Types and use of personal protective equipment (e.g., goggles or glasses with solid side shields and masks, or wrap-around face shields) should be determined with reference to the institutional risk assessment. Specifications of protective equipment must be balanced with the work to be performed so that the barriers selected do not increase work place risk by obscuring vision and contributing to increased risk of bites, needle sticks, scratches, or splashes.

Special Issues

Post-exposure prophylaxis with oral acyclovir or valacyclovir should be considered for significant exposures to B virus. Therapy with intravenous acyclovir and/or ganciclovir in documented B virus infections is also important in reduction of morbidity following B virus zoonotic infection.[36] In selected cases, IND permission has been granted for therapy with experimental antiviral drugs. Because of the seriousness of B virus infection, experienced medical and laboratory personnel should be consulted to develop individual case management. Barrier precautions should be observed with confirmed cases. B virus infection, as with all alphaherpesviruses, is lifelong in macaques.[40] There are no effective vaccines available.

Select Agent B virus is a select agent requiring registration with CDC and/or USDA for possession, use, storage and/or transfer. See Appendix F for additional information.

Transfer of Agent Importation of this agent may require CDC and/or USDA importation permits. Domestic transport of this agent may require a permit from USDA/APHIS/VS.

Human Herpes Virus

The herpesviruses are ubiquitous human pathogens and are commonly present in a variety of clinical materials submitted for virus isolation. Thus far, nine herpesviruses have been isolated from humans: herpes simplex virus-1(HSV-1), HSV-2, human cytomegalovirus (HCMV), varicella-zoster virus (VZV), Epstein-Barr virus (EBV), and human herpesviruses (HHV) 6A, 6B, 7, and 8.[41]

HSV infection is characterized by a localized primary lesion. Primary infection with HSV-1 may be mild and unapparent occurring in early childhood. In approximately 10% of infections, overt illness marked by fever and malaise occurs. HSV-1 is a common cause of meningoencephalitis. Genital infections, usually caused by HSV-2, generally occur in adults and are sexually transmissible. Neonatal infections are most frequently caused by HSV-2 but HSV-1 infections are also common. In the neonate, disseminated disease and encephalitis are often fatal. EBV is the cause of infectious mononucleosis. It is also associated with the pathogenesis of several lymphomas and nasopharyngeal cancer.[42] EBV is serologically distinct from the other herpesviruses; it infects and transforms B-lymphocytes. HCMV infection is common and often undiagnosed presenting as a nonspecific febrile illness. HCMV causes up to 10% of all cases of mononucleosis in young adults. The most severe form of the disease is seen in infants infected *in utero*. Children surviving infection may evidence mental retardation, microencephaly, motor disabilities and chronic liver disease.[42] HCMV is one of the most common congenital diseases.

VZV is the causative agent of chickenpox and herpes zoster. Chickenpox usually occurs in childhood and zoster occurs more commonly in adults. HHV-6 is the causative agent of exanthema subitum (roseola), a common childhood exanthem.[43] Nonspecific febrile illness and febrile seizures are also clinical manifestations of disease. HHV-6 may reactivate in immunocompetent individuals during pregnancy or during critical illness. Two distinct variants, HHV-6A and HHV-6B, exist, the latter causing roseola. HHV-7 is a constitutive inhabitant of adult human saliva.[44] Clinical manifestations are less well understood but the virus has also been associated with roseola. HHV-8, also known as Kaposi's sarcoma-associated virus, was first identified by Chang and co-workers in 1994.[42] HHV-8 is believed to be the causative agent of Kaposi's sarcoma and has been associated with primary effusion lymphoma.[45] The natural history of HHV-8 has not been completely elucidated. High risk groups for HHV-8 include HIV-infected men who have sex with men and individuals from areas of high endemicity, such as Africa or the Mediterranean.[45] The prevalence of HHV-8 is also higher among intravenous drug users than in the general population.[45] At least one report has provided evidence that in African children, HHV-8 infection may be transmitted from mother to child.[46] While few of the human herpesviruses have been demonstrated to cause laboratory-acquired infections, they are both primary and opportunistic pathogens, especially in immunocompromised hosts. Herpesvirus simiae (B-virus, Monkey B virus) is discussed separately in another agent summary statement in this section.

Occupational Infections

Few of the human herpesviruses have been documented as sources of laboratory acquired infections.

In a limited study, Gartner and co-workers have investigated the HHV-8 immunoglobulin G (IgG) seroprevalence rates for healthcare workers caring for patients with a high risk for HHV-8 infection in a non-endemic area. Healthcare workers in contact with risk group patients were infected more frequently than healthcare workers without contact with risk groups. Workers without contact with risk group patients were infected no more frequently than the control group.[53]

Although this diverse group of indigenous viral agents has not demonstrated a high potential hazard for laboratory-associated infection, frequent presence in clinical materials and common use in research warrant the application of appropriate laboratory containment and safe practices.

Natural Modes of Infection

Given the wide array of viruses included in this family, the natural modes of infection vary greatly, as does the pathogenesis of the various viruses. Some have wide host ranges, multiply effectively, and rapidly destroy the cells they infect (HSV-1, HSV-2). Others have restricted host ranges or long replicative cycles (HHV-6).[41] Transmission of human herpesviruses in nature are, in general, associated with close, intimate contact with a person excreting the virus in their saliva, urine, or other bodily fluids.[47] VZV is transmitted person-to-person through direct contact, through aerosolized vesicular fluids and respiratory secretions, and indirectly transmitted by fomites. Latency is a trait common to most herpesviruses, although the site and duration vary greatly. For example, EBV will persist in an asymptomatic, latent form in the host immune system, primarily in EBV-specific cytotoxic T cells[42] while latent HSV has been detected only in sensory neurons.[48,49] HHV-8 has been transmitted through organ transplantation[50] and blood transfusion;[51] some evidence suggests non-sexual horizontal transmission.[52]

Laboratory Safety and Containment Recommendations

Clinical materials and isolates of herpesviruses may pose a risk of infection following ingestion, accidental parenteral inoculation, and droplet exposure of the mucous membranes of the eyes, nose, or mouth, or inhalation of concentrated aerosolized materials. HHV-8 may be present in human blood or blood products and tissues or saliva. Aerosol transmission cannot be excluded as a potential route of transmission. Clinical specimens containing the more virulent Herpesvirus simiae (B-virus) may be inadvertently submitted for diagnosis of suspected herpes simplex infection. HCMV may pose a special risk during pregnancy because of potential infection of the fetus. All human herpesviruses pose an increased risk to persons who are immunocompromised.

BSL-2 practices, containment equipment, and facilities are recommended for activities utilizing known or potentially infectious clinical materials or cultures of indigenous viral agents that are associated or identified as a primary pathogen of human disease. Although there is little evidence that infectious aerosols are a significant source of LAI, it is prudent to avoid the generation of aerosols during the handling of clinical materials or isolates, or during the necropsy of animals. Primary containment devices (e.g., BSC) should be utilized to prevent exposure of workers to infectious aerosols. Additional containment and procedures, such as those described for BSL-3, should be considered when producing, purifying, and concentrating human herpesviruses, based on risk assessment.

Containment recommendations for herpesvirus simiae (B-virus, Monkey B virus) are described in the preceding agent summary statement.

Special Issues

Vaccine A live, attenuated vaccine for varicella zoster is licensed and available in the United States. In the event of a laboratory exposure to a non-immune individual, varicella vaccine is likely to prevent or at least modify disease.[47]

Treatment Antiviral medications are available for treatment of several of the herpesviruses.

Transfer of Agent Importation of this agent may require CDC and/or USDA importation permits. Domestic transport of this agent may require a permit from USDA/APHIS/VS. A DoC permit may be required for the export of this agent to another country. See Appendix C for additional information.

Influenza

Influenza is an acute viral disease of the respiratory tract. The most common clinical manifestations are fever, headache, malaise, sore throat and cough. GI tract manifestations (nausea, vomiting and diarrhea) are rare but may accompany the respiratory phase in children. The two most important features of influenza are the epidemic nature of illness and the mortality that arises from pulmonary complications of the disease.[54]

The influenza viruses are enveloped RNA viruses belonging to the Orthomyxoviridae. There are three serotypes of influenza viruses, A, B and C. Influenza A is further classified into subtypes by the surface glycoproteins that possess either hemagglutinin (H) or neuraminidase (N) activity. Emergence of completely new subtypes (antigenic shift) occurs at irregular intervals with Type A viruses. New subtypes are responsible for pandemics and can result from reassortment of human and avian influenza virus genes. Antigenic changes within a type or subtype (antigenic drift) of A and B viruses are ongoing processes that are responsible for frequent epidemics and regional outbreaks and make the annual reformulation of influenza vaccine necessary.

Influenza viral infections, with different antigenic subtypes, occur naturally in swine, horses, mink, seals and in many domestic and wild avian species. Interspecies transmission and reassortment of influenza A viruses have been reported to occur among humans and wild and domestic fowl. The human influenza viruses responsible for the 1918, 1957 and 1968 pandemics contained gene segments closely related to those of avian influenza viruses.[55] Swine influenza has also been isolated in human outbreaks.[56]

Control of influenza is a continuing human and veterinary public health concern.

Occupational Infections

LAI have not been routinely documented in the literature, but informal accounts and published reports indicate that such infections are known to have occurred, particularly when new strains showing antigenic shift or drift are introduced into a laboratory for diagnostic/research purposes.[56] Occupationally-acquired, nosocomial infections are documented.[57,58] Laboratory animal-associated infections have not been reported; however, there is possibility of human infection acquired from infected ferrets and vice versa.

Natural Modes of Infection

Airborne spread is the predominant mode of transmission especially in crowded, enclosed spaces. Transmission may also occur through direct contact since influenza viruses may persist for hours on surfaces particularly in the cold and under conditions of low humidity.[55] The incubation period is from one to three days. Recommendations for treatment and prophylaxis of influenza are available.[59]

Laboratory Safety and Containment Recommendations

The agent may be present in respiratory tissues or secretions of humans and most infected animals and birds. In addition, the agent may be present in the intestines and cloacae of many infected avian species. Influenza viruses may be disseminated in multiple organs in some infected animal species. The primary laboratory hazard is inhalation of virus from aerosols generated by infecting animals or by aspirating, dispensing, mixing, centrifuging or otherwise manipulating virus-infected samples. In addition, laboratory infection can result from direct inoculation of mucus membranes through virus-contaminated gloves following handling of tissues, feces or secretions from infected animals. Genetic manipulation has the potential for altering the host range, pathogenicity, and antigenic composition of influenza viruses. The potential for introducing influenza viruses with novel genetic composition into humans is unknown.

BSL-2 facilities, practices and procedures are recommended for diagnostic, research and production activities utilizing contemporary, circulating human influenza strains (e.g., H1/H3/B) and low pathogenicity avian influenza (LPAI) strains (e.g., H1-4, H6, H8-16), and equine and swine influenza viruses. ABSL-2 is appropriate for work with these viruses in animal models. All avian and swine influenza viruses require an APHIS permit. Based on economic ramifications and source of the virus, LPAI H5 and H7 and swine influenza viruses may have additional APHIS permit-driven containment requirements and personnel practices and/or restrictions.

Non-Contemporary Human Influenza (H2N2) Strains

Non-contemporary, wild-type human influenza (H2N2) strains should be handled with increased caution. Important considerations in working with these strains are the number of years since an antigenically related virus last circulated and the potential for presence of a susceptible population. BSL-3 and ABSL-3 practices, procedures and facilities are recommended with rigorous adherence to additional respiratory protection and clothing change protocols. Negative pressure, HEPA-filtered respirators or positive air-purifying respirators (PAPRs) are recommended for use. Cold-adapted, live attenuated H2N2 vaccine strains may continue to be worked with at BSL-2.

1918 Influenza Strain

Any research involving reverse genetics of the 1918 influenza strain should proceed with *extreme* caution. The risk to laboratory workers is unknown, but the pandemic potential is thought to be significant. Until further risk assessment data are available, the following practices and conditions are recommended for manipulation of reconstructed 1918 influenza viruses and laboratory animals infected with the viruses. These practices and procedures are considered minimum standards for work with the fully reconstructed virus.

- BSL-3 and ABSL-3 practices, procedures and facilities.

- Large laboratory animals such as NHP should be housed in primary barrier systems in ABSL-3 facilities.

- Rigorous adherence to additional respiratory protection and clothing change protocols.

- Use of negative pressure, HEPA-filtered respirators or PAPRs.

- Use of HEPA filtration for treatment of exhaust air.

- Amendment of personnel practices to include personal showers prior to exiting the laboratory.

Highly Pathogenic Avian Influenza (HPAI)

Manipulating HPAI viruses in biomedical research laboratories requires similar caution because some strains may pose increased risk to laboratory workers and have significant agricultural and economic implications. BSL-3 and ABSL-3 practices, procedures and facilities are recommended along with clothing change and personal showering protocols. Loose-housed animals infected with HPAI strains must be contained within BSL-3-Ag facilities. (See Appendix D.) Negative pressure, HEPA-filtered respirators or positive air-purifying respirators are recommended for HPAI viruses with potential to infect humans. The HPAI are agricultural select agents requiring registration of personnel and facilities with the lead agency for the institution (CDC or USDA-APHIS). An APHIS permit is also required. Additional containment requirements and personnel practices and/or restrictions may be added as conditions of the permit.

Other Influenza Recombinant or Reassortant Viruses

When considering the biocontainment level and attendant practices and procedures for work with other influenza recombinant or reassortant viruses, the local IBC should consider but not limit consideration to the following in the conduct of protocol-driven risk assessment.

- The gene constellation used.

- Clear evidence of reduced virus replication in the respiratory tract of appropriate animal models, compared with the level of replication of the wild-type parent virus from which it was derived.

- Evidence of clonal purity and phenotypic stability.

- The number of years since a virus that was antigenically related to the donor of the hemagglutinin and neuraminidase genes last circulated.

If adequate risk assessment data are not available, a more cautious approach utilizing elevated biocontainment levels and practices is warranted. There may be specific requirements regarding the setting of containment levels if your institution is subject to the *NIH Guidelines*.

Special Issues

Occupational Health Considerations Institutions performing work with HPAI and avian viruses that have infected humans; non-contemporary wild-type human influenza strains, including recombinants and reassortants; and viruses created by reverse genetics of the 1918 pandemic strain should develop and implement a specific medical surveillance and response plan. At the minimum these plans should: 1) require storage of baseline serum samples from individuals working with these influenza strains; 2) strongly recommend annual vaccination with the currently licensed influenza vaccine for such individuals; 3) provide employee counseling regarding disease symptoms including fever, conjunctivitis and respiratory symptoms; 4) establish a protocol for monitoring personnel for these symptoms; and 5) establish a clear medical protocol for responding to suspected laboratory-acquired infections. Antiviral drugs (e.g., oseltamivir, amantadine, rimantadine, zanamivir) should be available for treatment and prophylaxis, as necessary.[59] It is recommended that the sensitivities of the virus being studied to the antivirals be ascertained. All personnel should be enrolled in an appropriately constituted respiratory protection program.

Influenza viruses may require USDA and/or USPHS import permits depending on the host range and pathogenicity of the virus in question.

Select Agent Strains of HPAI and 1918 influenza virus are select agents requiring registration with CDC and/or USDA for possession, use, storage and/or transfer. See Appendix F for additional information.

Transfer of Agent Importation of this agent may require CDC and/or USDA importation permits. Domestic transport of this agent may require a permit from USDA/APHIS/VS. A DoC permit may be required for the export of this agent to another country. See Appendix C for additional information.

Lymphocytic Choriomeningitis Virus

Lymphocytic choriomeningitis (LCM) is a rodent-borne viral infectious disease that presents as aseptic meningitis, encephalitis, or meningoencephalitis. The causative agent is the LCM virus (LCMV) that was initially isolated in 1933. The virus is the protypical member of the family *Arenaviridae.*

Occupational Infections

LAI with LCM virus are well documented. Most infections occur when chronic viral infection exists in laboratory rodents, especially mice, hamsters and guinea pigs.[60-62] Nude and severe combined immune deficient (SCID) mice may pose a special risk of harboring silent chronic infections. Inadvertently infected cell cultures also represent a potential source of infection and dissemination of the agent.

Natural Modes of Infection

LCM and milder LCMV infections have been reported in Europe, the Americas, Australia, and Japan, and may occur wherever infected rodent hosts of the virus are found. Several serologic studies conducted in urban areas have shown that the prevalence of LCMV infection among humans ranges from 2% to 10%. Seroprevalence of 37.5% has been reported in humans in the Slovak Republic.[63]

The common house mouse, *Mus musculus*, naturally spreads LCMV. Once infected, these mice can become chronically infected as demonstrated by the presence of virus in blood and/or by persistently shedding virus in urine. Infections have also occurred in NHP in zoos, including macaques and marmosets. (*Callitrichid* hepatitis virus is a LCMV.)

Humans become infected by inhaling infectious aerosolized particles of rodent urine, feces, or saliva; by ingesting food contaminated with virus; by contamination of mucous membranes with infected body fluids; or by directly exposing cuts or other open wounds to virus-infected blood. Four recipients of organs from a donor who had unrecognized disseminated LCMV infection sustained severe disease and three succumbed. The source of donor infection was traced to a pet hamster that was not overtly ill.[64]

Laboratory Safety and Containment Recommendations

The agent may be present in blood, CSF, urine, secretions of the nasopharynx, feces and tissues of infected animal hosts and humans. Parenteral inoculation, inhalation, contamination of mucous membranes or broken skin with infectious tissues or fluids from infected animals are common hazards. Aerosol transmission is well documented.[60]

Of special note, tumors may acquire LCMV as an adventitious virus without obvious effects on the tumor. Virus may survive freezing and storage in liquid nitrogen for long periods. When infected tumor cells are transplanted, subsequent infection of the host and virus excretion may ensue. Pregnant women infected with LCMV have transmitted the virus to their fetuses with death or serious central nervous system malformation as a consequence.[65]

BSL-2 practices, containment equipment, and facilities are suitable for activities utilizing known or potentially infectious body fluids, and for cell culture passage of laboratory-adapted strains. BSL-3 is required for activities with high potential for aerosol production, work with production quantities or high concentrations of infectious materials, and for manipulation of infected transplantable tumors, field isolates and clinical materials from human cases. Strains of LCMV that are shown to be lethal in non-human primates should be handled at BSL-3. ABSL-2 practices, containment equipment, and facilities are suitable for studies in adult mice with mouse brain-passaged strains requiring BSL-2 containment. Work with infected hamsters also should be done at ABSL-3.

Special Issues

Vaccines Vaccines are not available for use in humans.

Transfer of Agent Importation of this agent may require CDC and/or USDA importation permits. Domestic transport of this agent may require a permit from USDA/APHIS/VS. A DoC permit may be required for the export of this agent to another country. See Appendix C for additional information.

Poliovirus

Poliovirus is the type species of the *Enterovirus* genus in the family *Picornaviridae*. Enteroviruses are transient inhabitants of the gastrointestinal tract, and are stable at acid pH. Picornaviruses are small, ether-insensitive viruses with an RNA genome.

There are three poliovirus serotypes (P1, P2, and P3). Immunity to one serotype does not produce significant immunity to the other serotypes.

Occupational Infections

Laboratory-associated poliomyelitis is uncommon. Twelve cases, including two deaths, were reported between 1941 and 1976.[62,66] No laboratory-associated poliomyelitis has been reported for nearly 30 years. Both inactivated poliovirus vaccine (IPV) and oral poliovirus vaccine (OPV) are highly effective in preventing disease, but neither vaccine provides complete protection against infection. Poliovirus infections among immunized laboratory workers are uncommon but remain undetermined in the absence of laboratory confirmation. An immunized laboratory worker may unknowingly be a source of poliovirus transmission to unvaccinated persons in the community.[67]

Natural Modes of Infection

At one time poliovirus infection occurred throughout the world. Transmission of wild poliovirus ceased in the United States in 1979, or possibly earlier. A polio eradication program conducted by the Pan American Health Organization led to elimination of polio from the Western Hemisphere in 1991. The Global Polio Eradication Program has dramatically reduced poliovirus transmission throughout the world.

Humans are the only known reservoir of poliovirus, which is transmitted most frequently by persons with unapparent infections. Person-to-person spread of poliovirus via the fecal-oral route is the most important route of transmission, although the oral-oral route may account for some cases.

Laboratory Safety and Containment Recommendations

The agent is present in the feces and in throat secretions of infected persons and in lymph nodes, brain tissue, and spinal cord tissue in fatal cases. For non-immunized persons in the laboratory, ingestion or parenteral inoculation are the primary routes of infection. For immunized persons, the primary risks are the same, except for parenteral inoculation, which likely presents a lower risk. The importance of aerosol exposure is unknown. Laboratory animal-associated infections have not been reported, but infected nonhuman primates should be considered to present a risk.

BSL-2 practices, containment equipment, and facilities are recommended for all activities utilizing wild poliovirus infectious culture fluids, environmental samples, and clinical materials. In addition, potentially infectious materials collected for any purpose should be handled at BSL-2. Laboratory personnel working with such materials must have documented polio vaccination. Persons who have had a primary series of OPV or IPV and who are at an increased risk can receive another dose of IPV, but available data do not indicate the need for more than a single lifetime IPV booster dose for adults.[68] ABSL-2 practices, containment equipment, and facilities are recommended for studies of virulent viruses in animals. Laboratories should use authentic Sabin OPV attenuated strains unless there are strong scientific reasons for working with wild polioviruses.

In anticipation of polio eradication, the WHO recommends destruction of all poliovirus stocks and potential infectious materials if there is no longer a programmatic or research need for such materials.[69] Institutions/laboratories in the United States that currently retain wild poliovirus infectious or potential infectious material should be on the United States National Inventory maintained by CDC. When one year has elapsed after detection of the last wild poliovirus worldwide, CDC will inform relevant institutions/laboratories about additional containment procedures. Safety recommendations are subject to change based on international polio eradication activities.

Special Issues

When OPV immunization stops, global control and biosafety requirements for wild as well as attenuated (Sabin) poliovirus materials are expected to become more stringent, consistent with the increased consequences of inadvertent transmission to a growing susceptible community.

Transfer of Agent Importation of this agent may require CDC and/or USDA importation permits. Domestic transport of this agent may require a permit from USDA/APHIS/VS. A DoC permit may be required for the export of this agent to another country. See Appendix C for additional information.

Poxviruses

Four genera of the subfamily *Chordopoxvirinae,* family *Poxviridae,* (*Orthopoxvirus, Parapoxvirus, Yatapoxvirus,* and *Molluscipoxvirus*) contain species that can cause lesions on human skin or mucous membranes with mild to severe systemic rash illness in laboratorians. Species within the first three genera mostly arise as zoonotic agents.[70,71] Laboratory-acquired poxvirus infections of most concern are from the orthopoxviruses that infect humans: variola virus (causes smallpox; human-specific), monkeypox virus (causes smallpox-like disease), *cowpox virus* (causes skin pustule, generalized rash), and vaccinia virus (causes skin pustule, systemic illness).[70-75]

Occupational Infections

Vaccinia virus, the leading agent of laboratory-acquired poxvirus infections, is used to make the current smallpox vaccine and may occur as a rare zoonosis.[70,71] Laboratory-acquired infections with standard, mutant, or bioengineered forms of vaccinia virus have occurred, even in previously vaccinated laboratorians. In addition, vaccination with live vaccinia virus sometimes has side effects, which range from mild events (e.g., fever, fatigue, swollen lymph nodes) to rare, severe, and at times fatal outcomes (e.g., generalized vaccinia, encephalitis, vaccinia necrosum, eczema vaccinatum, ocular keratitis, corneal infection, fetal infection of pregnancy, and possibly myocardial infarction, myopericarditis, or angina), thus vaccination contraindications should be carefully followed.[70,73-75]

Natural Modes of Infection

Smallpox has been eradicated from the world since 1980, but monkey pox virus is endemic in rodents in parts of Africa. Importation of African rodents into North America in 2003 resulted in an outbreak of monkeypox in humans.[72] Molluscum contagiosum, a disease due to *Molluscipoxvirus* infection, results in pearly white lesions that may persist for months in persons immunocompromised for various

218   Biosafety in Microbiological and Biomedical Laboratories

reasons, including chronic illness, AIDS, other infections, medications, cancer and cancer therapies, or pregnancy.[70]

Laboratory Safety and Containment Recommendations

Poxviruses are stable in a wide range of environmental temperatures and humidity and may be transmitted by fomites.[70] Virus may enter the body through mucous membranes, broken skin, or by ingestion, parenteral inoculation or droplet or fine-particle aerosol inhalation. Sources of laboratory-acquired infection include exposure to aerosols, environmental samples, naturally or experimentally infected animals, infectious cultures, or clinical samples, including vesiculopustular rash lesion fluid or crusted scabs, various tissue specimens, excretions and respiratory secretions.

Worldwide, all live variola virus work is to be done only within WHO approved BSL-4/ABSL-4 facilities; one is at the CDC in Atlanta and the other is at the State Research Center of Virology and Biotechnology (VECTOR) in Koltsovo, Russia.[76]

In general, all persons working in or entering laboratory or animal care areas where activities with vaccinia, monkey pox, or cowpox viruses are being conducted should have evidence of satisfactory vaccination. Vaccination is advised every three years for work with monkeypox virus and every 10 years for cowpox and vaccinia viruses (neither vaccination nor vaccinia immunoglobulin protect against poxviruses of other genera).[73-75]

ABSL-3 practices, containment equipment, and facilities are recommended for monkeypox work in experimentally or naturally infected animals. BSL-2 facilities with BSL-3 practices are advised if vaccinated personnel perform other work with monkeypox virus. These practices include the use of Class I or II BSCs and barriers, such as safety cups or sealed rotors, for all centrifugations. The *NIH Guidelines* have assessed the risk of manipulating attenuated vaccinia strains (modified virus Ankara [MVA], NYVAC, TROVAC, and ALVAC) in areas where no other human orthopoxviruses are being used and have recommended BSL-1.[76] However, higher levels of containment are recommended if these strains are used in work areas where other orthopoxviruses are manipulated. Vaccination is not required for individuals working only in laboratories where no other orthopoxviruses or recombinants are handled.[75] BSL-2 and ABSL-2 plus vaccination are recommended for work with most other poxviruses.

Special Issues

Other Considerations The CDC Web site *www.cdc.gov* provides information on poxviruses, especially variola and monkeypox viruses, smallpox vaccination, and reporting vaccination adverse events. Clinical and other laboratories using poxviruses and clinicians can phone the CDC Clinician Information Line (877-554-4625) and/or the CDC public information hotline (888-246-2675) concerning variola and other human poxvirus infections, smallpox vaccine, vaccinia immunoglobulin, poxvirus antiviral drugs, or other treatments or quarantine issues. Contact CDC regarding applications to transfer monkeypox viruses.

Transfer of Agent Importation of this agent may require CDC and/or USDA importation permits. Domestic transport of this agent may require a permit from USDA/APHIS/VS. A DoC permit may be required for the export of this agent to another country. See Appendix C for additional information.

Rabies Virus (and related lyssaviruses)

Rabies is an acute, progressive, fatal encephalitis caused by negative-stranded RNA viruses in the genus *Lyssavirus*, family *Rhabdoviridae*.[77] *Rabies virus* is the representative member (type species) of the genus. Members of the group include Australian bat lyssavirus, Duvenhage virus, European bat lyssavirus[1], European bat lyssavirus[2], Lagos bat virus, and Mokola virus.

Occupational Infections

Rabies LAI are extremely rare; two have been documented. Both resulted from presumed exposure to high concentrations of infectious aerosols, one generated in a vaccine production facility,[78] and the other in a research facility.[79] Naturally or experimentally infected animals, their tissues, and their excretions are a potential source of exposure for laboratory and animal care personnel.

Natural Modes of Infection

The natural hosts of rabies are many bat species and terrestrial carnivores, but most mammals can be infected. The saliva of infected animals is highly infectious, and bites are the usual means of transmission, although infection through superficial skin lesions or mucosa is possible.

Laboratory Safety and Containment Recommendations

When working with infected animals, the highest viral concentrations are present in central nervous system (CNS) tissue, salivary glands, and saliva, but rabies viral antigens may be detected in all innervated tissues. The most likely sources for exposure of laboratory and animal care personnel are accidental parenteral inoculation, cuts, or needle sticks with contaminated laboratory equipment, bites by infected animals, and exposure of mucous membranes or broken skin to infectious tissue or fluids. Infectious aerosols have not been a demonstrated hazard to personnel working with routine clinical materials or conducting diagnostic examinations. Fixed and attenuated strains of virus are presumed to be less hazardous, but the two recorded cases of laboratory-associated rabies resulted from presumed exposure to the fixed Challenge Virus Standard and Street Alabama Dufferin strains, respectively.

BSL-2 and/or ABSL-2 practices, containment equipment, and facilities are recommended for all activities utilizing known or potentially infectious materials or animals. Pre-exposure rabies vaccination is recommended for all individuals prior to working with lyssaviruses or infected animals, or engaging in diagnostic, production, or research activities with these viruses.[80] Rabies vaccination also is recommended for all individuals entering or working in the same room where lyssaviruses or infected animals are used. Prompt administration of postexposure booster vaccinations is recommended following recognized exposures in previously vaccinated individuals per current guidelines.[81] For routine diagnostic activities, it is not always feasible to open the skull or remove the brain of an infected animal within a BSC, but it is pertinent to use appropriate methods and personal protection equipment, including dedicated laboratory clothing, heavy protective gloves to avoid cuts or sticks from cutting instruments or bone fragments, and a face shield or PAPR to protect the skin and mucous membranes of the eyes, nose, and mouth from exposure to tissue fragments or infectious droplets.

If a Stryker saw is used to open the skull, avoid contacting brain tissue with the blade of the saw. Additional primary containment and personnel precautions, such as those described for BSL-3, are indicated for activities with a high potential for droplet or aerosol production, and for activities involving large production quantities or high concentrations of infectious materials.

Special Issues

Transfer of Agent Importation of this agent may require CDC and/or USDA importation permits. Domestic transport of this agent may require a permit from USDA/APHIS/VS. A DoC permit may be required for the export of this agent to another country. See Appendix C for additional information.

Retroviruses, including Human and Simian Immunodeficiency Viruses (HIV and SIV)

The family *Retroviridae* is divided into two subfamilies, the *Orthoretrovirinae* with six genera including the Lentivirus genus, which includes HIV-1 and HIV-2. Other important human pathogens are human T-lymphotropic viruses 1 and 2 (HTLV-1 and HTLV-2), members of the Deltaretrovirus genus. The Spumaretrovirinae, with one genus, Spumavirus, contains a variety of NHP viruses (foamy viruses) that can occasionally infect humans in close contact with NHPs.

Occupational Infections

Data on occupational HIV transmission in laboratory workers are collected through two CDC-supported national surveillance systems: surveillance for 1) AIDS, and 2) HIV-infected persons who may have acquired their infection through occupational exposures. For surveillance purposes, laboratory workers are defined as those persons, including students and trainees, who have worked in a clinical or HIV laboratory setting anytime since 1978. Cases reported in these two systems are classified as either documented or possible occupational transmission. Those classified as documented occupational transmission had evidence of HIV seroconversion (a negative HIV-antibody test at the time of the exposure which converted to positive) following a discrete percutaneous or mucocutaneous occupational exposure to blood, body fluids, or other clinical or laboratory specimens. As of June 1998, CDC had reports of 16 laboratory workers (all clinical) in the United States with documented occupational transmission.[82]

Workers have been reported to develop antibodies to simian immunodeficiency virus (SIV) following exposures. One case was associated with a needle-stick that occurred while the worker was manipulating a blood-contaminated needle after bleeding an SIV-infected macaque monkey.[83] Another case involved a laboratory worker who handled macaque SIV-infected blood specimens without gloves. Though no specific incident was recalled, this worker had dermatitis on the forearms and hands while working with the infected blood specimens.[84] A third worker[85] was exposed to SIV-infected primate blood through a needle-stick and subsequently developed antibodies to SIV. To date there is no evidence of illness or immunological incompetence in any of these workers.

Natural Modes of Infection

Retroviruses are widely distributed as infectious agents of vertebrates. Within the human population, spread is by close sexual contact or parenteral exposure through blood or blood products.

Laboratory Safety and Containment Recommendations

HIV has been isolated from blood, semen, saliva, tears, urine, CSF, amniotic fluid, breast milk, cervical secretion, and tissues of infected persons and experimentally infected nonhuman primates.[86]

Although the risk of occupationally-acquired HIV is primarily through exposure to infected blood, it is also prudent to wear gloves when manipulating other body fluids such as feces, saliva, urine, tears, sweat, vomitus, and human breast milk. This also reduces the potential for exposure to other microorganisms that may cause other types of infections.

In the laboratory, virus should be presumed to be present in all blood or clinical specimens contaminated with blood, in any unfixed tissue or organ (other than intact skin) from a human (living or dead), in HIV cultures, in all materials derived from HIV cultures, and in/on all equipment and devices coming into direct contact with any of these materials.

SIV has been isolated from blood, CSF, and a variety of tissues of infected nonhuman primates. Limited data exist on the concentration of virus in semen, saliva, cervical secretions, urine, breast milk, and amniotic fluid. Virus should be presumed to be present in all SIV cultures, in animals experimentally infected or inoculated with SIV, in all materials derived from SIV cultures, and in/on all equipment and devices coming into direct contact with any of these materials.[87]

The skin (especially when scratches, cuts, abrasions, dermatitis, or other lesions are present) and mucous membranes of the eye, nose, and mouth should be considered as potential pathways for entry of these retroviruses during laboratory activities. It is unknown whether infection can occur via the respiratory tract. The need for using sharps in the laboratory should be evaluated. Needles, sharp instruments, broken glass, and other sharp objects must be carefully handled and properly discarded. Care must be taken to avoid spilling and splashing infected cell-culture liquid and other potentially infected materials.[85]

BSL-2 practices, containment equipment, and facilities are recommended for activities involving blood-contaminated clinical specimens, body fluids and tissues. HTLV-1 and HTLV-2 should also be handled at this level. Activities such as producing research-laboratory-scale quantities of HIV or SIV, manipulating concentrated virus preparations, and conducting procedures that may produce droplets or aerosols, are performed in a BSL-2 facility, using BSL-3 practices. Activities involving large-scale volumes or preparation of concentrated HIV or SIV are conducted at BSL-3. ABSL-2 is appropriate for NHP and other animals infected with HIV or SIV. Human serum from any source that is used as a control or reagent in a test procedure should be handled at BSL-2.

In addition to the aforementioned recommendations, persons working with HIV, SIV, or other bloodborne pathogens should consult the OSHA Bloodborne Pathogen Standard.[88] Questions related to interpretation of this Standard should be directed to federal, regional or state OSHA offices.

Special Issues

It is recommended that all institutions establish written policies regarding the management of laboratory exposure to HIV and SIV, including treatment and prophylaxis protocols. (See Section VII.)

The risk associated with retroviral vector systems can vary significantly, especially lentiviral vectors. Because the risk associated with each gene transfer system can vary, no specific guideline can be offered other than to have all gene transfer protocols reviewed by an IBC.

Transfer of Agent Importation of this agent may require CDC and/or USDA importation permits. Domestic transport of this agent may require a permit from USDA/APHIS/VS. A DoC permit may be required for the export of this agent to another country. See Appendix C for additional information.

Severe Acute Respiratory Syndrome (SARS) Coronavirus

SARS is a viral respiratory illness caused by a previously undescribed coronavirus, SARS-associated coronavirus (SARS-CoV) within the family *Coronaviridae*. SARS was retrospectively recognized in China in November 2002. Over the next few months, the illness spread to other south-east.

Asian countries, North America, South America, and Europe following major airline routes. The majority of disease spread occurred in hospitals, among family members and contacts of hospital workers. From November 2002 through July 2003, when the global outbreak was contained, a total of 8,098 probable cases of SARS were reported to the WHO from 29 countries.[89]

In general, SARS patients present with fever (temperature greater than 100.4°F [>38.0°C]), malaise and myalgias quickly followed by respiratory symptoms including shortness of breath and cough. Ten to 20 percent of patients may have diarrhea. Review of probable cases indicates that the shortness of breath sometimes rapidly progresses to respiratory failure requiring ventilation. The case fatality rate is about 11%.

Occupational Infections

Healthcare workers are at increased risk of acquiring SARS from an infected patient especially if involved in pulmonary/respiratory procedures such as endotracheal intubation, aerosolization or nebulization of medications, diagnostic sputum induction, airway suctioning, positive pressure ventilation and high-frequency oscillatory ventilation.

Two confirmed episodes of SARS-CoV transmission to laboratory workers occurred in research laboratories in Singapore and Taiwan.[89,90] Both occurrences were linked to breaches in laboratory practices. Laboratory-acquired infections in China during 2004 demonstrated secondary and tertiary spread of the disease to close contacts and healthcare providers of one of the employees involved.[91] Although no laboratory-acquired cases have been associated with the routine processing of diagnostic specimens, SARS coronavirus represents an emerging infectious disease for which risk to the medical and laboratory community is not fully understood.

Natural Modes of Infection

The mode of transmission in nature is not well understood. It appears that SARS is transmitted from person-to-person through close contact such as caring for, living with, or having direct contact with respiratory secretions or body fluids of a suspect or probable case.[92] SARS is thought to be spread primarily through droplets, aerosols and possibly fomites. The natural reservoir for SARS CoV is unknown.

Laboratory Safety and Containment Recommendations

SARS-CoV may be detected in respiratory, blood, or stool specimens. The exact mode of transmission of SARS-CoV laboratory-acquired infection has not been established, but in clinical settings the primary mode of transmission appears through direct or indirect contact of mucous membranes with infectious respiratory droplets.[93,94]

In clinical laboratories, whole blood, serum, plasma and urine specimens should be handled using Standard Precautions, which includes use of gloves, gown, mask, and eye protection. Any procedure with the potential to generate aerosols (e.g., vortexing or sonication of specimens in an open tube) should be performed in a BSC. Use sealed centrifuge rotors or gasketed safety carriers for centrifugation. Rotors and safety carriers should be loaded and unloaded in a BSC. Procedures conducted outside a BSC must be performed in a manner that minimizes the risk of personnel exposure and environmental release.

The following procedures may be conducted in the BSL-2 setting: pathologic examination and processing of formalin-fixed or otherwise inactivated tissues, molecular analysis of extracted nucleic acid preparations, electron microscopic studies with glutaraldehyde-fixed grids, routine examination of bacterial and fungal cultures, routine staining and microscopic analysis of fixed smears, and final packaging of specimens for transport to diagnostic laboratories for additional testing (specimens should already be in a sealed, decontaminated primary container).

Activities involving manipulation of untreated specimens should be performed in BSL-2 facilities following BSL-3 practices. In the rare event that a procedure or process involving untreated specimens cannot be conducted in a BSC, gloves, gown, eye protection, and respiratory protection (acceptable methods of respiratory protection include: a properly fit-tested, National Institute for Occupational Safety and Health [NIOSH]-approved filter respirator [N-95 or higher level] or a PAPR equipped with HEPA filters) should be used. All personnel who use respiratory protective devices should be enrolled in an appropriately constituted respiratory protection program.

Work surfaces should be decontaminated upon completion of work with appropriate disinfectants. All waste must be decontaminated prior to disposal.

SARS-CoV propagation in cell culture and the initial characterization of viral agents recovered in cultures of SARS specimens must be performed in a BSL-3 facility using BSL-3 practices and procedures. Risk assessment may dictate the additional use of respiratory protection.

Inoculation of animals for potential recovery of SARS-CoV from SARS samples, research studies, and protocols involving animal inoculation for characterization of putative SARS agents must be performed in ABSL-3 facilities using ABSL-3 work practices. Respiratory protection should be used as warranted by risk assessment.

In the event of any break in laboratory procedure or accidents (e.g., accidental spillage of material suspected of containing SARS-CoV), procedures for emergency exposure management and/or environmental decontamination should be immediately implemented and the supervisor should be notified. The worker and the supervisor, in consultation with occupational health or infection control personnel, should evaluate the break in procedure to determine if an exposure occurred (see Special Issues, below).

Special Issues

Occupational Health Considerations Institutions performing work with SARS coronavirus should require storage of a baseline serum sample from individuals who work with the virus or virus-containing specimens. Personnel working with the virus or samples containing or potentially containing the virus should be trained regarding the symptoms of SARS-CoV infection and counseled to report any fever or respiratory symptoms to their supervisor immediately. They should be evaluated for possible exposure and the clinical features and course of their illness should be closely monitored. Institutions performing work with the SARS-CoV or handling specimens likely to contain the agent should develop and implement a specific occupational medical plan with respect to this agent. The plan, at a minimum, should contain procedures for managing:

- identifiable breaks in laboratory procedures;

- exposed workers without symptoms;

- exposed workers who develop symptoms within ten days of an exposure; and

- symptomatic laboratory workers with no recognized exposure.

Further information and guidance regarding the development of a personnel exposure response plan is available from the CDC.[95] Laboratory workers who are believed to have had a laboratory exposure to SARS-CoV should be evaluated, counseled about the risk of SARS-CoV transmission to others, and monitored for fever or lower respiratory symptoms as well as for any of the following: sore throat, rhinorrhea, chills, rigors, myalgia, headache, and diarrhea.

Local and/or state public health departments should be promptly notified of laboratory exposures and illness in exposed laboratory workers.

Transfer of Agent Importation of this agent may require CDC and/or USDA importation permits. Domestic transport of this agent may require a permit from USDA/APHIS/VS.

References

1. Tsai TF. Hemorrhagic fever with renal syndrome: mode of transmission to humans. Lab Animal Sci. 1987;37:428-30.
2. Umenai T, Lee HW, Lee PW, et al. Korean haemorrhagic fever in staff in an animal laboratory. Lancet. 1979;1:1314-6.
3. Desmyter J, LeDuc JW, Johnson KM, et al. Laboratory rat associated outbreak of haemorrhagic fever with renal syndrome due to Hantaan-like virus in Belgium. Lancet. 1083;2:1445-8.
4. Lloyd G, Bowen ET, Jones N, et al. HFRS outbreak associated with laboratory rats in UK. Lancet. 1984;1:1175-6.
5. Centers for Disease Control and Prevention. Laboratory management of agents associated with hantavirus pulmonary syndrome: interim biosafety guidelines. MMWR Morb Mortal Wkly Rep. 1994;43(RR-7):1-7.
6. Lopez N, Padula P, Rossi C, et al. Genetic identification of a new hantavirus causing severe pulmonary syndrome in Argentina. Virology. 1996;220:223-6.
7. Padula PJ, Edelstein A, Miguel SD, et al. Hantavirus pulmonary syndrome outbreak in Argentina: molecular evidence for person-to-person transmission of Andes virus. Virology. 1998;241:323-30.
8. Nichol ST, Spiropoulou CF, Morzunov S, et al. Genetic identification of a hantavirus associated with an outbreak of acute respiratory illness. Science. 1993;262:914-7.
9. Hjelle B, Spiropoulou CF, Torrez-Martinez N, et al. Detection of Muerto Canyon virus RNA in peripheral blood mononuclear cells from patients with hantavirus pulmonary syndrome. J Infecti Dis. 1994;170:1013-7.
10. Centers for Disease Control and Prevention. Outbreak of Hendra-like virus—Malaysia and Singapore, 1998-99. MMWR Morb Mortal Wkly Rep. 1999;48:265-9.
11. Centers for Disease Control and Prevention. MMWR. Update outbreak of Nipah virus—Malaysia and Singapore. MMWR Morb Mortal Wkly Rep. 1999;48:335-7.
12. Chua KB, Goh KJ, Wong KT, et al. Fatal encephalitis due to Nipah virus among pig-farmers in Malaysia. Lancet. 1999;354:1257-9.
13. Paton NI, Leo YS, Zaki SR, et al. Outbreak of Nipah-virus infection among abattoir workers in Singapore. Lancet. 1999;354:1253-6.
14. Chua KB, Bellini WJ, Rota PA, et al. Nipah virus, a recently emergent deadly paramyxovirus. Science. 2000;288:1432-5.

15. Selvey LA, Wells RM, McCormack JG, et al. Infection of humans and horses by a newly described morbillivirus. Med J Aust. 1995;162:642-5.
16. Hooper PT, Gould AR, Russell GM, et al. The retrospective diagnosis of a second outbreak of equine morbillivirus infection. Aust Vet J. 1996;74:244-5.
17. Murray K, Selleck P, Hooper P, et al. A morbillivirus that caused fatal disease in horses and humans. Science. 1995;268:94-7.
18. Rogers RJ, Douglas IC, Baldock FC, et al. Investigation of a second focus of equine morbillivirus infection in coastal Queensland. Aust Vet J. 1996;74:243-4.
19. Williamson MM, Hooper PT, Selleck PW, et al. Transmission studies of Hendra virus (equine morbillivirus) in fruit bats, horses and cats. Aust Vet J. 1998;76:813-8.
20. Yu M, Hansson E, Shiell B, et al. Sequence analysis of the Hendra virus nucleoprotein gene comparison with other members of the subfamily paramyxovirinae. J Gen Virol. 1998;791:775-80.
21. Johara MY, Field H, Mohd Rashdi A, et al. Nipah virus infection in bats (order Chiroptera) in peninsular Malaysia. Emerg Infect Dis. 2001;7:439-41.
22. Young P, Halpin K, Selleck PW, et al. Serologic evidence for the presence in Pteropus bats of a paramyxovirus related to equine morbillivirus. Emerg Infect Dis. 1996;2:239-40.
23. Olson JG, Rupprecht C, Rollin PE, et al. Antibodies to Nipah-like virus in bats (Pteropus lylei), Cambodia. Emerg Infect Dis. 2002;8:987-8.
24. Outbreaks of encephalitis due to Nipah/Hendra-like viruses, Western Bangladesh. Health and Science Bulletin. 2003;1:1-6.
25. Selvey LA, Taylor R, Arklay A, et al. Screening of bat carers for antibodies to equine morbillivirus. Comm Dis Intell. 1996;20:477-8.
26. Wong KT, Shieh WJ, Kumar S, et al. Nipah virus infection: pathology and pathogenesis of an emerging paramyxoviral zoonosis. Am J Pathol. 2002;161:2153-67.
27. Chua KB, Lam SK, Goh KJ, et al. The presence of Nipah virus in respiratory secretions and urine of patients during an outbreak of Nipah virus encephalitis in Malaysia. J Infect. 2001;42:40-3.
28. Mounts AW, Kaur H, Parashar UD, et al. A cohort study of health care workers to assess nosocomial transmissibility of Nipah virus, Malaysia, 1999. J Infect Dis. 2001;183:810-3.
29. Pike, RM. Laboratory-associated infections: incidence, fatalities, causes and prevention. Ann Rev Microbiol. 1979;33:41-66.
30. Evans MR, Henderson DK, Bennett JE. Potential for laboratory exposures to biohazardous agents found in blood. Am J Public Health. 1990;80:423-7.
31. Centers for Disease Control and Prevention. Recommendations for follow-up of healthcare workers after occupational exposure to hepatitis C virus. MMWR Morb Mortal Wkly Rep. 1997;46:603-6.
32. Chung H, Kudo M, Kumada T, et al. Risk of HCV transmission after needle-stick injury, and the efficacy of short-duration interferon administration to prevent HCV transmission to medical personnel. J Gastrol. 2003;38:877-9.

33. Buster EHCJ, van der Eijk AA, Schalm SW. Doctor to patient transmission of hepatitis B virus: implications of HBV DNA levels and potential new solutions. Antiviral Res. 2003;60:79-85.

34. Occupational Exposure to Bloodborne Pathogens—OSHA. Final Rule. Fed Register. 1991;56:64175-82.

35. Centers for Disease Control and Prevention. Recommendations of the Advisory Committee on Immunization Practices (ACIP). Inactivated hepatitis B virus vaccine. MMWR Morb Mortal Wkly Rep. 1982;31:317-22, 327-8.

36. Cohen JI, Davenport DS, Stewart JA, et al. Recommendations for prevention of and therapy for exposure to B virus (Cercopithecine herpesvirus 1). Clin Infect Dis. 2002;35(10):1191-1203.

37. Centers for Disease Control and Prevention. Fatal cercopithecine herpesvirus 1 (B virus) infection following a muccutaneous exposure and Interim Recommendations for worker protection. MMWR Morb Mortal Wkly Rep. 1998;47:1073-6,1083.

38. Committee on Occupational Health and Safety in the Care and Use of Non-Human Primates. Occupational health and safety in the care and use of nonhuman primates. Washington, DC: The National Academies Press; 2003.

39. The B Virus Working Group. Guidelines for prevention of Herpesvirus simiae (B virus) infection in monkey handlers. J Med Primatol. 1988;17(2):77-83.

40. Huff JL, Eberle R, Capitanio J, et al. Differential detection of B virus and rhesus cytomegalovirus in rhesus macaques. Journal of General Virology. 2003;84:83-92.

41. Roizman B, Pellett, P. The family Herpesviridae: a brief introduction. In: Knipe DM, Howley PM, editors. Fields virology. 4th ed. Volume 2. Philadelphia: Lippincott Williams and Wilkins. 2001. p. 2381-98.

42. Heymann, D, editor. Control of communicable diseases. 18th ed. Washington, DC: America Public Health Association. 2004.

43. Straus S. Human herpesvirus types 6 and 7. In: Mandell G, Bennett J, Dolin R, editors. Mandell, Douglas, and Bennett's principles and practice of infectious diseases. 6th ed. Volume 2. Philadelphia: Elsevier Inc. 2005. p. 1821-25.

44. Wyatt LS, Frenkel N. Human herpesvirus 7 is a constitutive inhabitant of adult human saliva. J Virol. 1992;66:3206-9.

45. Chang Y, Cesarman E, Pessin MS, et al. Identification of herpesvirus-like DNA sequences in AIDS-associated Kaposi's sarcoma. Science. 1994;266:1865-69.

46. Dukers NHTM, Reeza G. Human herpesvirus 8 epidemiology: what we know and do not know. AIDS. 2003;17:1177-30.

47. Plancoulaine S, Abel L, van Beveren M, et al. Human herpesvirus 8 transmission from mother to child in an endemic population. Lancet. 2000;357:307.

48. Yamashita N, Kimura H, Morishima T. Virological aspects of Epstein-Barr virus infections. Acta Med Okayama. 2005;59:239-46.

49. Khanna KM, Lepisto AJ, Decman V, et al. Immune control of herpes simplex virus during latency. Curr Opin Immunol. 2004;16:463-9.

50. Regamey N, Tamm M, Wernli M, et al. Transmission of human herpesvirus 8 infection from renal transplant donors to recipients. N Engl J Med. 1998; 339:1358-63.

51. Luppi M, Barozzi P, Guaraldi G, et al. Human herpesvirus 8-associated diseases in solid-organ transplantation: importance of viral transmission from the donor. Clin Infect Dis. 2003;37:606-7.

52. Mbulaiteye SM, Biggar RJ, Bakaki PM, et al. Human herpes 8 infection and transfusion history in children with sickle-cell disease in Uganda. J Natl Cancer Inst. 2003;95:1330-5.

53. Whitby D, Luppi M, Sabin C, et al. Detection of antibodies to human herpesvirus 8 in Italian children: evidence for horizontal transmission. Brit J Cancer. 2000;82:702-4.

54. Treanor, JJ. Influenza virus. In: Mandell GL, Bennett, JE, Dolin R, editors. Principles and practice of infectious diseases. 6th ed. New York: Churchill Livingstone; 2005.

55. Influenza. In: Chin J, editor. Control of communicable diseases manual. 17th ed. Washington, DC: American Public Health Association. 2000. p. 270-276.

56. Dowdle WR, Hattwick MA. Swine influenza virus infections in humans. J Infect Dis. 1977;136 Suppl:S386-5399.

57. Stott DJ, Kerr G, Carman WF. Nosocomial transmission of influenza. Occup Med. 2002;52:249-53.

58. Horcajada JP, Pumarola T, Martinez JA, et al. A nosocomial outbreak of influenza during a period without influenza epidemic activity. Eur Respir J. 2003;21:303-7.

59. Centers for Disease Control and Prevention. Guidelines for preventing healthcare-associated pneumonia, 2003: recommendations of CDC and the Healthcare Infection Control Practices Advisory Committee. MMWR Recomm Rep. 2004;53(RR-3):1-36.

60. Bowen GS, Calisher CH, Winkler WG, et al. Laboratory studies of a lymphocytic choriomeningitis virus outbreak in man and laboratory animals. Am J Epidemiol. 1975;102:233-40.

61. Jahrling PB, Peters CJ. Lymphocytic choriomeningitis virus: a neglected pathogen of man. Arch Pathol Lab Med. 1992;116:486-8.

62. Pike RM. Laboratory-associated infections: summary and analysis of 3,921 cases. Hlth Lab Sci. 1976;13:105-14.

63. Reiserova L, Kaluzova M, Kaluz S, et al. Identification of MaTu-MX agent as a new strain of lymphocytic choriomeningitis virus (LCMV) and serological indication of horizontal spread of LCMV in human population. Virology. 1999;257:73-83.

64. Centers for Disease Control and Prevention. Lymphocytic choriomeningitis virus infection in organ transplant recipients: Massachusetts, Rhode Island, 2005. MMWR. 2005;54:537-9.

65. Wright R, Johnson D, Neumann M, et al. Congenital lymphocytic choriomeningitis virus syndrome: a disease that mimics congenital toxoplasmosis or Cytomegalovirus infection. Pediatrics. 1997;100:E9.

66. Dowdle WR, Gary HE, Sanders R, et al. Can post-eradication laboratory containment of wild polioviruses be achieved? Bull World Health Organ. 2002;80:311-6.
67. Mulders MN, Reimerink JHJ, Koopmans MPG, et al. Genetic analysis of wild type poliovirus importation into The Netherlands (1979-1995). J Infect Dis. 1997;176:617-24.
68. Centers for Disease Control and Prevention. Poliomyelitis prevention in the United States. Updated recommendations of the Advisory Committee on Immunization Practices (ACIP). 2002;49(RR-5):1-22.
69. Uirusu. WHO global action plan for laboratory containment of wild poliovirus. 2nd ed. 2005;55:161-78.
70. Esposito JJ, Fenner F. Poxviruses. In: Knipe DM, Howley PM, Griffin DE, editors. Fields virology. 4th ed. Philadelphia: Lippincott, Williams and Wilkins, Philadelphia. 2002. p. 2885-921.
71. Lewis-Jones S. Zoonotic poxvirus infections in humans. Curr Opin Infect Dis. 2004;17:81-9.
72. Reed KD, Melski JW, Graham MB, et al. The detection of monkeypox in humans in the Western Hemisphere. N Engl J Med. 2004;350:342-50.
73. Wharton M, Strikas RA, Harpaz R, et al. Recommendations for using smallpox vaccine in a pre-event vaccination program. Supplemental recommendations of the Advisory Committee on Immunization Practices (ACIP) and the Healthcare Infection Control Practices Advisory Committee (HICPAC). MMWR Recomm Rep. 2003;52(RR-7):1-16.
74. Centers for Disease Control and Prevention. Supplemental recommendations on adverse events following smallpox vaccine in the pre-event vaccination program: recommendations of the Advisory Committee on Immunization Practices (ACIP). MMWR Morb Mortal Wkly Rep. 2003;52:282-4.
75. Centers for Disease Control and Prevention. Vaccinia (smallpox) vaccine: recommendations of the Advisory Committee on Immunization Practices (ACIP), 2001. MMWR Morb Mortal Wkly Rep. 2001;50(RR-10):1-25.
76. NIH Guidelines for Research Involving Recombinant DNA Molecules. Bethesda: The National Institutes of Health (US), Office of Biotechnology Activities; 2002, April.
77. Rupprecht CE, Hanlon CA, Hemachudha T. Rabies re-examined. Lancet Infect Dis. 2002;2:327-43.
78. Winkler WG, Fashinell TR, Leffingwell L, et al. Airborne rabies transmission in a laboratory worker. JAMA. 1973;226:1219-21.
79. Centers for Disease Control and Prevention. Rabies in a laboratory worker, New York. MMWR Morb Mortal Wkly Rep. 1977;26:183-4.
80. Centers for Disease Control and Prevention. Human rabies prevention—United States, 1999. MMWR Morb Mortal Wkly Rep. 1999;48(RR-1):1-29.
81. Rupprecht CE, Gibbons RV. Prophylaxis against rabies. New Engl J Med. 2004;351:2626-35.

82. Centers for Disease Control and Prevention. HIV/AIDS surveillance report, June 1998. Atlanta, GA; 1998.

83. Khabbaz RF, Rowe T, Murphey-Corb M, et al. Simian immunodeficiency virus needlestick accident in a laboratory worker. Lancet. 1992;340:271-3.

84. Centers for Disease Control and Prevention. Seroconversion to simian immunodeficiency virus in two laboratory workers. MMWR Morb Mortal Wkly Rep. 1992;41:678-81.

85. Sotir M, Switzer W, Schable C, et al. Risk of occupational exposure to potentially infectious nonhuman primate materials and to simian immunodeficiency virus. J Med Primatol. 1997;26:233-40.

86. Schochetman G, George JR. AIDS testing: methodology and management issues. New York: Springer-Verlag; 1991.

87. Centers for Disease Control and Prevention. Update: universal precautions for prevention of transmission of human immunodeficiency virus, hepatitis B virus and other bloodborne pathogens in healthcare settings. MMWR Morb Mortal Wkly Rep. 1988; 37:377-82, 387, 388.

88. Occupational exposure to bloodborne pathogens. Final Rule. Fed. Register 1991;56:64175-82.

89. Center for Disease Control, Taiwan. A Report on the Laboratory-Acquired SARS Case in Taiwan [cited 2004 Jan 7]. [about one screen]. Available from: *http://www.cdc.gov.tw/mp.asp?mp=5*

90. Singapore Ministry of Health. Singapore: Singapore Ministry of Health. Biosafety and SARS Incident in Singapore, September 2003. Report of the review panel on new SARS case and biosafety; [about 31 screens]. Available from: *http://www.moh.gov.sg/mohcorp/default.aspx*.

91. Centers for Disease Control and Prevention. (*www.cdc.gov*). Atlanta, GA; [updated 2004 May 19]. Severe Acute Respiratory Syndrome (SARS); [about one screen]. Available from: *http://www.cdc.gov/ncidod/sars/situation/may19.htm*.

92. Severe acute respiratory syndrome. In: Heymann, D, editor. Control of communicable diseases. 18th ed. Washington, DC: America Public Health Association. 2004. p. 480-487.

93. Chow PK, Ooi, EE, Tan, HK, et al. Healthcare worker seroconversion in SARS outbreak. Emerg Infect Dis. 2004;10:225-31.

94. Loeb M, McGeer A, Henry B, et al. SARS among critical care nurses, Toronto. Emerg Infect Dis. 2004;10:251-5.

95. Centers for Disease Control and Prevention. (*www.cdc.gov*). Atlanta, GA; [updated 2004 May 19]. Supplement F: Laboratory Guidance; [about two screens]. Available from: *http://www.cdc.gov/ncidod/sars/guidance/f/index.htm*.

Section VIII-F: Arboviruses and Related Zoonotic Viruses

In 1979, the American Committee on Arthropod-Borne Viruses (ACAV) Subcommittee on Arbovirus Laboratory Safety (SALS) first provided biosafety recommendations for each of the 424 viruses then registered in the International Catalogue of Arboviruses, including Certain Other Viruses of Vertebrates.[1] Working together, SALS, the CDC and the NIH have periodically updated the catalogue by providing recommended biosafety practices and containment for arboviruses registered since 1979. These recommendations are based, in part, on risk assessments derived from information provided by a worldwide survey of laboratories working with arboviruses, new published reports on the viruses, as well as discussions with scientists working with each virus.

Table 6, located at the end of this Section, provides an alphabetical listing of 597 viruses and includes common name, virus family or genus, acronym, BSL recommendation, the basis for the rating, the antigenic group[2] (if known), HEPA filtration requirements, and regulatory requirements (i.e., import/export permits from either the CDC or the USDA). In addition, many of the organisms are classified as select agents and require special security measures to possess, use, or transport. (See Appendix F.) Table 4 provides a key for the SALS basis for assignment of viruses listed in Table 6.

Agent summary statements have been included for certain arboviruses. They were submitted by a panel of experts for more detailed consideration due to one or more of the following factors:

- at the time of writing this edition, the organism represented an emerging public health threat in the United States;

- the organism presented unique biocontainment challenge(s) that required further detail; and

- the organism presented a significant risk of laboratory-acquired infection.

These recommendations were made in August 2005; requirements for biosafety, shipping, and select agent registration can change. Please be sure to confirm the requirements with the appropriate Federal agency. If the pathogen of interest is one listed in Appendix D, contact the USDA for additional biosafety requirements. USDA guidance may supersede the information found in this Chapter.

Recommendations for the containment of infected arthropod vectors were drafted by a subcommittee of the American Committee on Medical Entomology (ACME), and circulated widely among medical entomology professionals. (See Appendix E.)

Some commonly used vaccine strains for which attenuation has been firmly established are recognized by SALS. These vaccine strains may be handled safely at BSL-2 (Table 5). The agents in Table 4 and 5 may require permits from USDA/DOC/DHHS.

Table 4. Explanation of Symbols Used in Table 6 to Define Basis for Assignment of Viruses to Biosafety Levels

Symbol	Definition
S	Results of SALS survey and information from the Catalog.[1]
IE	Insufficient experience with virus in laboratory facilities with low biocontainment.
A	Additional criteria.
A1	Disease in sheep, cattle or horses.
A2	Fatal human laboratory infection—probably aerosol.
A3	Extensive laboratory experience and mild nature of aerosol laboratory infections justifies BSL-2.
A4	Placed in BSL-4 based on the close antigenic relationship with a known BSL-4 agent plus insufficient experience.
A5	BSL-2 arenaviruses are not known to cause serious acute disease in humans and are not acutely pathogenic for laboratory animals including primates. In view of reported high frequency of laboratory aerosol infection in workers manipulating high concentrations of Pichinde virus, it is strongly recommended that work with high concentrations of BSL-2 arenaviruses be done at BSL-3.
A6	Level assigned to prototype or wild-type virus. A lower level may be recommended for variants with well-defined reduced virulence characteristics.
A7	Placed at this biosafety level based on close antigenic or genetic relationship to other viruses in a group of 3 or more viruses, all of which are classified at this level.
A8	BSL-2 hantaviruses are not known to cause laboratory infections, overt disease in humans, or severe disease in experimental primates. Because of antigenic and biologic relationships to highly pathogenic hantaviruses and the likelihood that experimentally infected rodents may shed large amounts of virus, it is recommended that work with high concentrations or experimentally infected rodents be conducted at BSL-3.

Table 5. Vaccine Strains of BSL-3 and BSL-4 Viruses that May Be Handled as BSL-2

Virus	Vaccine Strain
Chikungunya	181/25
Junin	Candid #1
Rift Valley fever	MP-12
Venezuelan equine encephalomyelitis	TC83 & V3526
Yellow fever	17-D
Japanese encephalitis	14-14-2

Based on the recommendations listed with the tables, the following guidelines should be adhered to where applicable.

Viruses with BSL-2 Containment Recommended

The recommendation for conducting work with the viruses listed in Table 6 at BSL-2 are based on the existence of historical laboratory experience adequate to assess the risks when working with this group of viruses. This indicates a) no overt laboratory-associated infections are reported, b) infections resulted from exposures other than by infectious aerosols, or c) if disease from aerosol exposure is documented, it is uncommon.

Laboratory Safety and Containment Recommendations

Agents listed in this group may be present in blood, CSF, various tissues, and/or infected arthropods, depending on the agent and the stage of infection. The primary laboratory hazards comprise accidental parenteral inoculation, contact of the virus with broken skin or mucous membranes, and bites of infected laboratory rodents or arthropods. Properly maintained BSCs, preferable Class II, or other appropriate personal protective equipment or physical containment devices are used whenever procedures with a potential for creating infectious aerosols or splashes are conducted.

BSL-2 practices, containment equipment, and facilities are recommended for activities with potentially infectious clinical materials and arthropods and for manipulations of infected tissue cultures, embryonate hen's eggs, and rodents.

Large quantities and/or high concentrations of any virus have the potential to overwhelm both innate immune mechanisms and vaccine-induced immunity. When a BSL-2 virus is being produced in large quantities or in high concentrations, additional risk assessment is required. This might indicate BSL-3 practices, including additional respiratory protection, based on the risk assessment of the proposed experiment.

Viruses with BSL-3 Containment Recommended

The recommendations for viruses listed in Table 6 that require BSL-3 containment are based on multiple criteria. SALS considered the laboratory experience for some viruses to be inadequate to assess risk, regardless of the available information regarding disease severity. In some cases, SALS recorded overt LAI transmitted by the aerosol route in the absence or non-use of protective vaccines, and considered that the natural disease in humans is potentially severe, life threatening, or causes residual damage.[1] Arboviruses also were classified as requiring BSL-3 containment if they caused diseases in domestic animals in countries outside of the United States.

Laboratory Safety and Containment Recommendations

The agents listed in this group may be present in blood, CSF, urine, and exudates, depending on the specific agent and stage of disease. The primary laboratory hazards are exposure to aerosols of infectious solutions and animal bedding, accidental parenteral inoculation, and contact with broken skin. Some of these agents (e.g., VEE virus) may be relatively stable in dried blood or exudates.

BSL-3 practices, containment equipment, and facilities are recommended for activities using potentially infectious clinical materials and infected tissue cultures, animals, or arthropods.

A licensed attenuated live virus is available for immunization against yellow fever. It is recommended for all personnel who work with this agent or with infected animals, and those entering rooms where the agents or infected animals are present.

Junin virus has been reclassified to BSL-3, provided that all at-risk personnel are immunized and the laboratory is equipped with HEPA-filtered exhaust. SALS also has reclassified Central European tick-borne encephalitis (CETBE) viruses to BSL-3, provided all at-risk personnel are immunized. CETBE is not a registered name in *The International Catalogue of Arboviruses* (1985). Until the registration issue is resolved taxonomically, CETBE refers to the following group of very closely related, if not essentially identical, tick-borne flaviviruses isolated from Czechoslovakia, Finland and Russia: Absettarov, Hanzalova, Hypr, and Kumlinge viruses. While there is a vaccine available that confers immunity to the CETBE group of genetically (>98%) homogeneous viruses, the efficacy of this vaccine against Russian spring-summer encephalitis (RSSE) virus infections has not been established. Thus, the CETBE group of viruses has been reclassified as BSL-3 when personnel are immunized with CETBE vaccine, while RSSE remains classified as BSL-4. It should be noted that CETBE viruses are currently listed as select agents and require special security and permitting considerations. (See Appendix F.)

Investigational vaccines for eastern equine encephalomyelitis (EEE) virus, Venezuelan equine encephalitis (VEE), western equine encephalomyelitis (WEE) virus, and Rift Valley fever viruses (RVFV), may be available in limited quantities and administered on-site at the Special Immunization Program of USAMRIID, located at Ft. Detrick, Frederick, MD. Details are available at the end of this section.

The use of investigational vaccines for laboratory personnel should be considered if the vaccine is available. Initial studies have shown the vaccine to be effective in producing an appropriate immunologic response, and the adverse effects of vaccination are within acceptable parameters. The decision to recommend vaccines for laboratory personnel must be carefully considered and based on an risk assessment which includes a review of the characteristics of the agent and the disease, benefits versus the risk of vaccination, the experience of the laboratory personnel, laboratory procedures to be used with the agent, and the contraindications for vaccination including the health status of the employee.

If the investigational vaccine is contraindicated, does not provide acceptable reliability for producing an immune response, or laboratory personnel refuse vaccination, the use of appropriate personal protective equipment may provide an alternative. Respiratory protection, such as use of a PAPR, should be considered in areas using organisms with a well-established risk of aerosol infections in the laboratory, such as VEE viruses.

Any respiratory protection equipment must be provided in accordance with the institution's respiratory protection program. Other degrees of respiratory protection may be warranted based on an assessment of risk as defined in Chapter 2 of this manual. All personnel in a laboratory with the infectious agent must use comparable personal protective equipment that meets or exceeds the requirements, even if they are not working with the organism. Sharps precautions as described under BSL-2 and BSL-3 requirements must be continually and strictly reinforced, regardless of whether investigational vaccines are used.

Non-licensed vaccines are available in limited quantities and administered on-site at the Special Immunization Program of USAMRIID. IND vaccines are administered under a cooperative agreement between the U.S. Army and the individual's requesting organization. Contact the Special Immunization Program by telephone at (301) 619-4653.

Enhanced BSL-3 Containment

Situations may arise for which enhancements to BSL-3 practices and equipment are required; for example, when a BSL-3 laboratory performs diagnostic testing on specimens from patients with hemorrhagic fevers thought to be due to dengue or yellow fever viruses. When the origin of these specimens is Africa, the Middle East, or South America, such specimens might contain etiologic agents, such as arenaviruses, filoviruses, or other viruses that are usually manipulated in a BSL-4 laboratory. Examples of enhancements to BSL-3 laboratories might include: 1) enhanced respiratory protection of personnel against aerosols; 2) HEPA filtration of dedicated exhaust air from the laboratory; and 3) personal body shower. Additional appropriate training for all animal care personnel should be considered.

Viruses with BSL-4 Containment Recommended

The recommendations for viruses assigned to BSL-4 containment are based on documented cases of severe and frequently fatal naturally occurring human infections and aerosol-transmitted laboratory infections. SALS recommends that certain agents with a close antigenic relationship to agents assigned to BSL-4 also be provisionally handled at this level until sufficient laboratory data indicates that work with the agent may be assigned to a lower biosafety level.

Laboratory Safety and Containment Recommendations

The infectious agents may be present in blood, urine, respiratory and throat secretions, semen, and other fluids and tissues from human or animal hosts, and in arthropods, rodents, and NHPs. Respiratory exposure to infectious aerosols, mucous membrane exposure to infectious droplets, and accidental parenteral inoculation are the primary hazards to laboratory or animal care personnel.[3,4]

BSL-4 practices, containment equipment, and facilities are recommended for all activities utilizing known or potentially infectious materials of human, animal, or arthropod origin. Clinical specimens from persons suspected of being infected with one of the agents listed in this summary should be submitted to a laboratory with a BSL-4 maximum containment facility.[5]

Dealing with Unknown Arboviruses

The ACAV has published reports documenting laboratory workers who acquired arbovirus infections during the course of their duties.[6] In the first such document, it was recognized that these laboratory infections typically occurred by unnatural routes such as percutaneous or aerosol exposure, that "lab adapted" strains were still pathogenic for humans, and that as more laboratories worked with newly identified agents, the frequency of laboratory-acquired infections was increasing. Therefore, to assess the risk of these viruses and provide safety guidelines to those working with them, ACAV appointed SALS to evaluate the hazards of working with arboviruses in the laboratory setting.[7,8]

The SALS committee made a series of recommendations, published in 1980, describing four levels of laboratory practices and containment guidelines that were progressively more restrictive. These levels were determined after widely-distributed surveys evaluated numerous criteria for each particular virus including: 1) past occurrence of laboratory-acquired infections correlated with facilities and practices used; 2) volume of work performed as a measure of potential exposure risk; 3) immune status of laboratory personnel; 4) incidence and severity of naturally-acquired infections in adults; and 5) incidence of disease in animals outside the United States (to assess import risk).

While these criteria are still important factors to consider in any risk assessment for manipulating arboviruses in the laboratory, it is important to note that there have been many modifications to personal laboratory practices (e.g., working in BSC while wearing extensive personal protective equipment in contrast to working with viruses on an open bench top) and significant changes in laboratory equipment and facilities (e.g., BSC, PAPR) available since the initial SALS evaluation. Clearly, when dealing with a newly recognized arbovirus, there is insufficient previous experience with it; thus, the virus should be assigned a higher biosafety level. However, with increased ability to safely characterize viruses, the relationship to other disease-causing arboviruses can be established with reduced exposure to the investigators. Therefore, in addition to those established by SALS, additional assessment criteria should be considered.

One criterion for a newly identified arbovirus is a thorough description of how the virus will be handled and investigated. For example, experiments involving pure genetic analysis could be handled differently than those where the virus will be put into animals or arthropods.[9] Additionally, an individual risk assessment should consider the fact that not all strains of a particular virus exhibit the same degree of pathogenicity or transmissibility. While variable pathogenicity occurs frequently with naturally identified strains, it is of particular note for strains that are modified in the laboratory. It may be tempting to assign biosafety levels to hybrid or chimeric strains based on the parental types but due to possible altered biohazard potential, assignment to a different biosafety level may be justified.[10] A clear description of the strains involved should accompany any risk assessment.

Most of the identified arboviruses have been assigned biosafety levels; however, a number of those that are infrequently studied, newly identified, or have only single isolation events may not have been evaluated by SALS, ACAV, CDC, or the NIH (Table 6). Thorough risk assessment is important for all arboviral research and it is of particular importance for work involving unclassified viruses. A careful assessment by the laboratory director, institutional biosafety officer and safety committee, and as necessary, outside experts is necessary to minimize the risk of human, animal, and environmental exposure while allowing research to progress.

Chimeric Viruses

The ability to construct cDNA clones encoding a complete RNA viral genome has led to the generation of recombinant viruses containing a mixture of genes from two or more different viruses. Chimeric, full-length viruses and truncated replicons have been constructed from numerous alphaviruses and flaviviruses. For example, alphavirus replicons encoding foreign genes have been used widely as immunogens against bunyavirus, filovirus, arenavirus, and other antigens. These replicons have been safe and usually immunogenic in rodent hosts leading to their development as candidate human vaccines against several virus groups including retroviruses.[11-14]

Because chimeric viruses contain portions of multiple viruses, the IBC, in conjunction with the biosafety officer and the researchers, must conduct a risk assessment that, in addition to standard criteria, includes specific elements that need to be considered before assigning appropriate biosafety levels and containment practices. These elements include: 1) the ability of the chimeric virus to replicate in cell culture and animal model systems in comparison with its parental strains;[15] 2) altered virulence characteristics or attenuation compared with the parental viruses in animal models;[16] 3) virulence or attenuation patterns by intracranial routes using large doses for agents affecting the CNS;[17,18] and 4) demonstration of lack of reversion to virulence or parental phenotype.

Many patterns of attenuation have been observed with chimeric flaviviruses and alphaviruses using the criteria described above. Additionally, some of these chimeras are in phase II testing as human vaccines.[19]

Chimeric viruses may have some safety features not associated with parental viruses. For example, they are generated from genetically stable cDNA clones without the need for animal or cell culture passage. This minimizes the possibility of mutations that could alter virulence properties. Because some chimeric strains incorporate genomic segments lacking gene regions or genetic elements critical for virulence, there may be limited possibility of laboratory recombination to generate strains exhibiting wild-type virulence.

Ongoing surveillance and laboratory studies suggest that many arboviruses continue to be a risk to human and animal populations. The attenuation of all chimeric strains should be verified using the most rigorouscontainment requirements of the parental strains. The local IBC should evaluate containment recommendations for each chimeric virus on a case-by-case basis, using virulence data from an appropriate animal model. Additional guidance from the NIH Office of Biotechnology Activities and/or the Recombinant DNA Advisory Committee (RAC) may be necessary.

West Nile Virus (WNV)

WNV has emerged in recent years in temperate regions of Europe and North America, presenting a threat to public and animal health. This virus belongs to the family *Flaviviridae* and the genus *Flavivirus*, Japanese encephalitis virus antigenic complex. The complex currently includes Alfuy, Cacipacore, Japanese encephalitis, Koutango, Kunjin, Murray Valley encephalitis, St. Louis encephalitis,

Rocio, Stratford, Usutu, West Nile, and Yaounde viruses. Flaviviruses share a common size (40-60nm), symmetry (enveloped, icosahedral nucleocapsid), nucleic acid (positive-sense, single stranded RNA approximately 10,000-11,000 bases) and virus morphology. The virus was first isolated from a febrile adult woman in the West Nile District of Uganda in 1937.[20] The ecology was characterized in Egypt in the 1950s; equine disease was first noted in Egypt and France in the early 1960s.[21,22] It first appeared in North America in 1999 as encephalitis reported in humans and horses.[23] The virus has been detected in Africa, Europe, the Middle East, west and central Asia, Oceania (subtype Kunjin virus), and most recently, North America.

Occupational Infections

LAI with WNV have been reported in the literature. SALS reported 15 human infections from laboratory accidents in 1980. One of these infections was attributed to aerosol exposure. Two parenteral inoculations have been reported recently during work with animals.[24]

Natural Modes of Infections

In the United States, infected mosquitoes, primarily members of the *Culex* genus, transmit WNV. Virus amplification occurs during periods of adult mosquito blood-feeding by continuous transmission between mosquito vectors and bird reservoir hosts. People, horses, and most other mammals are not known to develop infectious viremias very often, and thus are probably "dead-end" or incidental hosts.

Laboratory Safety and Containment Recommendations

WNV may be present in blood, serum, tissues, and CSF of infected humans, birds, mammals, and reptiles. The virus has been found in oral fluids and feces of birds. Parenteral inoculation with contaminated materials poses the greatest hazard; contact exposure of broken skin is a possible risk. Sharps precautions should be strictly adhered to when handling potentially infectious materials. Workers performing necropsies on infected animals may be at higher risk of infection.

BSL-2 practices, containment equipment, and facilities are recommended for activities with human diagnostic specimens, although it is unusual to recover virus from specimens obtained from clinically ill patients. BSL-2 is recommended for processing field collected mosquito pools whereas BSL-3 and ABSL-3 practices, containment equipment, and facilities are recommended for all manipulations of WNV cultures and for experimental animal and vector studies, respectively.

Dissection of field collected dead birds for histopathology and culture is recommended at BSL-3 containment due to the potentially high levels of virus found in such samples. Non-invasive procedures performed on dead birds (such as oropharyngeal or cloacal swabs) can be conducted at BSL-2.

Transfer of Agent Importation of this agent may require CDC and/or USDA importation permits. Domestic transport of this agent may require a permit from USDA/APHIS/VS. A DoC permit may be required for the export of this agent to another country. See Appendix C for additional information.

Eastern Equine Encephalitis (EEE) Virus, Venezuelan Equine Encephalitis (VEE) Virus, and Western Equine Encephalitis (WEE) Virus

VEE, EEE, and WEE viruses are members of the genus *Alphavirus* in the family *Togaviridae*. They are small, enveloped viruses with a genome consisting of a single strand of positive-sense RNA. All three viruses can cause encephalitis often accompanied by long-term neurological sequelae. Incubation period ranges from 1-10 days and the duration of acute illness is typically days to weeks depending upon severity of illness. Although not the natural route of transmission, the viruses are highly infectious by the aerosol route; laboratory acquired infections have been documented.[25]

Occupational Infections

These alphaviruses, especially VEE virus, are infectious by aerosol in laboratory studies and more than 160 EEE virus, VEE virus, or WEE virus laboratory-acquired infections have been documented. Many infections were due to procedures involving high virus concentrations and aerosol-generating activities such as centrifugation and mouth pipetting. Procedures involving animals (e.g., infection of newly hatched chicks with EEE virus and WEE virus) and mosquitoes also are particularly hazardous.

Natural Modes of Infection

Alphaviruses are zoonoses maintained and amplified in natural transmission cycles involving a variety of mosquito species and either small rodents or birds. Humans and equines are accidental hosts with naturally acquired alphavirus infections resulting from the bites of infected mosquitoes.

EEE virus occurs in focal locations along the eastern seaboard, the Gulf Coast and some inland Midwestern locations of the United States, in Canada, some Caribbean Islands, and Central and South America.[26] Small outbreaks of human disease have occurred in the United States, the Dominican Republic, Cuba, and Jamaica. In the United States, equine epizootics are common occurrences during the summer in coastal regions bordering the Atlantic and Gulf of Mexico, in other eastern and Midwestern states, and as far north as Quebec, Ontario, and Alberta in Canada.

In Central and South America, focal outbreaks due to VEE virus occur periodically with rare large regional epizootics involving thousands of equine cases and deaths in predominantly rural settings. These epizootic/epidemic viruses are theorized to emerge periodically from mutations occurring in the continuously circulating enzootic VEE viruses in northern South America. The classical epizootic varieties of the virus are not present in the United States. An enzootic subtype, Everglades virus (VEE antigenic complex subtype II virus), exists naturally in southern Florida, while endemic foci of Bijou Bridge virus (VEE antigenic complex subtype III-B virus), have been described in the western United States.[27]

The WEE virus is found mainly in western parts of the United States and Canada. Sporadic infections also occur in Central and South America.

Laboratory Safety and Containment Recommendations

Alphaviruses may be present in blood, CSF, other tissues (e.g., brain), or throat washings. The primary laboratory hazards are parenteral inoculation, contact of the virus with broken skin or mucus membranes, bites of infected animals or arthropods, or aerosol inhalation.

Diagnostic and research activities involving clinical material, infectious cultures, and infected animals or arthropods should be performed under BSL-3 practices, containment equipment, and facilities. Due to the high risk of aerosol infection, additional personal protective equipment, including respiratory protection, should be considered for non-immune personnel. Animal work with VEE virus, EEE virus and WEE virus should be performed under ABSL-3 conditions. HEPA filtration is required on the exhaust system of laboratory and animal facilities using VEE virus.

Special Issues

Vaccines Two strains of VEE virus (TC-83 and V3526) are highly attenuated in vertebrate studies and have been either exempted (strain TC-83) or excluded (strain V3526) from select agent regulations. Because of the low level of pathogenicity, these strains may be safely handled under BSL-2 conditions without vaccination or additional personal protective equipment.

Investigational vaccine protocols have been developed to immunize at-risk laboratory or field personnel against these alphaviruses, however, the vaccines are available only on a limited basis and may be contraindicated for some personnel. Therefore, additional personal protective equipment may be warranted in lieu of vaccination. For personnel who have no neutralizing antibody titer (either by previous vaccination or natural infection), additional respiratory protection is recommended for all procedures.

Select Agent VEE virus and EEE virus are select agents requiring registration with CDC and/or USDA for possession, use, storage and/or transfer. See Appendix F for additional information.

Transfer of Agent Importation of this agent may require CDC and/or USDA importation permits. Domestic transport of this agent may require a permit from USDA/APHIS/VS.

Rift Valley Fever Virus (RVFV)

RVFV was first isolated in Kenya in 1936 and subsequently shown to be endemically present in almost all areas of sub-Saharan Africa.[28] In periods of heavy rainfall, large epizootics occur involving primarily sheep, cattle, and human disease, although many other species are infected. The primordial vertebrate reservoir is unknown, but the introduction of large herds of highly susceptible domestic breeds in the last few decades has provided a substrate for massive virus amplification. The virus has been introduced into Egypt, Saudi Arabia, and Yemen and caused epizootics and epidemics in those countries. The largest of these was in 1977 to 1979 in Egypt with many thousands of human cases and 610 reported deaths.[29]

Most human infections are symptomatic and the most common syndrome consists of fever, myalgia, malaise, anorexia, and other non-specific symptoms. Recovery within one to two weeks is usual but hemorrhagic fever, encephalitis, or retinitis also occurs. Hemorrhagic fever develops as the primary illness proceeds and is characterized by disseminated intravascular coagulation and hepatitis. Perhaps 2% of cases will develop this complication and the mortality is high. Encephalitis follows an apparent recovery in <1% of cases and results in a substantial mortality and sequelae. Retinal vasculitis occurs in convalescence of a substantial but not precisely known proportion of cases. The retinal lesions are often macular and permanent, leading to substantial loss of visual acuity.

Infected sheep and cattle suffer a mortality rate of 10-35%, and spontaneous abortion occurs virtually in all pregnant females. Other animals studied have lower viremia and lesser mortality but may abort. This virus is an OIE List A disease and triggers export sanctions.

Occupational Infections

The potential for infection of humans by routes other than arthropod transmission was first recognized in veterinarians performing necropsies. Subsequently, it became apparent that contact with infected animal tissues and infectious aerosols were dangerous; many infections were documented in herders, slaughterhouse workers, and veterinarians. Most of these infections resulted from exposure to blood and other tissues including aborted fetal tissues of sick animals.

There have been 47 reported laboratory infections; before modern containment and vaccination became available virtually every laboratory that began work with the virus suffered infections suggestive of aerosol transmission.[30,31]

Natural Modes of Infection

Field studies show RVFV to be transmitted predominantly by mosquitoes, although other arthropods may be infected and transmit. Mechanical transmission also has been documented in the laboratory. Floodwater *Aedes* species are the primary vector and transovarial transmission is an important part of the maintenance cycle.[32] However, many different mosquito species are implicated in horizontal transmission in field studies, and laboratory studies have shown a large number of mosquito species worldwide to be competent vectors, including North American mosquitoes.

It is currently believed that the virus passes dry seasons in the ova of flood-water *Aedes* mosquitoes. Rain allows infectious mosquitoes to emerge and feed on vertebrates. Several mosquito species can be responsible for horizontal spread, particularly in epizootic/epidemic situations. The vertebrate amplifiers are usually sheep and cattle, with two caveats; as yet undefined native African vertebrate amplifier is thought to exist and very high viremias in humans are thought to play some role in viral amplifications.[33]

Transmission of diseases occurs between infected animals but is of low efficiency and virus titers in throat swabs are low. Nosocomial infection rarely if ever occurs. There are no examples of latency with RVFV, although virus may be isolated from lymphoid organs of mice and sheep for four to six weeks post-infection.

Laboratory Safety and Containment Recommendations

Concentrations of RVFV in blood and tissues of sick animals are often very high. Placenta, amniotic fluid, and fetuses from aborted domestic animals are highly infectious. Large numbers of infectious virus also are generated in cell cultures and laboratory animals.

BSL-3 practices, containment equipment and facilities are recommended for processing human or animal material in endemic zones or in non-endemic areas in emergency circumstances. Particular care should be given to stringent aerosol containment practices, autoclaving waste, decontamination of work areas, and control of egress of material from the laboratory. Other cultures, cells, or similar biological material that could potentially harbor RVFV should not be used in a RVFV laboratory and subsequently removed.

Diagnostic or research studies outside endemic areas should be performed in a BSL-3 laboratory. Personnel also must have additional respiratory protection (such as a PAPR) or be vaccinated for RVFV. In addition, the USDA may require full BSL-3-Ag containment for research conducted in non-endemic areas in loose-housed animals. (See Appendix D.)

Special Issues

Vaccines Two apparently effective vaccines have been developed by the Department of Defense (DoD) and have been used in volunteers, laboratory staff, and field workers under investigational protocols, but neither vaccine is available at this time.

Select Agent RVFV is a select agent requiring registration with CDC and/or USDA for possession, use, storage and/or transfer. See Appendix F for additional information.

The live-attenuated MP-12 vaccine strain is specifically exempted from the Select Agent rules. In general, BSL-2 containment is recommended for working with this strain.

The USDA may require enhanced ABSL-3, ABSL-3, or BSL-3-Ag facilities and practices for working with RVFV in the United States. (See Appendix D.) Investigators should contact the USDA for further guidance before initiating research.

Transfer of Agent Importation of this agent may require CDC and/or USDA importation permits. Domestic transport of this agent may require a permit from USDA/APHIS/VS.

Table 6. Alphabetic Listing of 597 Arboviruses and Hemorrhagic Fever Viruses*

Name	Acronym	Taxonomic Status (Family or Genus)	Recommended Biosafety Level	Basis of Rating	Antigenic Group	HEPA Filtration on Lab Exhaust
Abras	ABRV	*Orthobunvavarus*	2	A7	Patois	No
Absettarov	ABSV	*Flavivirus*	4	A4	B^f	Yes
Abu Hammad	AHV	*Nairovirus*	2	S	Dera Ghazi Khan	No
Acado	ACDV	*Orbivirus*	2	S	Corriparta	No
Acara	ACAV	*Orthobunyavirus*	2	S	Capim	No
Adelaide River	ARV	*Lyssavirus*	2	IE	Bovine Ephemeral Fever	No
African Horse sickness	AHSV	*Orbivirus*	3^c	A1	African Horsesickness	Yes
African Swine Fever	ASFV	*Asfivirus*	3^c	IE	Asfivirus	Yes

Name	Acronym	Taxonomic Status (Family or Genus)	Recom- mended Biosafety Level	Basis of Rating	Antigenic Group	HEPA Filtration on Lab Exhaust
Aguacate	AGUV	*Phlebovirus*	2	S	Phlebotomus Fever	No
Aino	AINOV	*Orthobunyavirus*	2	S	Simbu	No
Akabane	AKAV	*Orthobunyavirus*	3ᶜ	S	Simbu	Yes
Alenquer	ALEV	*Phlebovirus*	2	IE	Phlebotomus Fever	No
Alfuy	ALFV	*Flavivirus*	2	S	Bᶠ	No
Alkhumra	ALKV	*Flavivirus*	4	A4	Bᶠ	Yes
Allpahuayo	ALLPV	*Arenavirus*	3	IE	Tacaribe	No
Almeirim	ALMV	*Orbivirus*	2	IE	Changuinola	No
Almpiwar	ALMV	*Rhabdoviridae*	2	S		No
Altamira	ALTV	*Orbivirus*	2	IE	Changuinola	No
Amapari	AMAV	*Arenavirus*	2	A5	Tacaribe	No
Ambe	AMBEV	*Phlebovirus*	2	IE		No
Ananindeua	ANUV	*Orthobunyavirus*	2	A7	Guama	No
Andasibe	ANDV	*Orbivirus*	2	A7		No
Andes	ANDV	*Hantavirus*	3ᵃ	IE	Hantaan	No
Anhanga	ANHV	*Phlebovirus*	2	S	Phlebotomus Fever	No
Anhembi	AMBV	*Orthobunyavirus*	2	S	Bunyamwera	No
Anopheles A	ANAV	*Orthobunyavirus*	2	S	Anopheles A	No
Anopheles B	ANBV	*Orthobunyavirus*	2	S	Anopheles B	No
Antequera	ANTV	*Bunyaviridae*	2	IE	Resistencia	No
Apeu	APEUV	*Orthobunyavirus*	2	S	Cᶠ	No
Apoi	APOIV	*Flavivirus*	2	S	Bᶠ	No
Araguari	ARAV	Unassigned	3	IE		No
Aransas Bay	ABV	*Bunyaviridae*	2	IE	UPOLU	No
Arbia	ARBV	*Phlebovirus*	2	IE	Phlebotomus Fever	No
Arboledas	ADSV	*Phlebovirus*	2	A7	Phlebotomus Fever	No
Aride	ARIV	*Unassigned*	2	S		No
Ariquemes	ARQV	*Phlebovirus*	2	A7	Phlebotomus Fever	No
Arkonam	ARKV	*Orbivirus*	2	S	Ieri	No
Armero	ARMV	*Phlebovirus*	2	A7	Phlebotomus Fever	No
Aroa	AROAV	*Flavivirus*	2	S	Bᶠ	No
Aruac	ARUV	*Rhabdoviridae*	2	S		No

Name	Acronym	Taxonomic Status (Family or Genus)	Recom- mended Biosafety Level	Basis of Rating	Antigenic Group	HEPA Filtration on Lab Exhaust
Arumateua	ARMTV	*Orthobunyavirus*	2	A7		No
Arumowot	AMTV	*Phlebovirus*	2	S	Phlebotomus Fever	No
Aura	AURAV	*Alphavirus*	2	S	Aᶠ	No
Avalon	AVAV	*Nairovirus*	2	S	Sakhalin	No
Babahoyo	BABV	*Orthobunyavirus*	2	A7	Patois	No
Babanki	BBKV	*Alphavirus*	2	A7	Aᶠ	No
Bagaza	BAGV	*Flavivirus*	2	S	Bᶠ	No
Bahig	BAHV	*Orthobunyavirus*	2	S	Tete	No
Bakau	BAKV	*Orthobunyavirus*	2	S	Bakau	No
Baku	BAKUV	*Orbivirus*	2	S	Kemerovo	No
Bandia	BDAV	*Nairovirus*	2	S	Qalyub	No
Bangoran	BGNV	*Rhabdoviridae*	2	S		No
Bangui	BGIV	*Bunyaviridae*	2	S		No
Banzi	BANV	*Flavivirus*	2	S	Bᶠ	No
Barmah Forest	BFV	*Alphavirus*	2	A7	Aᶠ	No
Barranqueras	BQSV	*Bunyaviridae*	2	IE	Resistencia	No
Barur	BARV	*Rhabdoviridae*	2	S	Kern Canyon	No
Batai	BATV	*Orthobunyavirus*	2	S	Bunyamwera	No
Batama	BMAV	*Orthobunyavirus*	2	A7	Tete	No
Batken	BKNV	*Thogotovirus*	2	IE		No
Bauline	BAUV	*Orbivirus*	2	S	Kemerovo	No
Bear Canyon	BRCV	*Arenavirus*	3	A7		No
Bebaru	BEBV	*Alphavirus*	2	S	Aᶠ	No
Belem	BLMV	*Bunyaviridae*	2	IE		No
Belmont	ELV	*Bunyaviridae*	2	S		No
Belterra	BELTV	*Phlebovirus*	2	A7	Phlebotomus Fever	No
Benevides	BENV	*Orthobunyavirus*	2	A7	Capim	No
Benfica	BENV	*Orthobunyavirus*	2	A7	Capim	No
Bermejo	BMJV	*Hantavirus*	3	IE	Hantaan	No
Berrimah	BRMV	*Lyssavirus*	2	IE	Bovine Ephem- eral Fever	No
Beritoga	BERV	*Orthobunyavirus*	2	S	Guama	No
Bhanja	BHAV	*Bunyaviridae*	3	S	Bhanja	No
Bimbo	BBOV	*Rhabdoviridae*	2	IE		No

Name	Acronym	Taxonomic Status (Family or Genus)	Recommended Biosafety Level	Basis of Rating	Antigenic Group	HEPA Filtration on Lab Exhaust
Bimitti	BIMV	*Orthobunyavirus*	2	S	Guama	No
Birao	BIRV	*Orthobunyavirus*	2	S	Bunyamwera	No
Bluetoungue (exotic serotypes)	BTV	*Orbivirus*	3^c	S	Bluetongue	No
Bluetoungue (non-exotic)	BTV	*Orbivirus*	2^c	S	Bluetongue	No
Bobaya	BOBV	*Bunyaviridae*	2	IE		No
Bobia	BIAV	*Orthobunyavirus*	2	IE	Olifantsylei	No
Boraceia	BORV	*Orthobunyavirus*	2	S	Anopheles B	No
Botambi	BOTV	*Orthobunyavirus*	2	S	Olifantsylei	No
Boteke	BTKV	*Vesiculovirus*	2	S	Vesicular Stomatitis	No
Bouboui	BOUV	*Flavivirus*	2	S	B^f	No
Bovine Ephemeral Fever	BEFV	*Lyssavirus*	3^c	A1	Bovine Ephemeral Fever	No
Bozo	BOZOV	*Orthobunyavirus*	2	A7	Bunyamwera	No
Breu Branco	BRBV	*Orbivirus*	2	A7		No
Buenaventura	BUEV	*Phlebovirus*	2	IE	Phlebotomus Fever	No
Bujaru	BUJV	*Phlebovirus*	2	S	Phlebotomus Fever	No
Bunyamwera	BUNV	*Orthobunyavirus*	2	S	Bunyamwera	No
Bunyip Creek	BCV	*Orbivirus*	2	S	Palyam	No
Burg El Arab	BEAV	*Rhabdoviridae*	2	S	Matariva	No
Bushbush	BSBV	*Orthobunyavirus*	2	S	Capim	No
Bussuquara	BSQV	*Flavivirus*	2	S	B^f	No
Buttonwillow	BUTV	*Orthobunyavirus*	2	S	Simbu	No
Bwamba	BWAV	*Orthobunyavirus*	2	S	Bwamba	No
Cabassou	CABV	*Alphavirus*	3	IE	A^f	Yes
Cacao	CACV	*Phlebovirus*	2	S	Phlebotomus Fever	No
Cache Valley	CVV	*Orthobunyavirus*	2	S	Bunyamwera	No
Cacipacore	CPCV	*Flavivirus*	2	IE	B^f	No
Caimito	CAIV	*Phlebovirus*	2	S	Phlebotomus Fever	No
Calchaqui	CQIV	*Vesiculovirus*	2	A7	Vesicular Stomatitis	No
California Encephalitis	CEV	*Orthobunyavirus*	2	S	California	No
Calovo	CVOV	*Orthobunyavirus*	2	S	Bunyamwera	No
Cananeia	CNAV	*Orthobunyavirus*	2	IE	GUAMA	No

Name	Acronym	Taxonomic Status (Family or Genus)	Recommended Biosafety Level	Basis of Rating	Antigenic Group	HEPA Filtration on Lab Exhaust
Candiru	CDUV	*Phlebovirus*	2	S	Phlebotomus Fever	No
Caninde	CANV	*Orbivirus*	2	IE	Changuinola	No
Cano Delgadito	CADV	*Hantavirus*	3ª	IE	Hantaan	No
Cape Wrath	CWV	*Orbivirus*	2	S	Kemerovo	No
Capim	CAPV	*Orthobunyavirus*	2	S	Capim	No
Caraipe	CRPV	*Orthobunyavirus*	2	A7		No
Carajas	CRJV	*Vesiculovirus*	2	A7	Vesicular Stomatitis	No
Caraparu	CARV	*Orthobunyavirus*	2	S	Cᶠ	No
Carey Island	CIV	*Flavivirus*	2	S	Bᶠ	No
Catu	CATUV	*Orthobunyavirus*	2	S	Guama	No
Chaco	CHOV	*Rhabdoviridae*	2	S	Timbo	No
Chagres	CHGV	*Phlebovirus*	2	S	Phlebotomus Fever	No
Chandipura	CHPV	*Vesiculovirus*	2	S	Vesicular Stomatitis	No
Changuinola	CGLV	*Orbivirus*	2	S	Changuinola	No
Charleville	CHVV	*Lyssavirus*	2	S	Rab	No
Chenuda	CNUV	*Orbivirus*	2	S	Kmerovo	No
Chikungunya	CHIKV	*Alphavirus*	3	S	Aᶠ	Yes
Chilibre	CHIV	*Phlebovirus*	2	S	Phlebotomus Fever	No
Chim	CHIMV	*Bunyaviridae*	2	IE		No
Chobar Gorge	CGV	*Orbivirus*	2	S	Chobar Gorge	No
Clo Mor	CMV	*Nairovirus*	2	S	Sakhalin	No
Coastal Plains	CPV	*Lyssavirus*	2	IE	Tibrogargan	No
Cocal	COCV	*Vesiculovirus*	2	A3	Vesicular Stomatitis	No
Codajas	CDJV	*Orbivirus*	2	A7		No
Colorado Tick Fever	CTFV	*Coltivirus*	2	S	Colorado Tick Fever	No
Congo-Crimean Hemorrhagic Fever	CCHFV	*Nairovirus*	4	A6	CCHF	Yes
Connecticut	CNTV	*Rhabdoviridae*	2	IE	Sawgrass	No
Corfou	CFUV	*Phlebovirus*	2	A7	Phlebotomus Fever	No
Corriparta	CORV	*Orbivirus*	2	S	Corriaparta	No
Cotia	CPV	*Poxviridae*	2	S		No
Cowbone Ridge	CRV	*Flavivirus*	2	S	Bᶠ	No

Name	Acronym	Taxonomic Status (Family or Genus)	Recommended Biosafety Level	Basis of Rating	Antigenic Group	HEPA Filtration on Lab Exhaust
Csiro Village	CVGV	*Orbivirus*	2	S	Palyam	No
Cuiaba	CUIV	*Rhabdoviridae*	2	S		No
Curionopolis	CRNPV	*Rhabdoviridae*	2	A7		No
Dabakala	DABV	*Orthobunyavirus*	2	A7	Olifantsylei	No
D'Aguilar	DAGV	*Orbivirus*	2	S	Palyam	No
Dakar Bat Virus	DBV	*Flavivirus*	2	S	B[f]	No
Deer Tick Virus	DRTV	*Flavivirus*	3	A7		No
Dengue Virus Type 1	DENV-1	*Flavivirus*	2	S	B[f]	No
Dengue Virus Type 2	DENV-2	*Flavivirus*	2	S	B[f]	No
Dengue Virus Type 3	DENV-3	*Flavivirus*	2	S	B[f]	No
Dengue Virus Type 4	DENV-4	*Flavivirus*	2	S	B[f]	No
Dera Ghazi Khan	DGKV	*Nairovirus*	2	S	Dera Ghazi Khan	No
Dobrava-Belgrade	DOBV	*Hantavirus*	3[a]	IE		No
Dhori	DHOV	*Orthomyxoviridae*	2	S		No
Douglas	DOUV	*Orthobunyavirus*	3	IE	Simbu	No
Durania	DURV	*Phlebovirus*	2	A7	Phlebotomus Fever	No
Dugbe	DUGV	*Nairovirus*	3	S	Nairobi Sheep Disease	No
Eastern Equine Encephalitis	EEEV	*Alphavirus*	3[c]	S	A[f]	No
Ebola (Including Reston)	EBOV	*Filovirus*	4	S	EBO	Yes
Edge Hill	EHV	*Flavivirus*	2	S	B[f]	No
Enseada	ENSV	*Bunyaviridae*	3	IE		No
Entebbe Bat	ENTV	*Flavivirus*	2	S	B[f]	No
Epizootic Hemorrhagic Disease	EHDV	*Orbivirus*	2	S	Epizootic Hemorrhagic Disease	No
Erve	ERVEV	*Bunyaviridae*	2	S	Thiafora	No
Estero Real	ERV	*Orthobunyavirus*	2	IE	Patois	No
Eubenangee	EUBV	*Orbivirus*	2	S	Eubenangee	No
Everglades	EVEV	*Alphavirus*	3	S	A[f]	Yes
Eyach	EYAV	*Coltivirus*	2	S	Colorado Tick Fever	No
Farmington	FRMV	*Vesiculovirus*	2	A7		No
Flanders	FLAV	*Rhabdoviridae*	2	S	Hart Park	No

Name	Acronym	Taxonomic Status (Family or Genus)	Recommended Biosafety Level	Basis of Rating	Antigenic Group	HEPA Filtration on Lab Exhaust
Flexal	FLEV	*Arenavirus*	3	S	Tacaribe	No
Fomede	FV	*Orbivirus*	2	A7	Chobar Gorge	No
Forecariah	FORV	*Bunyaviridae*	2	A7	Bhanja	No
Fort Morgan	FMV	*Alphavirus*	2	S	A^f	No
Fort Sherman	FSV	*Orthobunyavirus*	2	A7	Bunyamwera	No
Frijoles	FRIV	*Phlebovirus*	2	S	Phlebotomus Fever	No
Gabek Forest	GFV	*Phlebovirus*	2	A7	Phlebotomus Fever	No
Gadgets Gully	GGYV	*Flavivirus*	2	IE	B^f	No
Gamboa	GAMV	*Orthobunyavirus*	2	S	Gamboa	No
Gan Gan	GGV	*Bunyaviridae*	2	A7	Mapputta	No
Garba	GARV	*Rhabdoviridae*	2	IE	Matariva	No
Garissa	GRSV	*Orthobunyavirus*	3	A7	Bunyamwera	No
Germiston	GERV	*Orthobunyavirus*	3		Bunyamwera	Yes
Getah	GETV	*Alphavirus*	2	A1	A^f	No
Gomoka	GOMV	*Orbivirus*	2	S	Ieri	No
Gordil	GORV	*Phlebovirus*	2	IE	Phlebotomus Fever	No
Gossas	GOSV	*Rhabdoviridae*	2	S		No
Grand Arbaud	GAV	*Phlebovirus*	2	S	Uukuniemi	No
Gray Lodge	GLOV	*Vesiculovirus*	2	IE	Vesicular Stomatitis	No
Great Island	GIV	*Orbivirus*	2	S	Kemerovo	No
Guajara	GJAV	*Orthobunyavirus*	2	S	Capim	No
Guama	GMAV	*Orthobunyavirus*	2	S	Guama	No
Guanarito	GTOV	*Arenavirus*	4	A4	Tacaribe	Yes
Guaratuba	GTBV	*Orthobunyavirus*	2	A7	Guama	No
Guaroa	GROV	*Orthobunyavirus*	2	S	California	No
Gumbo Limbo	GLV	*Orthobunyavirus*	2	S	C^f	No
Gurupi	GURV	*Orbivirus*	2	IE	Changuinola	No
Hantaan	HTNV	*Hantavirus*	3^a	S	Hantaan	No
Hanzalova	HANV	*Flavivirus*	4	A4	B^f	Yes
Hart Park	HPV	*Rhabdoviridae*	2	S	Hart Park	No
Hazara	HAZV	*Nairovirus*	2	S	CHF-Congo	No
Highlands J	HJV	*Alphavirus*	2	S	A^f	No
Huacho	HUAV	*Orbivirus*	2	S	Kemerovo	No

Name	Acronym	Taxonomic Status (Family or Genus)	Recommended Biosafety Level	Basis of Rating	Antigenic Group	HEPA Filtration on Lab Exhaust
Hughes	HUGV	*Nairovirus*	2	S	Hughes	No
Hypr	HYPRV	*Flavivirus*	4	S	B^f	Yes
Iaco	IACOV	*Orthobunyavirus*	2	IE	Bunyamwera	No
Ibaraki	IBAV	*Orbivirus*	2	IE	Epizootic Hemorrhagic Disease	Yes
Icoaraci	ICOV	*Phlebovirus*	2	S	Phlebotomus Fever	No
Ieri	IERIV	*Orbivirus*	2	S	Ieri	No
Ife	IFEV	*Orbivirus b*	2	IE		No
Iguape	IGUV	*Flavivirus*	2	A7	B^f	No
Ilesha	ILEV	*Orthobunyavirus*	2	S	Bunyamwera	No
Ilheus	ILHV	*Flavivirus*	2	S	B^f	No
Ingwavuma	INGV	*Orthobunyavirus*	2	S	Simbu	No
Inhangapi	INHV	*Rhabdoviridae*	2	IE		No
Inini	INIV	*Orthobunyavirus*	2	IE	Simbu	No
Inkoo	INKV	*Orthobunyavirus*	2	S	California	No
Ippy	IPPYV	*Arenavirus*	2	S	Tacaribe	No
Iriri	IRRV	*Rhabdoviridae*	2	A7		No
Irituia	IRIV	*Orbivirus*	2	S	Changuinola	No
Isfahan	ISFV	*Vesiculovirus*	2	S	Vesicular Stomatitis	No
Israel Turkey Meningitis	ITV	*Flavivirus*	2 with 3 practices	S	B^f	No
Issyk-Kul	ISKV	*Bunyaviridae*	3	IE		No
Itacaiunas	ITCNV	*Rhabdoviridae*	2	A7		No
Itaituba	ITAV	*Phlebovirus*	2	IE	Phlebotomus Fever	No
Itaporanga	ITPV	*Phlebovirus*	2	S	Phlebotomus Fever	No
Itaqui	ITQV	*Orthobunyavirus*	2	S	C^f	No
Itimirim	ITIV	*Orthobunyavirus*	2	IE	Guama	No
Itupiranga	ITUV	*Orbivirus b*	2	IE		No
Ixcanal	IXCV	*Phlebovirus*	2	A7	Phlebotomus Fever	No
Jacareacanga	JACV	*Orbivirus*	2	IE	Corriparta	No
Jacunda	JCNV	*Phlebovirus*	2	A7	Phlebotomus Fever	No
Jamanxi	JAMV	*Orbivirus*	2	IE	Changuinola	No
Jamestown Canyon	JCV	*Orthobunyavirus*	2	S	California	No

Name	Acronym	Taxonomic Status (Family or Genus)	Recom- mended Biosafety Level	Basis of Rating	Antigenic Group	HEPA Filtration on Lab Exhaust
Japanaut	JAPV	*Orbivirus b*	2	S		No
Japanese Encephalitis	JEV	*Flavivirus*	3c	S	Bf	No
Jari	JARIV	*Orbivirus*	2	IE	Changuinola	No
Jatobal	JTBV	*Orthobunyavirus*	2	A7		No
Jerry Slough	JSV	*Orthobunyavirus*	2	S	California	No
Joa	JOAV	*Phlebovirus*	2	A7		No
Johnston Atoll	JAV	Unassigned	2	S	Quaranfil	No
Joinjakaka	JOIV	*Rhabdoviridae*	2	S		No
Juan Diaz	JDV	*Orthobunyavirus*	2	S	Capim	No
Jugra	JUGV	*Flavivirus*	2	S	Bf	No
Junin	JUNV	*Arenavirus*	4	A6	Tacaribe	Yes
Jurona	JURV	*Vesiculovirus*	2	S	Vesicular Stomatitis	No
Juruaca	JRCV	*Picornavirus b*	2	A7		No
Jutiapa	JUTV	*Flavivirus*	2	S	Bf	No
Kadam	KADV	*Flavivirus*	2	S	Bf	No
Kaeng Khoi	KKV	*Orthobunyavirus b*	2	S		No
Kaikalur	KAIV	*Orthobunyavirus*	2	S	Simbu	No
Kairi	KRIV	*Orthobunyavirus*	2	A1	Bunyamwera	No
Kaisodi	KSOV	*Bunyaviridae*	2	S	Kaisodi	No
Kamese	KAMV	*Rhabdoviridae*	2	S	Hart Park	No
Kamiti River	KRV	*Flavivirus*	2	A7		No
Kammavanpettai	KMPV	*Orbivirus*	2	S		No
Kannamangalam	KANV	*Rhabdoviridae*	2	S		No
Kao Shuan	KSV	*Nairovirus*	2	S	Dera Ghazi Khan	No
Karimabad	KARV	*Phlebovirus*	2	S	Phlebotomus Fever	No
Karshi	KSIV	*Flavivirus*	2	S	Bf	No
Kasba	KASV	*Orbivirus*	2	S	Palyam	No
Kedougou	KEDV	*Flavivirus*	2	A7	Bf	No
Kemerovo	KEMV	*Orbivirus*	2	S	Kemerovo	No
Kern Canyon	KCV	*Rhabdoviridae*	2	S	Kern Canyon	No
Ketapang	KETV	*Orthobunyavirus*	2	S	Bakau	No
Keterah	KTRV	*Bunyaviridae*	2	S		No
Keuraliba	KEUV	*Rhabdoviridae*	2	S	Le Dantec	No

Name	Acronym	Taxonomic Status (Family or Genus)	Recom- mended Biosafety Level	Basis of Rating	Antigenic Group	HEPA Filtration on Lab Exhaust
Keystone	KEYV	*Orthobunyavirus*	2	S	California	No
Khabarovsk	KHAV	*Hantavirus*	3ª	IE	Hantaan	No
Khasan	KHAV	*Nairovirus*	2	IE	CCHF	No
Kimberley	KIMV	*Lyssavirus*	2	A7	Bovine Ephem- eral Fever	No
Kindia	KINV	*Orbivirus*	2	A7	Palyam	No
Kismayo	KISV	*Bunyaviridae*	2	S	Bhanja	No
Klamath	KLAV	*Vesiculovirus*	2	S	Vesicular Stomatitis	No
Kokobera	KOKV	*Flavivirus*	2	S	Bᶠ	No
Kolongo	KOLV	*Lyssavirus*	2	S	Rab	No
Koongol	KOOV	*Orthobunyavirus*	2	S	Koongol	No
Kotonkan	KOTV	*Lyssavirus*	2	S	Rab	No
Koutango	KOUV	*Flavivirus*	3	S	Bᶠ	No
Kowanyama	KOWV	*Bunyaviridae*	2	S		No
Kumlinge	KUMV	*Flavivirus*	4	A4	Bᶠ	Yes
Kunjin	KUNV	*Flavivirus*	2	S	Bᶠ	No
Kununurra	KNAV	*Rhabdoviridae*	2	S		No
Kwatta	KWAV	*Vesiculovirus*	2	S	Vesicular Stomatitis	No
Kyasanur Forest Disease	KFDV	*Flavivirus*	4	S	Bᶠ	Yes
Kyzylagach	KYZV	*Alphavirus*	2	IE	Aᶠ	No
La Crosse	LACV	*Orthobunyavirus*	2	S	California	No
Lagos Bat	LBV	*Lyssavirus*	2	S	Rab	No
Laguna Negra	LANV	*Hantavirus*	3ª	IE		No
La Joya	LJV	*Vesiculovirus*	2	S	Vesicular Stomatitis	No
Lake Clarendon	LCV	*Orbivirus b*	2	IE		No
Landjia	LJAV	*Rhabdoviridae*	2	S		No
Langat	LGTV	*Flavivirus*	2	S	Bᶠ	No
Lanjan	LJNV	*Bunyaviridae*	2	S	Kaisodi	No
Las Maloyas	LMV	*Orthobunyavirus*	2	A7	Anopheles A	No
Lassa	LASV	*Arenavirus*	4	S	Tacaribe	Yes
Latino	LATV	*Arenavirus*	2	A5	Tacaribe	No
Lebombo	LEBV	*Orbivirus*	2	S		No
Lechiguanas	LECHV	*Hantavirus*	3ª	IE	Hantaan	No

Name	Acronym	Taxonomic Status (Family or Genus)	Recommended Biosafety Level	Basis of Rating	Antigenic Group	HEPA Filtration on Lab Exhaust
Le Dantec	LDV	*Rhabdoviridae*	2	S	Le Dantec	No
Lednice	LEDV	*Orthobunyavirus*	2	A7	Turlock	No
Lipovnik	LIPV	*Orbivirus*	2	S	Kemerovo	No
Llano Seco	LLSV	*Orbivirus*	2	IE	Umatilla	No
Lokern	LOKV	*Orthobunyavirus*	2	S	Bunyamwera	No
Lone Star	LSV	*Bunyaviridae*	2	S		No
Louping Ill	LIV	*Flavivirus*	3ᶜ	S	Bᶠ	Yes
Lukuni	LUKV	*Orthobunyavirus*	2	S	Anopheles A	No
Macaua	MCAV	*Orthobunyavirus*	2	IE	Bunyamwera	No
Machupo	MACV	*Arenavirus*	4	S	Tacaribe	Yes
Madrid	MADV	*Orthobunyavirus*	2	S	Cᶠ	No
Maguari	MAGV	*Orthobunyavirus*	2	S	Bunyamwera	No
Mahogany Hammock	MHV	*Orthobunyavirus*	2	S	Guama	No
Main Drain	MDV	*Orthobunyavirus*	2	S	Bunyamwera	No
Malakal	MALV	*Lyssavirus*	2	S	Bovine Ephemeral	No
Manawa	MWAV	*Phlebovirus*	2	S	Uukumiemi	No
Manitoba	MNTBV	*Rhabdoviridae*	2	A7		No
Manzanilla	MANV	*Orthobunyavirus*	2	S	Simbu	No
Mapputta	MAPV	*Bunyaviridae*	2	S	Mapputta	No
Maporal	MPRLV	*Hantavirus*	3ᵃ	IE	Hantaan	No
Maprik	MPKV	*Bunyaviridae*	2	S	Mapputta	No
Maraba	MARAV	*Vesiculovirus*	2	A7		No
Marajo	MRJV	Unassigned	2	IE		No
Marburg	MARV	*Filovirus*	4	S	Marburg	Yes
Marco	MCOV	*Rhabdoviridae*	2	S		No
Mariquita	MRQV	*Phlebovirus*	2	A7	Phlebotomus Fever	No
Marituba	MTBV	*Orthobunyavirus*	2	S	Cᶠ	No
Marrakai	MARV	*Orbivirus*	2	S	Palyam	No
Matariya	MTYV	*Rhabdoviridae*	2	S	Matariva	No
Matruh	MTRV	*Orthobunyavirus*	2	S	Tete	No
Matucare	MATV	*Orbivirus*	2	S		No
Mayaro	MAYV	*Alphavirus*	2	S	Aᶠ	No
Mboke	MBOV	*Orthobunyavirus*	2	A7	Bunyamwera	No
Meaban	MEAV	*Flavivirus*	2	IE	Bᶠ	No

Name	Acronym	Taxonomic Status (Family or Genus)	Recommended Biosafety Level	Basis of Rating	Antigenic Group	HEPA Filtration on Lab Exhaust
Melao	MELV	*Orthobunyavirus*	2	S	California	No
Mermet	MERV	*Orthobunyavirus*	2	S	Simbu	No
Middelburg	MIDV	*Alphavirus*	2	A1	Aᶠ	No
Minatitlan	MNTV	*Orthobunyavirus*	2	S	Minatitlan	No
Minnal	MINV	*Orbivirus*	2	S	Umatilla	No
Mirim	MIRV	*Orthobunyavirus*	2	S	Guama	No
Mitchell River	MRV	*Orbivirus*	2	S		No
Mobala	MOBV	*Arenavirus*	3	A7	Tacaribe	No
Modoc	MODV	*Flavivirus*	2	S	Bᶠ	No
Moju	MOJUV	*Orthobunyavirus*	2	S	Guama	No
Mojui Dos Campos	MDCV	*Orthobunyavirus*	2	IE		No
Mono Lake	MLV	*Orbivirus*	2	S	Kemerovo	No
Mont. Myotis Leukemia	MMLV	*Flavivirus*	2	S	Bᶠ	No
Monte Dourado	MDOV	*Orbivirus*	2	IE	Changuinola	No
Mopeia	MOPV	*Arenavirus*	3	A7		No
Moriche	MORV	*Orthobunyavirus*	2	S	Capim	No
Morro Bay	MBV	*Orthobunyavrius*	2	IE	California	No
Morumbi	MRMBV	*Phlebovirus*	2	A7	Phlebotomus Fever	No
Mosqueiro	MQOV	*Rhabdoviridae*	2	A7	Hart Park	No
Mossuril	MOSV	*Rhabdoviridae*	2	S	Hart Park	No
Mount Elgon Bat	MEBV	*Vesiculovirus*	2	S	Vesicular Stomatitis	No
M'Poko	MPOV	*Orthobunyavirus*	2	S	Turlock	No
Mucambo	MUCV	*Alphavirus*	3	S	Aᶠ	Yes
Mucura	MCRV	*Phlebovirus*	2	A7	Phlebotomus Fever	No
Munguba	MUNV	*Phlebovirus*	2	IE	Phlebotomus Fever	No
Murray Valley Encephalitis	MVEV	*Flavivirus*	3	S	Bᶠ	No
Murutucu	MURV	*Orthobunyavirus*	2	S	Cᶠ	No
Mykines	MYKV	*Orbivirus*	2	A7	Kemerovo	No
Nairobi Sheep Disease	NSDV	*Nairovirus*	3ᶜ	A1	Nairobi Sheep Disease	No
Naranjal	NJLV	*Flavivirus*	2	IE	Bᶠ	No
Nariva	NARV	*Paramyxoviridae*	2	IE		No
Nasoule	NASV	*Lyssavirus*	2	A7	Rab	No

Name	Acronym	Taxonomic Status (Family or Genus)	Recom- mended Biosafety Level	Basis of Rating	Antigenic Group	HEPA Filtration on Lab Exhaust
Navarro	NAVV	*Rhabdoviridae*	2	S		No
Ndelle	NDEV	*Orthoreovirus*	2	A7	Ndelle	No
Ndumu	NDUV	*Alphavirus*	2	A1	A^f	No
Negishi	NEGV	*Flavivirus*	3	S	B^f	No
Nepuyo	NEPV	*Orthobunyavirus*	2	S	C^f	No
Netivot	NETV	*Orbivirus*	2	A7		No
New Minto	NMV	*Rhabdoviridae*	2	IE	Sawgrass	No
Ngaingan	NGAV	*Lyssavirus*	2	S	Tibrogargan	No
Ngari d	NRIV	*Orthobunyavirus*	3	A7	Bunyamera	No
Ngoupe	NGOV	*Orbivirus*	2	A7	Eubenangee	No
Nique	NIQV	*Phlebovirus*	2	S	Phlebotomus Fever	No
Nkolbisson	NKOV	*Rhabdoviridae*	2	S	Kern Canyon	No
Nodamura	NOV	*Alphanodavirus*	2	IE		No
Nola	NOLAV	*Orthobunyavirus*	2	S	Bakau	No
Northway	NORV	*Orthobunyavirus*	2	IE	Bunyamwera	No
Ntaya	NTAV	*Flavivirus*	2	S	B^f	No
Nugget	NUGV	*Orbivirus*	2	S	Kemerovo	No
Nyamanini	NYMV	Unassigned	2	S	Nyamanini	No
Nyando	NDV	*Orthobunyavirus*	2	S	Nyando	No
Oak Vale	OVV	*Rhabdoviridae*	2	A7		No
Odrenisrou	ODRV	*Phlebovirus*	2	A7	Phlebotomus Fever	No
Okhotskiy	OKHV	*Orbivirus*	2	S	Kemerovo	No
Okola	OKOV	*Bunyaviridae*	2	S	Tanga	No
Olifantsvlei	OLIV	*Orthobunyavirus*	2	S	Olifantsylei	No
Omo	OMOV	*Nairovirus*	2	A7	Qalyub	No
Omsk Hemorrhagic	OHFV	*Flavivirus*	4	S	B^f	Yes
O'Nyong-Nyong	ONNV	*Alphavirus*	2	S	A^f	Yes
Oran	ORANV	*Hantavirus*	3^a	IE	Hantaan	No
Oriboca	ORIV	*Orthobunyavirus*	2	S	C^f	No
Oriximina	ORXV	*Phlebovirus*	2	IE	Phlebotomus Fever	No
Oropouche	OROV	*Orthobunyavirus*	3	S	Simbu	Yes
Orungo	ORUV	*Orbivirus*	2	S	Orungo	No
Ossa	OSSAV	*Orthobunyavirus*	2	S	C^f	No

Name	Acronym	Taxonomic Status (Family or Genus)	Recom- mended Biosafety Level	Basis of Rating	Antigenic Group	HEPA Filtration on Lab Exhaust
Ouango	OUAV	*Rhabdoviridae*	2	IE		No
Oubangui	OUBV	*Poxviridae*	2	IE		No
Oubi	OUBIV	*Orthobunyavirus*	2	A7	Olifantsylei	No
Ourem	OURV	*Orbivirus*	2	IE	Changuinola	No
Pacora	PCAV	*Bunyaviridae*	2	S		No
Pacui	PACV	*Phlebovirus*	2	S	Phlebotomus Fever	No
Pahayokee	PAHV	*Orthobunyavirus*	2	S	Patois	No
Palma	PMAV	*Bunyaviridae*	2	IE	Bhanja	No
Palestina	PLSV	*Orthobunyavirus*	2	IE	Minatitlan	No
Palyam	PALV	*Orbivirus*	2	S	Palyam	No
Para	PARAV	*Orthobunyavirus*	2	IE	Simbu	No
Paramushir	PMRV	*Nairovirus*	2	IE	Sakhalin	No
Parana	PARV	*Arenavirus*	2	A5	Tacaribe	No
Paroo River	PRV	*Orbivirus*	2	IE		No
Pata	PATAV	*Orbivirus*	2	S		No
Pathum Thani	PTHV	*Nairovirus*	2	S	Dera Ghazi Khan	No
Patois	PATV	*Orthobunyavirus*	2	S	Patois	No
Peaton	PEAV	*Orthobunyavirus*	2	A1	Simbu	No
Pergamino	PRGV	*Hantavirus*	3[a]	IE		No
Perinet	PERV	*Vesiculovirus*	2	A7	Vesicular Stomatitis	No
Petevo	PETV	*Orbivirus*	2	A7	Palyam	No
Phnom-Penh Bat	PPBV	*Flavivirus*	2	S	Bf	No
Pichinde	PICV	*Arenavirus*	2	A5	Tacaribe	No
Picola	PIAV	*Orbivirus*	2	IE	Wongorr	No
Pirital	PIRV	*Arenavirus*	3	IE		No
Piry	PIRYV	*Vesiculovirus*	3	S	Vesicular Stomatitis	No
Pixuna	PIXV	*Alphavirus*	2	S	A[f]	No
Playas	PLAV	*Orthobunyavirus*	2	IE	Bunyamwera	No
Pongola	PGAV	*Orthobunyavirus*	2	S	Bwamba	No
Ponteves	PTVV	*Phlebovirus*	2	A7	Uukuniemi	No
Potosi	POTV	*Orthobunyavirus*	2	IE	Bunyamwera	No
Powassan	POWV	*Flavivirus*	3	S	B[f]	No
Precarious Point	PPV	*Phlebovirus*	2	A7	Uukuniemi	No

Name	Acronym	Taxonomic Status (Family or Genus)	Recommended Biosafety Level	Basis of Rating	Antigenic Group	HEPA Filtration on Lab Exhaust
Pretoria	PREV	*Nairovirus*	2	S	Dera Ghazi Khan	No
Prospect Hill	PHV	*Hantavirus*	2	A8	Hantaan	No
Puchong	PUCV	*Lyssavirus*	2	S	Bovine Ephemeral ever	No
Pueblo Viejo	PVV	*Orthobunyavirus*	2	IE	Gamboa	No
Punta Salinas	PSV	*Nairovirus*	2	S	Hughes	No
Punta Toro	PTV	*Phlebovirus*	2	S	Phlebotomus Fever	No
Purus	PURV	*Orbivirus*	2	IE	Changuinola	No
Puumala	PUUV	*Hantavirus*	3ª	IE	Hantaan	No
Qalyub	QYBV	*Nairovirus*	2	S	Qalyub	No
Quaranfil	QRFV	Unassigned	2	S	Quaranfil	No
Radi	RADIV	*Vesiculovirus*	2	A7	Vesicular Stomatitis	No
Razdan	RAZV	*Bunyaviridae*	2	IE		No
Resistencia	RTAV	*Bunyaviridae*	2	IE	Resistencia	No
Restan	RESV	*Orthobunyavirus*	2	S	Cᶠ	No
Rhode Island	RHIV	*Rhabdoviridae*	2	A7		No
Rift Valley Fever	RVFV	*Phlebovirus*	3ᶜ	S	Phlebotomus Fever	Yes
Rio Bravo	RBV	*Flavivirus*	2	S	Bᶠ	No
Rio Grande	RGV	*Phlebovirus*	2	S	Phlebotomus Fever	No
Rio Preto	RIOPV	Unassigned	2	IE		No
Rochambeau	RBUV	*Lyssavirus*	2	IE	Rab	No
Rocio	ROCV	*Flavivirus*	3	S	Bᶠ	Yes
Ross River	RRV	*Alphavirus*	2	S	Aᶠ	No
Royal Farm	RFV	*Flavivirus*	2	S	Bᶠ	No
Russian Spring-Summer Encephalitis	RSSEV	*Flavivirus*	4	S	Bᶠ	Yes
Saaremaa	SAAV	*Hantavirus*	3ª	IE	Hantaan	No
Sabia	SABV	*Arenavirus*	4	A4		Yes
Sabo	SABOV	*Orthobunyavirus*	2	S	Simbu	No
Saboya	SABV	*Flavivirus*	2	S	Bᶠ	No
Sagiyama	SAGV	*Alphavirus*	2	A1	Aᶠ	No
Saint-Floris	SAFV	*Phlebovirus*	2	S	Phlebotomus Fever	No
Sakhalin	SAKV	*Nairovirus*	2	S	Sakhalin	No

Name	Acronym	Taxonomic Status (Family or Genus)	Recommended Biosafety Level	Basis of Rating	Antigenic Group	HEPA Filtration on Lab Exhaust
Salanga	SGAV	*Poxviridae*	2	IE	SGA	No
Salehabad	SALV	*Phlebovirus*	2	S	Phlebotomus Fever	No
Salmon River	SAVV	*Coltivirus*	2	IE	Colorado Tick Fever	No
Sal Vieja	SVV	*Flavivirus*	2	A7	Bᶠ	No
San Angelo	SAV	*Orthobunyavirus*	2	S	California	No
Sandfly Fever, Naples	SFNV	*Phlebovirus*	2	S	Phlebotomus Fever	No
Sandfly Fever, Sicilian	SFSV	*Phlebovirus*	2	S	Phlebotomus Fever	No
Sandjimba	SJAV	*Lyssavirus*	2	S	Rab	No
Sango	SANV	*Orthobunyavirus*	2	S	Simbu	No
San Juan	SJV	*Orthobunyavirus*	2	IE	Gamboa	No
San Perlita	SPV	*Flavivirus*	2	A7	Bᶠ	No
Santarem	STMV	*Bunyaviridae*	2	IE		No
Santa Rosa	SARV	*Orthobunyavirus*	2	IE	Bunyamwera	No
Saraca	SRAV	*Orbivirus*	2	IE	Changuinola	No
Sathuperi	SATV	*Orthobunyavirus*	2	S	Simbu	No
Saumarez Reef	SREV	*Flavivirus*	2	IE	Bᶠ	No
Sawgrass	SAWV	*Rhabdoviridae*	2	S	Sawgrass	No
Sebokele	SEBV	Unassigned	2	S		No
Sedlec	SEDV	*Bunyaviridae*	2	A7		No
Seletar	SELV	*Orbivirus*	2	S	Kemerovo	No
Sembalam	SEMV	Unassigned	2	S		No
Semliki Forest	SFV	*Alphavirus*	3	A2	Aᶠ	No
Sena Madureira	SMV	*Rhabdoviridae*	2	IE	Timbo	No
Seoul	SEOV	*Hantavirus*	3ª	IE	Hantaan	No
Sepik	SEPV	*Flavivirus*	2	IE	Bᶠ	No
Serra Do Navio	SDNV	*Orthobunyavirus*	2	A7	California	No
Serra Norte	SRNV	*Phlebovirus*	2	A7		No
Shamonda	SHAV	*Orthobunyavirus*	2	S	Simbu	No
Shark River	SRV	*Orthobunyavirus*	2	S	Patois	No
Shokwe	SHOV	*Orthobunyavirus*	2	IE	Bunyamwera	No
Shuni	SHUV	*Orthobunyavirus*	2	S	Simbu	No
Silverwater	SILV	*Bunyaviridae*	2	S	Kaisodi	No
Simbu	SIMV	*Orthobunyavirus*	2	S	Simbu	No

Name	Acronym	Taxonomic Status (Family or Genus)	Recommended Biosafety Level	Basis of Rating	Antigenic Group	HEPA Filtration on Lab Exhaust
Simian Hemorrhagic Fever	SHFV	*Arterivirus*	2	A2	Simian Hemorrhagic Fever	No
Sindbis	SINV	*Alphavirus*	2	S	A^f	No
Sin Nombre	SNV	*Hantavirus*	3ª	IE	Hantaan	No
Sixgun City	SCV	*Orbivirus*	2	S	Kemerovo	No
Slovakia	SLOV	Unassigned	3	IE		No
Snowshoe Hare	SSHV	*Orthobunyavirus*	2	S	California	No
Sokoluk	SOKV	*Flavivirus*	2	S	B^f	No
Soldado	SOLV	*Nairovirus*	2	S	Hughes	No
Somone	SOMV	Unassigned	3	IE	Somone	No
Sororoca	SORV	*Orthobunyavirus*	2	S	Bunyamwera	No
Spondweni	SPOV	*Flavivirus*	2	S	B^f	No
Sripur	SRIV	*Rhabdoviridae*	3	IE		No
St. Louis Encephalitis	SLEV	*Flavivirus*	3	S	B^f	No
Stratford	STRV	*Flavivirus*	2	S	B^f	No
Sunday Canyon	SCAV	*Bunyaviridae*	2	S		No
Tacaiuma	TCMV	*Orthobunyavirus*	2	S	Anopheles A	No
Tacaribe	TCRV	*Arenavirus*	2	A5	Tacaribe	No
Taggert	TAGV	*Nairovirus*	2	S	Sakhalin	No
Tahyna	TAHV	*Orthobunyavirus*	2	S	California	No
Tai	TAIV	*Bunyaviridae*	2	A7	Bunyamwera	No
Tamdy	TDYV	*Bunyaviridae*	2	IE		No
Tamiami	TAMV	*Arenavirus*	2	A5	Tacaribe	No
Tanga	TANV	*Bunyaviridae*	2	S	Tanga	No
Tanjong Rabok	TRV	*Orthobunyavirus*	2	S	Bakau	No
Tapara	TAPV	*Phlebovirus*	2	A7		No
Tataguine	TATV	*Bunyaviridae*	2	S		No
Tehran	THEV	*Phlebovirus*	2	A7	Phlebotomus Fever	No
Telok Forest	TFV	*Orthobunyavirus*	2	IE	Bakau	No
Tembe	TMEV	*Orbivirus b*	2	S		No
Tembusu	TMUV	*Flavivirus*	2	S	B^f	No
Tensaw	TENV	*Orthobunyavirus*	2	S	Bunyamwera	No
Termeil	TERV	*Bunyavirus b*	2	IE		No
Tete	TETEV	*Orthobunyavirus*	2	S	Tete	No

Name	Acronym	Taxonomic Status (Family or Genus)	Recommended Biosafety Level	Basis of Rating	Antigenic Group	HEPA Filtration on Lab Exhaust
Thiafora	TFAV	*Bunyaviridae*	2	A7	Thiafora	No
Thimiri	THIV	*Orthobunyavirus*	2	S	Simbu	No
Thogoto	THOV	*Orthomyxoviridae*	2	S	Thogoto	No
Thottapalayam	TPMV	*Hantavirus*	2	S	Hantaan	No
Tibrogargan	TIBV	*Lyssavirus*	2	S	Tibrogargan	No
Tilligerry	TILV	*Orbivirus*	2	IE	Eubenangee	No
Timbo	TIMV	*Rhabdoviridae*	2	S	Timbo	No
Timboteua	TBTV	*Orthobunyavirus*	2	A7	Guama	No
Tinaroo	TINV	*Orthobunyavirus*	2	IE	Simbu	No
Tindholmur	TDMV	*Orbivirus*	2	A7	Kemerovo	No
Tlacotalpan	TLAV	*Orthobunyavirus*	2	IE	Bunyamwera	No
Tonate	TONV	*Alphavirus*	3	IE	A^f	Yes
Topografov	TOPV	*Hantavirus*	3^a	IE	Hantaan	No
Toscana	TOSV	*Phlebovirus*	2	S	Phlebotomus Fever	No
Toure	TOUV	Unassigned	2	S		No
Tracambe	TRCV	*Orbivirus*	2	A7		No
Tribec	TRBV	*Orbivirus*	2	S	Kemerovo	No
Triniti	TNTV	*Togaviridae*	2	S		No
Trivittatus	TVTV	*Orthobunyavirus*	2	S	California	No
Trocara	TROCV	*Alphavirus*	2	IE	A^f	No
Trombetas	TRMV	*Orthobunyavirus*	2	A7		No
Trubanaman	TRUV	*Bunyaviridae*	2	S	Mapputta	No
Tsuruse	TSUV	*Orthobunyavirus*	2	S	Tete	No
Tucurui	TUCRV	*Orthobunyavirus*	2	A7		No
Tula	TULV	*Hantavirus*	2	A8		No
Tunis	TUNV	*Phlebovirus*	2	A7	Phlebotomus Fever	No
Turlock	TURV	*Orthobunyavirus*	2	S	Turlock	No
Turuna	TUAV	*Phlebovirus*	2	IE	Phlebotomus Fever	No
Tyuleniy	TYUV	*Flavivirus*	2	S	B^f	No
Uganda S	UGSV	*Flavivirus*	2	S	B^f	No
Umatilla	UMAV	*Orbivirus*	2	S	Umatilla	No
Umbre	UMBV	*Orthobunyavirus*	2	S	Turlock	No
Una	UNAV	*Alphavirus*	2	S	A^f	No
Upolu	UPOV	*Bunyaviridae*	2	S	Upolu	No

Name	Acronym	Taxonomic Status (Family or Genus)	Recommended Biosafety Level	Basis of Rating	Antigenic Group	HEPA Filtration on Lab Exhaust
Uriurana	UURV	*Phlebovirus*	2	A7	Phlebotomus Fever	No
Urucuri	URUV	*Phlebovirus*	2	S	Phlebotomus Fever	No
Usutu	USUV	*Flavivirus*	2	S	B[f]	No
Utinga	UTIV	*Orthobunyavirus*	2	IE	Simbu	No
Uukuniemi	UUKV	*Phlebovirus*	2	S	Uukuniemi	No
Vellore	VELV	*Orbivirus*	2	S	Palyam	No
Venezuelan Equine Encephalitis	VEEV	*Alphavirus*	3[c]	S	A[f]	Yes
Venkatapuram	VKTV	Unassigned	2	S		No
Vinces	VINV	*Orthobunyavirus*	2	A7	C[f]	No
Virgin River	VRV	*Orthobunyavirus*	2	A7	Anopheles A	No
Vesicular Stomatitis-Alagoas	VSAV	*Vesiculovirus*	2[c]	S	Vesicular Stomatitis	No
Vesicular Stomatitis-Indiana	VSIV	*Vesiculovirus*	2[c]	A3	Vesicular Stomatitis	No
Vesicular Stomatitis-New Jersey	VSNJV	*Vesiculovirus*	2[c]	A3	Vesicular Stomatitis	No
Wad Medani	WMV	*Orbivirus*	2	S	Kemerovo	No
Wallal	WALV	*Orbivirus*	2	S	Wallal	No
Wanowrie	WANV	*Bunyaviridae*	2	S		No
Warrego	WARV	*Orbivirus*	2	S	Warrego	No
Wesselsbron	WESSV	*Flavivirus*	3[c]	S	B[f]	Yes
Western Equine Encephalitis	WEEV	*Alphavirus*	3	S	A[f]	No
West Nile	WNV	*Flavivirus*	3	S	B[f]	No
Whataroa	WHAV	*Alphavirus*	2	S	A[f]	No
Whitewater Arroyo	WWAV	*Arenavirus*	3	IE	Tacaribe	No
Witwatersrand	WITV	*Bunyaviridae*	2	S		No
Wongal	WONV	*Orthobunyavirus*	2	S	Koongol	No
Wongorr	WGRV	*Orbivirus*	2	S	Wongorr	No
Wyeomyia	WYOV	*Orthobunyavirus*	2	S	Bunyamwera	No
Xiburema	XIBV	*Rhabdoviridae*	2	IE		No
Xingu	XINV	*Orthobunyavirus*	3			No
Yacaaba	YACV	*Bunyaviridae*	2	IE		No
Yaounde	YAOV	*Flavivirus*	2	A7	B[f]	No

Name	Acronym	Taxonomic Status (Family or Genus)	Recommended Biosafety Level	Basis of Rating	Antigenic Group	HEPA Filtration on Lab Exhaust
Yaquina Head	YHV	*Orbivirus*	2	S	Kemerovo	No
Yata	YATAV	*Rhabdoviridae*	2	S		No
Yellow Fever	YFV	*Flavivirus*	3	S	B[f]	Yes
Yogue	YOGV	*Bunyaviridae*	2	S	Yogue	No
Yoka	YOKA	*Poxviridae*	2	IE		No
Yug Bogdanovac	YBV	*Vesiculovirus*	2	IE	Vesicular Stomatitis	No
Zaliv Terpeniya	ZTV	*Phlebovirus*	2	S	Uukuniemi	No
Zegla	ZEGV	*Orthobunyavirus*	2	S	Patois	No
Zika	ZIKV	*Flavivirus*	2	S	B[f]	No
Zirqa	ZIRV	*Nairovirus*	2	S	Hughes	No

* Federal regulations, import/export requirements, and taxonomic status are subject to changes. Check with the appropriate federal agency to confirm regulations.

[a] Containment requirements will vary based on virus concentration, animal species, or virus type. See the Hantavirus agent summary statement in the viral agent chapter.

[b] Tentative placement in the genus.

[c] These organisms are considered pathogens of significant agricultural importance by the USDA (see Appendix D) and may require additional containment (up to and including BSL-3-Ag containment). Not all strains of each organism are necessarily of concern to the USDA. Contact USDA for more information regarding exact containment/permit requirements before initiating work.

[d] Alternate name for Ganjam virus.

[e] Garissa virus is considered an isolate of this virus, so same containment requirements apply.

[f] Antigenic groups designated A, B, and C refer to the original comprehensive and unifying serogroups established by Casals, Brown, and Whitman based on cross-reactivity among known arboviruses (2,21). Group A viruses are members of the genus *Alphavirus*, group B belong to the family *Flaviviridae*, and Group C viruses are members of the family *Bunyaviridae*.

References

1. American Committee on Arthropod-borne Viruses. Subcommittee on Information Exchange International catalogue of arboviruses including certain other viruses of vertebrates. 3rd ed. San Antonio (TX): American Society of Tropical Medicine and Hygiene; 1985.
2. Casals J, Brown LV. Hemagglutinations with arthropod-borne viruses. J Exp Med. 1954;99:429-49.
3. Leifer E, Gocke DJ, Bourne H. Lassa fever, a new virus disease of man from West Africa. II. Report of a laboratory acquired infection treated with plasma from a person recently recovered from the disease. Am J Trop Med Hyg. 1970;19:667-9.
4. Weissenbacher MC, Grela ME, Sabattini MS, et al. Inapparent infections with Junin virus among laboratory workers. J Infect Dis. 1978;137:309-13.

5. Centers for Disease Control and Prevention, Office of Biosafety. Classification of etiologic agents on the basis of hazard. 4th edition. US Department of Health, Education and Welfare; US Public Health Service. 1979.

6. Hanson RP, Sulkin SE, Beuscher EL, et al. Arbovirus infections of laboratory workers. Extent of problem emphasizes the need for more effective measures to reduce hazards. Science. 1967;158:1283-86.

7. Karabatsos N. Supplement to international catalogue of arboviruses, including certain other viruses of vertebrates. Am J Trop Med Hyg. 1978;27:372-440.

8. Department of Health, Education, and Welfare. International catalogue of arboviruses including certain other viruses of vertebrates. Berge TO, editor. Washington, DC. 1975. Pub No. (CDC) 75-8301.

9. Hunt GJ, Tabachnick WJ. Handling small arbovirus vectors safely during biosafety level 3 containment: Culicoides variipennis sonorensis (Diptera: Ceratopogonidae) and exotic bluetongue viruses. J Med Entomol. 1996;33:271-7.

10. Warne SR. The safety of work with genetically modified viruses. In: Ring CJA, Blair ED, editors. Genetically engineered viruses: development and applications. Oxford: BIOS Scientific Publishers; 2001. p. 255-73.

11. Berglund P, Quesada-Rolander M, Putkonen P, et al. Outcome of immunization of cynomolgus monkeys with recombinant Semliki Forest virus encoding human immunodeficiency virus type 1 envelope protein and challenge with a high dose of SHIV-4 virus. AIDS Res Hum Retroviruses. 1997;13:1487-95.

12. Davis NL, Caley IJ, Brown KW, et al. Vaccination of macaques against pathogenic simian immunodeficiency virus with Venezuelan equine encephalitis virus replicon particles. J Virol. 2000;74:371-8.

13. Fernandez IM, Golding H, Benaissa-Trouw BJ, et al. Induction of HIV-1 IIIb neutralizing antibodies in BALB/c mice by a chimaeric peptide consisting of a T-helper cell epitope of Semliki Forest virus and a B-cell epitope of HIV. Vaccine. 1998;16:1936-40.

14. Notka F, Stahl-Hennig C, Dittmer U, et al. Construction and characterization of recombinant VLPs and Semliki-Forest virus live vectors for comparative evaluation in the SHIV monkey model. Biol Chem. 1999;380:341-52.

15. Kuhn RJ, Griffin DE, Owen KE, et al. Chimeric Sindbis-Ross River viruses to study interactions between alphavirus nonstructural and structural regions. J Virol. 1996;70:7900-9.

16. Schoepp RJ, Smith JF, Parker MD. Recombinant chimeric western and eastern equine encephalitis viruses as potential vaccine candidates. Virology. 2002;302:299-309.

17. Paessler S, Fayzulin RZ, Anishchenko M, et al. Recombinant Sindbis/ Venezuelan equine encephalitis virus is highly attenuated and immunogenic. J Virol. 2003;77:9278-86.

18. Arroyo J, Miller CA, Catalan J, et al. Yellow fever vector live-virus vaccines: West Nile virus vaccine development. Trends Mol Med. 2001;7:350-4.

19. Monath TP, McCarthy K, Bedford P, et al. Clinical proof of principle for ChimeriVax: recombinant live attenuated vaccines against flavivirus infections. Vaccine. 2002;20:1004-18.

20. Smithburn KC, Hughes TP, Burke AW, et al. A neurotropic virus isolated from the blood of a native of Uganda. Am J Trop Med Hyg. 1940;20:471-92.

21. Melnick JL, Paul JR, Riordan JT, et al. Isolation from human sera in Egypt of a virus apparently identical to West Nile virus. Proc Soc Exp Biol Med. 1951;77:661-5.

22. Taylor RM, Work TH, Hurlbut HS, Rizk F. A study of the ecology of West Nile virus in Egypt. Am J Trop Med Hyg. 1956;5:579-620.

23. Gerhardt R. West Nile virus in the United States (1999-2005). J Am Anim Hosp Assoc. 2006;42:170-7.

24. Centers for Disease Control and Prevention. Laboratory-acquired West Nile virus infections—United States, 2002. MMWR Morb Mortal Wkly Rep. 2002;51:1133-5.

25. Rusnak JM, Kortepeter MG, Hawley RJ, et al. Risk of occupationally acquired illnesses from biological threat agents in unvaccinated laboratory workers. Biosecur Bioterror. 2004;2:281-93.

26. Morris CD. Eastern equine encephalitis. In: Monath TP editor. The arboviruses: epidemiology and ecology. Vol III. Boca Raton: CRC Press; 1988. p. 2-20.

27. Kinney RM, Trent DW, France JK. Comparative immunological and biochemical analyses of viruses in the Venezuelan equine encephalitis complex. J Gen Virol. 1983;64:135-47.

28. Flick R, Bouloy M. Rift Valley fever virus. Curr Mol Med. 2005;5:827-34

29. Imam IZE, Darwish MA. A preliminary report on an epidemic of Rift Valley fever (RVF) in Egypt. J Egypt Public Health Assoc. 1977;52:417-8.

30. Francis T Jr., Magill TP. Rift valley fever: a report of three cases of laboratory infection and the experimental transmission of the disease to ferrets. J Exp Med. 1935;62:433-48.

31. Smithburn KC, Haddow AJ, Mahaffy AF. Rift valley fever: accidental infections among laboratory workers. J Immunol. 1949;62:213-27.

32. Linthicum KJ, Anyamba A, Tucker CJ, et al. Climate and satellite indicators to forecast Rift Valley epidemics in Kenya. Science. 1999;285:397-400.

33. Weaver SC. Host range, amplification and arboviral disease emergence. Arch Virol Suppl. 2005;19:33-44.

Section VIII-G: Toxin Agents

Botulinum Neurotoxin

Seven immunologically distinct serotypes of Botulinum neurotoxin (BoNT) have been isolated (A, B, C1, D, E, F and G). Each BoNT holotoxin is a disulfide-bonded heterodimer composed of a zinc metallo-protease "light chain" (approximately 50 kD) and a receptor binding "heavy chain" (approximately 100 kD). The heavy chain enhances cell binding and translocation of the catalytic light chain across the vesicular membrane.[1] There are also a number of important accessory proteins that can stabilize the natural toxin complex in biological systems or in buffer.

Four of the serotypes (A, B, E and, less commonly, F) are responsible for most human poisoning through contaminated food, wound infection, or infant botulism, whereas livestock may be at greater risk for poisoning with serotypes B, C1 and D.[2,3] It is important to recognize, however, that all BoNT serotypes are highly toxic and lethal by injection or aerosol delivery. BoNT is one of the most toxic proteins known; absorption of less than one microgram (μg) of BoNT can cause severe incapacitation or death, depending upon the serotype and the route of exposure.

Diagnosis of Laboratory Exposures

Botulism is primarily clinically diagnosed through physician observations of signs and symptoms that are similar for all serotypes and all routes of intoxication.[4] There typically is a latency of several hours to days, depending upon the amount of toxin absorbed, before the signs and symptoms of BoNT poisoning occur. The first symptoms of exposure generally include blurred vision, dry mouth and difficulty swallowing and speaking. This is followed by a descending, symmetrical flaccid paralysis, which can progress to generalized muscle weakness and respiratory failure. Sophisticated tests such as nerve conduction studies and single-fiber electromyography can support the diagnosis and distinguish it from similar neuromuscular conditions. Routine laboratory tests are of limited value because of the low levels of BoNT required to intoxicate, as well as the delay in onset of symptoms.

Laboratory Safety and Containment Recommendations

Solutions of sodium hypochlorite (0.1%) or sodium hydroxide (0.1N) readily inactivate the toxin and are recommended for decontamination of work surfaces and for spills. Additional considerations for the safe use and inactivation of toxins of biological origin are found in Appendix I. Because neurotoxin producing Clostridia species requires an anaerobic environment for growth and it is essentially not transmissible among individuals, exposure to pre-formed BoNT is the primary concern for laboratory workers. Two of the most significant hazards in working with BoNT or growing neurotoxin producing Clostridia species cultures are unintentional aerosol generation, especially during centrifugation, and accidental needle-stick. Although BoNT does not penetrate intact skin, proteins can be absorbed through broken or lacerated skin and, therefore, BoNT samples or contaminated material should be handled with gloves.

Workers in diagnostic laboratories should be aware that neurotoxin producing Clostridia species or its spores can be stable for weeks or longer in a variety of food products, clinical samples (e.g., serum, feces) and environmental samples (e.g., soil). Stability of the toxin itself will depend upon the sterility, temperature, pH and ionic strength of the sample matrix, but useful comparative data are available from the food industry. BoNT retains its activity for long periods (at least 6-12 months) in a variety of frozen foods, especially under acidic conditions (pH 4.5-5.0) and/or high ionic strength, but the toxin is readily inactivated by heating.[5]

A documented incident of laboratory intoxication with BoNT occurred in workers who were performing necropsies on animals that had been exposed 24 h earlier to aerosolized BoNT serotype A; the laboratory workers presumably inhaled aerosols generated from the animal fur. The intoxications were relatively mild, and all affected individuals recovered after a week of hospitalization.[6] Despite the low incidence of laboratory-associated botulism, the remarkable toxicity of BoNT necessitates that laboratory workers exercise caution during all experimental procedures.

BSL-2 practices, containment equipment, and facilities are recommended for routine dilutions, titrations or diagnostic studies with materials known to contain or have the potential to contain BoNT. Additional primary containment and personnel precautions, such as those recommended for BSL-3, should be implemented for activities with a high potential for aerosol or droplet production, or for those requiring routine handling of larger quantities of toxin.

Personnel not directly involved in laboratory studies involving botulinum toxin, such as maintenance personnel, should be discouraged from entering the laboratory when BoNT is in use until after the toxin and all work surfaces have been decontaminated. Purified preparations of toxin components, e.g. isolated BoNT "light chains" or "heavy chains," should be handled as if contaminated with holotoxin unless proven otherwise by toxicity bioassays.

Special Issues

Vaccines A pentavalent (A, B, C, D and E) botulinum toxoid vaccine (PBT) is available through the CDC as an IND. Vaccination is recommended for all personnel working in direct contact with cultures of neurotoxin producing Clostridia species or stock solutions of BoNT. Due to a possible decline in the immunogenicity of available PBT stocks for some toxin serotypes, the immunization schedule for the PBT recently has been modified to require injections at 0, 2, 12, and 24 weeks, followed by a booster at 12 months and annual boosters thereafter. Since there is a possible decline in vaccine efficacy, the current vaccine contains toxoid for only 5 of the 7 toxin types, this vaccine should not be considered as the sole means of protection and should not replace other worker protection measures.

Select Agent Botulinum toxin is a select agent requiring registration with CDC and/or USDA for possession, use, storage and/or transfer if quantities are above the minimum exemption level. See Appendix F for additional information.

Transfer of Agent Importation of this agent may require CDC and/or USDA importation permits. Domestic transport of this agent may require a permit from USDA/APHIS/VS. A DoC permit may be required for the export of this agent to another country. See Appendix C for additional information.

Staphylococcal Enterotoxins (SE)

SE are a group of closely related extracellular protein toxins of 23 to 29 kD molecular weight that are produced by distinct gene clusters found in a wide variety of *S. aureus* strains.[8,9] SE belong to a large family of homologous pyrogenic exotoxins from staphylococci, streptococci and mycoplasma which are capable of causing a range of illnesses in man through pathological amplification of the normal T-cell receptor response, cytokine/lymphokine release, immunosuppression and endotoxic shock.[9,10]

SE serotype A (SEA) is a common cause of severe gastroenteritis in humans.[11] It has been estimated from accidental food poisoning that exposure to as little as 0.05 to 1 µg SEA by the gastric route causes incapacitating illness.[12-15] Comparative human toxicity for different serotypes of SE is largely unknown, but human volunteers exposed to 20-25 µg SE serotype B (SEB) in distilled water experienced enteritis similar to that caused by SEA.[16]

SE are highly toxic by intravenous and inhalation routes of exposure. By inference from accidental exposure of laboratory workers and controlled experiments with NHP, it has been estimated that inhalation of less than 1 ng/kg SEB can incapacitate more than 50% of exposed humans, and that the inhalation LD_{50} in humans may be as low as 20 ng/kg SEB.[17]

Exposure of mucous membranes to SE in a laboratory setting has been reported to cause incapacitating gastrointestinal symptoms, conjunctivitis and localized cutaneous swelling.[18]

Diagnosis of Laboratory Exposures

Diagnosis of SE intoxication is based on clinical and epidemiologic features. Gastric intoxication with SE begins rapidly after exposure (1-4 h) and is characterized by severe vomiting, sometimes accompanied by diarrhea, but without a high fever. At higher exposure levels, intoxication progresses to hypovolemia, dehydration, vasodilatation in the kidneys, and lethal shock.[11]

While fever is uncommon after oral ingestion, inhalation of SE causes a marked fever and respiratory distress. Inhalation of SEB causes a severe, incapacitating illness of rapid onset (3-4 h) lasting 3 to 4 days characterized by high fever, headache, and a nonproductive cough; swallowing small amounts of SE during an inhalation exposure may result in gastric symptoms as well.[19]

Differential diagnosis of SE inhalation may be unclear initially because the symptoms are similar to those caused by several respiratory pathogens such as influenza, adenovirus, and mycoplasma. Naturally occurring pneumonias or influenza, however, would typically involve patients presenting over a more prolonged interval of time, whereas SE intoxication tends to plateau rapidly, within a few hours. Nonspecific laboratory findings of SE inhalation include a neutrophilic leukocytosis, an elevated erythrocyte sedimentation rate, and chest X-ray abnormalities consistent with pulmonary edema.[19]

Laboratory confirmation of intoxication includes SE detection by immunoassay of environmental and clinical samples, and gene amplification to detect staphylococcal genes in environmental samples. SE may be undetectable in the serum at the time symptoms occur; nevertheless, a serum specimen should be drawn as early as possible after exposure. Data from animal studies suggest the presence of SE in the serum or urine is transient. Respiratory secretions and nasal swabs may demonstrate the toxin early (within 24 h of inhalation exposure). Evaluation of neutralizing antibody titers in acute and convalescent sera of exposed individuals can be undertaken, but may yield false positives resulting from pre-existing antibodies produced in response to natural SE exposure.

Laboratory Safety and Containment Recommendations

General considerations for the safe use and inactivation of toxins of biological origin are found in Appendix I. Accidental ingestion, parenteral inoculation, and droplet or aerosol exposure of mucous membranes are believed to be the primary hazards of SE for laboratory and animal-care personnel. SE are relatively stable, monomeric proteins, readily soluble in water, and resistant to proteolytic degradation and temperature fluctuations. The physical/chemical stability of SE suggests that additional care must be taken by laboratory workers to avoid exposure to residual toxin that may persist in the environment.

Active SE toxins may be present in clinical samples, lesion fluids, respiratory secretions, or tissues of exposed animals. Additional care should be taken during necropsy of exposed animals or in handling clinical stool samples because SE toxins retain toxic activity throughout the digestive tract.

Accidental laboratory exposures to SE serotype B have been reviewed.[18] Documented accidents included inhalation of SE aerosols generated from pressurized equipment failure, as well as re-aerosolization of residual toxin from the fur of exposed animals. The most common cause of laboratory intoxication with SE is expected to result from accidental self-exposure via the mucous membranes by touching contaminated hands to the face or eyes.

BSL-2 practices and containment equipment and facilities should be used when handling SE or potentially contaminated material. Because SE is highly active by the oral or ocular exposure route, the use of a laboratory coat, gloves and safety glasses is mandatory when handling toxin or toxin-contaminated solutions. Frequent and careful hand-washing and laboratory decontamination should be strictly enforced when working with SE. Depending upon a risk assessment of the laboratory operation, the use of a disposable face mask may be required to avoid accidental ingestion.

BSL-3 facilities, equipment, and practices are indicated for activities with a high potential for aerosol or droplet production and those involving the use of large quantities of SE.

Special Issues

Vaccines No approved vaccine or specific antidote is currently available for human use, but experimental, recombinant vaccines are under development.

Select Agent SE is a select agent requiring registration with CDC and/or USDA for possession, use, storage and/or transfer. See Appendix F for additional information.

Transfer of Agent Importation of this agent may require CDC and/or USDA importation permits. Domestic transport of this agent may require a permit from USDA/APHIS/VS. A DoC permit may be required for the export of this agent to another country. See Appendix C for additional information.

Ricin Toxin

Ricin is produced in maturing seeds of the castor bean, *Ricinus communis,* which has been recognized for centuries as a highly poisonous plant for humans and livestock.[20] Ricin belongs to a family of ribosome inactivating proteins from plants, including abrin, modeccin, and viscumin, that share a similar overall structure and mechanism of action.[21] The ricin holotoxin is a disulfide-bonded heterodimer composed of an A-chain (approximately 34 kD polypeptide) and a B-chain (approximately 32 kD). The A-chain is an N-glycosidase enzyme and a potent inhibitor of protein synthesis, whereas the B-chain is a relatively non-toxic lectin that facilitates toxin binding and internalization to target cells.[20]

Ricin is much less toxic by weight than is BoNT or SE, and published case reports suggest that intramuscular or gastric ingestion of ricin is rarely fatal in adults.[22] Animal studies and human poisonings suggest that the effects of ricin depend upon the route of exposure, with inhalation and intravenous exposure being the most toxic. In laboratory mice, for example, the LD_{50} by intravenous injection is about 5 µg/kg, whereas it is 20 mg/kg by intragasteric route.[23,24] The ricin aerosol LD_{50} for NHP is estimated to be 10-15 µg/kg.[17] The human lethal dose has not been established rigorously, but may be as low as 1-5 mg of ricin by injection 25 or by the aerosol route (extropolation from two species of NGP).

Diagnosis of Laboratory Exposures

The primary diagnosis is through clinical manifestations that vary greatly depending upon the route of exposure. Following inhalation exposure of NHP, there is typically a latency period of 24-72 h that may be characterized by loss of appetite and listlessness. The latency period progresses rapidly to severe pulmonary distress, depending upon the exposure level. Most of the pathology occurs in the lung and upper respiratory tract, including inflammation, bloody sputum, and pulmonary edema. Toxicity from ricin inhalation would be expected to progress despite treatment with antibiotics, as opposed to an infectious process. There would be no mediastinitis as seen with inhalation anthrax. Ricin patients would not be expected to plateau clinically as occurs after inhalation of SEB.

Gastric ingestion of ricin causes nausea, vomiting, diarrhea, abdominal cramps and dehydration. Initial symptoms may appear more rapidly following gastric ingestion (1-5 h), but generally require exposure to much higher levels of toxin compared with the inhalation route. Following intramuscular injection of ricin, symptoms may persist for days and include nausea, vomiting, anorexia, and high fever. The site of ricin injection typically shows signs of inflammation with marked swelling and induration. One case of poisoning by ricin injection resulted in fever, vomiting, irregular blood pressure, and death by vascular collapse after a period of several days; it is unclear in this case if the toxin was deposited intramuscularly or in the bloodstream.[25]

Specific immunoassay of serum and respiratory secretions or immunohistochemical stains of tissue may be used where available to confirm a diagnosis. Ricin is an extremely immunogenic toxin, and paired acute and convalescent sera should be obtained from survivors for measurement of antibody response. Polymerase chain reaction (PCR) can detect residual castor bean DNA in most ricin preparations. Additional supportive clinical or diagnostic features, after aerosol exposure to ricin, may include the following: bilateral infiltrates on chest radiographs, arterial hypoxemia, neutrophilic leukocytosis, and a bronchial aspirate rich in protein.[24]

Laboratory Safety and Containment Recommendations

General considerations for the safe use and inactivation of toxins of biological origin are found in Appendix I. Precautions should be extended to handling potentially contaminated clinical, diagnostic and post-mortem samples because ricin may retain toxicity in the lesion fluids, respiratory secretions, or unfixed tissues of exposed animals.

When the ricin A-chain is separated from the B-chain and administered parenterally to animals, its toxicity is diminished by >1,000-fold compared with ricin holotoxin.[26] However, purified preparations of natural ricin A-chain or B-chain, as well as crude extracts from castor beans, should be handled as if contaminated by ricin until proven otherwise by bioassay.

BSL-2 practices, containment equipment and facilities are recommended, especially a laboratory coat, gloves, and respiratory protection, when handling ricin toxin or potentially contaminated materials.

Ricin is a relatively non-specific cytotoxin and irritant that should be handled in the laboratory as a non-volatile toxic chemical. A BSC (Class II, Type B1 or B2) or a chemical fume hood equipped with an exhaust HEPA filter and charcoal filter are indicated for activities with a high potential for aerosol, such as powder samples, and the use of large quantities of toxin. Laboratory coat, gloves, and full-face respirator should be worn if there is a potential for creating a toxin aerosol.

Special Issues

Vaccines No approved vaccine or specific antidote is currently available for human use, but experimental, recombinant vaccines are under development.

Select Agent Ricin toxin is a select agent requiring registration with CDC and/or USDA for possession, use, storage and/or transfer. See Appendix F for additional information.

Transfer of Agent Importation of this agent may require CDC and/or USDA importation permits. Domestic transport of this agent may require a permit from USDA/APHIS/VS. A DoC permit may be required for the export of this agent to another country. See Appendix C for additional information.

Selected Low Molecular Weight (LMW) Toxins

LMW toxins comprise a structurally and functionally diverse class of natural poisons, ranging in size from several hundred to a few thousand daltons, that includes complex organic structures, as well as disulfide cross-linked and cyclic polypeptides. Tremendous structural diversity may occur within a particular type of LMW toxin, often resulting in incomplete toxicological or pharmacological characterization of minor isoforms. Grouping LMW toxins together has primarily been a means of distinguishing them from protein toxins with respect to key biophysical characteristics. Compared with proteins, the LMW toxins are of smaller size, which alters their filtration and biodistribution properties, are generally more stable and persistent in the environment, and may exhibit poor water-solubility necessitating the use of organic solvent; these characteristics pose special challenges for safe handling, containment, and decontamination of LMW toxins within the laboratory.

The set of LMW toxins selected for discussion herein are employed routinely as laboratory reagents, and/or have been designated as potential public health threats by the CDC, including: T-2 mycotoxin produced by *Fusarium* fungi;[27,28] saxitoxin and related paralytic shellfish poisons produced by dinoflagellates of the *Gonyaulax* family;[29] tetrodotoxin from a number of marine animals,[30] brevetoxin from the dinoflagellate *Ptychodiscus brevis*;[31] palytoxin from marine coelenterates belonging to the genus *Palythoa*,[32] polypeptide conotoxins α-GI (includes GIA) and α-MI from the *Conus* genus of gastropod mollusks;[33] and the monocyclic polypeptide, microcystin-LR from freshwater cyanobacteria *Microcystis aeruginosa*.[34]

Trichothecene mycotoxins comprise a broad class of structurally complex, non-volatile sesquiterpene compounds that are potent inhibitors of protein synthesis.[27,28] Mycotoxin exposure occurs by consumption of moldy grains, and at least one of these toxins, designated "T-2," has been implicated as a potential biological warfare agent.[27] T-2 is a lipid-soluble molecule that can be absorbed into the body rapidly through exposed mucosal surfaces.[35] Toxic effects are most pronounced in metabolically active target organs and include emesis, diarrhea, weight loss, nervous disorder, cardiovascular alterations, immunodepression, hemostatic derangement, bone marrow damage, skin toxicity, decreased reproductive capacity, and death.[27] The LD_{50} for T-2 in laboratory animals ranges from 0.2 to 10 mg/kg, depending on the route of exposure, with aerosol toxicity estimated to be 20 to 50 times greater than parenteral exposure.[17,27] Of special note, T-2 is a potent vesicant capable of directly damaging skin or corneas. Skin lesions, including frank blisters, have been observed in animals with local, topical application of 50 to 100 ng of toxin.[27,35]

Saxitoxin and tetrodotoxin are paralytic marine toxins that interfere with normal function of the sodium channel in excitable cells of heart, muscle and neuronal tissue.[36] Animals exposed to 1-10 μg/kg toxin by parenteral routes typically develop a rapid onset of excitability, muscle spasm, and respiratory distress; death may occur within 10-15 minutes from respiratory paralysis.[29,37] Humans ingesting seafood contaminated with saxitoxin or tetrodotoxin show similar signs of toxicity, typically preceded by paresthesias of the lips, face and extremities.[36,38]

Brevetoxins are cyclic-polyether, paralytic shellfish neurotoxins produced by marine dinoflagellates that accumulate in filter-feeding mollusks and may cause human intoxication from ingestion of contaminated seafood, or by irritation from sea spray containing the toxin.[36] The toxin depolarizes and opens voltage-gated sodium ion channels, effectively making the sodium channel of affected nerve or muscle cells hyper-excitable. Symptoms of human ingestion are expected to include paresthesias of the face, throat and fingers or toes, followed by dizziness, chills, muscle pains, nausea, gastroenteritis, and reduced heart rate. Brevetoxin has a parenteral LD_{50} of 200 µg/kg in mice and guinea pigs.[31] Guinea pigs exposed to a slow infusion of brevetoxin develop fatal respiratory failure within 30 minutes of exposure to 20 µg/kg toxin.[37]

Palytoxin is a structurally complex, articulated fatty acid associated with soft coral *Palythoa vestitus* that is capable of binding and converting the essential cellular Na+/K+ pump into a non-selective cation channel.[32,39] Palytoxin is among the most potent coronary vasoconstrictors known, killing animals within minutes by cutting off oxygen to the myocardium.[40] The LD_{50} for intravenous administration ranges from 0.025 to 0.45 µg/kg in different species of laboratory animals.[40] Palytoxin is lethal by several parenteral routes, but is about 200-fold less toxic if administered to the alimentary tract (oral or rectal) compared with intravenous administration.[40] Palytoxin disrupts normal corneal function and causes irreversible blindness at topically applied levels of approximately 400 ng/kg, despite extensive rinsing after ocular instillation.[40]

Conotoxins are polypeptides, typically 10-30 amino acids long and stabilized by distinct patterns of disulfide bonds, that have been isolated from the toxic venom of marine snails and shown to be neurologically active or toxic in mammals.[33] Of the estimated >105 different polypeptides (conopeptides) present in venom of over 500 known species of *Conus*, only a few have been rigorously tested for animal toxicity. Of the isolated conotoxin subtypes that have been analyzed, at least two post-synaptic paralytic toxins, designated α-GI (includes GIA) and α-MI, have been reported to be toxic in laboratory mice with LD_{50} values in the range of 10-100 µg/kg depending upon the species and route of exposure.

Workers should be aware, however, that human toxicity of whole or partially fractionated *Conus* venom, as well as synthetic combinations of isolated conotoxins, may exceed that of individual components. For example, untreated cases of human poisoning with venom of C. *geographus* result in an approximately 70% fatality rate, probably as a result of the presence of mixtures of various α- and µ-conotoxins with common or synergistic biological targets.[33,41] The α-conotoxins act as potent nicotinic antagonists and the µ-conotoxins block the sodium channel.[33] Symptoms of envenomation depend upon the *Conus* species involved, generally occur rapidly after exposure (minutes), and range from severe pain to spreading numbness.[42] Severe intoxication results in muscle paralysis, blurred or double vision, difficulty breathing and swallowing, and respiratory or cardiovascular collapse.[42]

Microcystins (also called cyanoginosins) are monocyclic heptapeptides composed of specific combinations of L-, and D-amino acids, some with uncommon side chain structures, that are produced by various freshwater cyanobacteria.[43] The toxins are potent inhibitors of liver protein phosphatase type 1 and are capable of causing massive hepatic hemorrhage and death.[43]

One of the more potent toxins in this family, microcystin-LR, has a parenteral LD_{50} of 30 to 200 µg/kg in rodents.[34] Exposure to microcystin-LR causes animals to become listless and prone in the cage; death occurs in 16 to 24 h. The toxic effects of microcystin vary depending upon the route of exposure and may include hypotension and cardiogenic shock, in addition to hepatotoxicity.[34,44]

Diagnosis of Laboratory Exposures

LMW toxins are a diverse set of molecules with a correspondingly wide range of signs and symptoms of laboratory exposure, as discussed above for each toxin. Common symptoms can be expected for LMW toxins with common mechanisms of action. For example, several paralytic marine toxins that interfere with normal sodium channel function cause rapid paresthesias of the lips, face and digits after ingestion. The rapid onset of illness or injury (minutes to hours) generally supports a diagnosis of chemical or LMW toxin exposure. Painful skin lesions may occur almost immediately after contact with T-2 mycotoxin, and ocular irritation or lesions will occur in minutes to hours after contact with T-2 or palytoxin.

Specific diagnosis of LMW toxins in the form of a rapid diagnostic test is not presently available in the field. Serum and urine should be collected for testing at specialized reference laboratories by methods including antigen detection, receptor-binding assays, or liquid chromatographic analyses of metabolites. Metabolites of several marine toxins, including saxitoxin, tetrodotoxin, and brevetoxins, are well-studied as part of routine regulation of food supplies.[36] Likewise, T-2 mycotoxin absorption and biodistribution has been studied, and its metabolites can be detected as late as 28 days after exposure.[27] Pathologic specimens include blood, urine, lung, liver, and stomach contents. Environmental and clinical samples can be tested using a gas liquid chromatography-mass spectrometry technique.

Laboratory Safety and Containment Recommendations

General considerations for the safe use and inactivation of toxins of biological origin are found in Appendix I. Ingestion, parenteral inoculation, skin and eye contamination, and droplet or aerosol exposure of mucous membranes are the primary hazards to laboratory and animal care personnel. LMW toxins also can contaminate food sources or small-volume water supplies. Additionally, the T-2 mycotoxin is a potent vesicant and requires additional safety precautions to prevent contact with exposed skin or eyes. Palytoxin also is highly toxic by the ocular route of exposure.

In addition to their high toxicity, the physical/chemical stability of the LMW toxins contribute to the risks involved in handling them in the laboratory environment. Unlike many protein toxins, the LMW toxins can contaminate surfaces as a stable, dry film that may pose an essentially indefinite contact threat to laboratory workers. Special emphasis, therefore, must be placed upon proper decontamination of work surfaces and equipment.[45]

When handling LMW toxins or potentially contaminated material, BSL-2 practices, containment, equipment and facilities are recommended, especially the wearing of a laboratory coat, safety glasses and disposable gloves; the gloves must be impervious to organic solvents or other diluents employed with the toxin.

A BSC (Class II, Type B1 or B2) or a chemical fume hood equipped with exhaust HEPA filters and a charcoal filter are indicated for activities with a high potential for aerosol, such as powder samples, and the use of large quantities of toxin. Laboratory coat and gloves should be worn if potential skin contact exists. The use of respiratory protection should be considered if potential aerosolization of toxin exists.

For LMW toxins that are not easily decontaminated with bleach solutions, it is recommended to use pre-positioned, disposable liners for laboratory bench surfaces to facilitate clean up and decontamination.

Special Issues

Vaccines No approved vaccines are currently available for human use. Experimental therapeutics for LMW toxins have been reviewed.[46]

Select Agent Some LMW toxins are a select agent requiring registration with CDC and/or USDA for possession, use, storage and/or transfer. See Appendix F for additional information.

Transfer of Agent Importation of this agent may require CDC and/or USDA importation permits. Domestic transport of this agent may require a permit from USDA/APHIS/VS. A DoC permit may be required for the export of this agent to another country. See Appendix C for additional information.

References

1. Simpson LL. Identification of the major steps in botulinum toxin action. Annu Rev Pharmacol Toxicol. 2004;44:167-93.
2. Gangarosa EJ, Donadio JA, Armstrong RW, et al. Botulism in the United States, 1899-1969. Am J Epidemiol. 1971;93:93-101.
3. Hatheway C. Botulism. In: Balows A, Hausler W, Ohashi M, et al., editors. Laboratory diagnosis of infectious diseases: principles and practice. Vol. 1. New York: Springer-Verlag; 1988. p. 111-33.
4. Shapiro RL, Hatheway C, Swerdlow DL. Botulism in the United States: a clinical and epidemiologic review. Ann Intern Med. 1998;129:221-8.

5. Woolford A, Schantz EJ, Woodburn M. Heat inactivation of botulinum toxin type A in some convenience foods after frozen storage. J Food Sci. 1978;43:622-4.
6. Holzer E. Botulismus durch inhalation. Med Klin. 1962;57:1735-8.
7. Franz DR, Pitt LM, Clayton MA, et al. Efficacy of prophylactic and therapeutic administration of antitoxin for inhalation botulism. In: DasGupta BR, editor. Botulinum and tetanus neurotoxins: neurotransmission and biomedical aspects. New York: Plenum Press; 1993. p. 473-6.
8. Llewelyn M, Cohen J. Superantigens: microbial agents that corrupt immunity. Lancet Infect Dis. 2002;2:156-62.
9. Jarraud S, Peyrat MA, Lim A, et al. egc, a highly prevalent operon of enterotoxin gene, forms a putative nursery of superantigens in *Staphylococcus aureus*. J Immunol. 2001;166:669-77. Erratum in: J Immunol. 2001;166:following 4259.
10. Marrack P, Kappler J. The *Staphylococcal enterotoxins* and their relatives. Science. 1990;248:705-11. Erratum in: Science. 1990;248:1066.
11. Jett M, Ionin B, Das R, et al. The *Staphylococcal enterotoxins*. In: Sussman M, editor. Molecular medical microbiology. Vol. 2. San Diego: Academic Press; 2001. p. 1089-1116.
12. Bergdoll MS. Enterotoxins. In: Montie TC, Kadis S, Ajl SJ, editors. Microbial toxins: bacterial protein toxins. Vol. 3. New York: Academic Press;1970. p. 265-326.
13. Evenson ML, Hinds MW, Bernstein RS, et al. Estimation of human dose of staphylococcal enterotoxin A from a large outbreak of staphylococcal food poisoning involving chocolate milk. Int J Food Microbiol. 1988;7:311-16.
14. Asao T, Kumeda Y, Kawai T, et al. An extensive outbreak of staphylococcal food poisoning due to low-fat milk in Japan: estimation of enterotoxin A in the incriminated milk and powdered skim milk. Epidemiol Infect. 2003;130:33-40.
15. Do Carmo LS, Cummings C, Linardi VR, et al. A case study of a massive staphylococcal food poisoning incident. Foodborne Pathog Dis. 2004;1:241-46.
16. Raj HD, Bergdoll MS. Effect of enterotoxin B on human volunteers. J Bacteriol. 1969;98:833-34.
17. LeClaire RD, Pitt MLM. Biological weapons defense: effect levels. In: Lindler LE, Lebeda FJ, Korch GW, editors. Biological weapons defense: infectious diseases and counter bioterrorism. Totowa, New Jersey: Humana Press, Inc.; 2005. p. 41-61.
18. Rusnak JM, Kortepeter M, Ulrich R, et al. Laboratory exposures to *Staphylococcal enterotoxin* B. Emerg Infect Dis. 2004;10:1544-9.
19. Ulrich R, Sidell S, Taylor T, et al. Staphylococcal enterotoxin B and related pyrogenic toxins. In: Sidell FR, Takafuji ET, Franz DR, editors. Medical aspects of chemical and biological warfare. Vol. 6. Textbook of military medicine, part 1: warfare, weaponry, and the casualty. Washington, DC: Office of the Surgeon General at TMM Publications, Borden Institute, Walter Reed Army Medical Center; 1997. p. 621-30.

20. Olsnes S. The history of ricin, abrin and related toxins. Toxicon. 2004;44:361-70.
21. Hartley MR, Lord JM. Cytotoxic ribosome-inactivating lectins from plants. Biochim Biophys Acta. 2004;1701:1-14.
22. Doan LG. Ricin: mechanism of toxicity, clinical manifestations, and vaccine development. A review. J Toxicol Clin Toxicol. 2004;42:201-8.
23. Cumber AJ, Forrester JA, Foxwell BM, et al. Preparation of antibody-toxin conjugates. Methods Enzymol. 1985;112:207-25.
24. Franz D, Jaax N. Ricin toxin. In: Sidell FR, Takafuji ET, Franz DR, editors. Medical aspects of chemical and biological warfare. Vol. 6. Textbook of military medicine, part 1: warfare, weaponry, and the casualty. Washington, DC. Office of the Surgeon General at TMM Publications, Borden Institute, Walter Reed Army Medical Center; 1997;631-642.
25. Crompton R, Gall D. Georgi Markov-death in a pellet. Med Leg J. 1980;48:51-62.
26. Soler-Rodriguez AM, Uhr JW, Richardson J, et al. The toxicity of chemically deglycosylated ricin A-chain in mice. Int J Immunopharmacol. 1992;14:281-91.
27. Wannemacher R, Wiener SL. Trichothecene mycotoxins. In: Sidell FR, Takafuji ET, Franz DR, editors. Medical aspects of chemical and biological warfare. Vol. 6. Textbook of military medicine, part 1: warfare, weaponry, and the casualty. Washington, DC: Office of the Surgeon General at TMM Publications, Borden Institute, Walter Reed Army Medical Center; 1997. p. 655-76.
28. Bamburg J. Chemical and biochemical studies of the trichothecene mycotoxins. In: JV Rodricks, editor. Mycotoxins and other fungal related food problems. Vol. 149. Advances in chemistry. Washington, DC: American Chemical Society, Diviosion of Agricultural and Food Chemistry; 1976. p. 144-162.
29. Schantz EJ. Chemistry and biology of saxitoxin and related toxins. Ann N Y Acad Sci. 1986;479:15-23.
30. Yasumoto T, Nagai H, Yasumura D, et al. Interspecies distribution and possible origin of tetrodotoxin. Ann N Y Acad Sci 1986;479:44-51.
31. Baden DG, Mende TJ, Lichter W, et al. Crystallization and toxicology of T34: a major toxin from Florida's red tide organism (Ptychodiscus brevis). Toxicon. 1981;19:455-62.
32. Moore RE, Scheuer PJ. Palytoxin: a new marine toxin from a coelenterate. Science. 1971;172:495-8.
33. Olivera BM, Cruz LJ. Conotoxins, in retrospect. Toxicon. 2001;39:7-14.
34. Carmichael W. Algal toxins. In: Callow J, editor. Advances in botanical research. Vol. 12. London: Academic Press; 1986. p. 47-101.
35. Bunner B, Wannemacher RW Jr, Dinterman RE, et al. Cutaneous absorption and decontamination of [3H]T-2 toxin in the rat model. J Toxicol Environ Health. 1989;26:413-423.

36. Poli M. Foodborne Marine biotoxins. In: Miliotis MD, Bier JW, editors. International handbook of foodborne pathogens. New York: Marcel Dekker; 2003;445-58.

37. Franz DR, LeClaire RD. Respiratory effects of brevetoxin and saxitoxin in awake guinea pigs. Toxicon. 1989;27:647-54.

38. Kao CY. Tetrodotoxin, saxitoxin and their significance in the study of excitation phenomena. Pharmacol Rev. 1966;18:997-1049.

39. Artigas P, Gadsby DC. Na+/K+-pump ligands modulate gating of palytoxin-induced ion channels. Proc Natl Acad Sci USA. 2003;100:501-5.

40. Wiles JS, Vick JA, Christensen MK. Toxicological evaluation of palytoxin in several animal species. Toxicon. 1974;12:427-33.

41. Cruz LJ, White J. Clinical Toxicology of *Conus* Snail Stings. In: Meier J, White J, editors. Handbook of clinical toxicology of animal venoms and poisons. Boca Raton: CRC Press; 1995. p. 117-28.

42. McIntosh JM, Jones RM. Cone venom-from accidental stings to deliberate injection. Toxicon. 2001;39:1447-51.

43. Dawson RM. The toxicology of microcystins. Toxicon. 1998;36:953-62.

44. LeClaire RD, Parker GW, Franz DR. Hemodynamic and calorimetric changes induced by microcystin-LR in the rat. J Appl Toxicol. 1995;15:303-11.

45. Wannemacher RW. Procedures for inactivation and safety containment of toxins. In: Proceedings for the symposium on agents of biological origin. 1989; Aberdeen Proving Ground, MD. Aberdeen, Maryland: U.S. Army Chemical Research, Development and Engineering Center; 1989. p. 115-22.

46. Paddle BM. Therapy and prophylaxis of inhaled biological toxins. J Appl Toxicol. 2003;23:139-70.

Section VIII-H: Prion Diseases

Transmissible spongiform encephalopathies (TSE) or prion diseases are neurodegenerative diseases which affect humans and a variety of domestic and wild animal species (Tables 7 and 8).[1,2] A central biochemical feature of prion diseases is the conversion of normal prion protein (PrP) to an abnormal, misfolded, pathogenic isoform designated PrP^{Sc} (named for "scrapie," the prototypic prion disease). The infectious agents that transmit prion diseases are resistant to inactivation by heat and chemicals and thus require special biosafety precautions. Prion diseases are transmissible by inoculation or ingestion of infected tissues or homogenates, and infectivity is present at high levels in brain or other central nervous system tissues, and at slightly lower levels in lymphoid tissues including spleen, lymph nodes, gut, bone marrow, and blood. Although the biochemical nature of the infectious TSE agent, or prion, is not yet proven, the infectivity is strongly associated with the presence of PrP^{Sc}, suggesting that this material may be a major component of the infectious agent.

A chromosomal gene encodes PrP^{C} (the cellular isoform of PrP) and no PrP genes are found in purified preparations of prions. PrP^{Sc} is derived from PrP^{C} by a posttranslational process whereby PrP^{Sc} acquires a high *beta*-sheet content and a resistance to inactivation by normal disinfection processes. The PrPSc is less soluble in aqueous buffers and, when incubated with protease (proteinase K), the PrP^{C} is completely digested (sometimes indicated by the "sensitive" superscript, PrP^{sen}) while PrP^{Sc} is resistant to protease (PrP^{res}). Neither PrP-specific nucleic acids nor virus-like particles have been detected in purified, infectious preparations.

Occupational Infections

No occupational infections have been recorded from working with prions. No increased incidence of Creutzfeldt-Jakob disease (CJD) has been found amongst pathologists who encounter cases of the disease post-mortem.

Natural Modes of Infection

The recognized diseases caused by prions are listed under Table 7 (human diseases) and Table 8 (animal diseases). The only clear risk factor for disease transmission is the consumption of infected tissues such as human brain in the case of kuru, and meat including nervous tissue in the case of bovine spongiform encephalopathy and related diseases such as feline spongiform encephalopathy. It is also possible to acquire certain diseases such as familial CJD by inheritance through the germ line.

Most TSE agents, or prions, have a preference for infection of the homologous species, but cross-species infection with a reduced efficiency is also possible. After cross-species infection there is often a gradual adaptation of specificity for the new host; however, infectivity for the original host may also be propagated for several passages over a time-span of years. The process of cross-species adaptation can also vary among individuals in the same species and the rate of adaptation and the final species specificity is difficult to predict with accuracy. Such considerations help to form the basis for the biosafety classification of different prions.

Table 7. The Human Prion Diseases

Disease	Abbreviation	Mechanism of Pathogenesis
Kuru		Infection through ritualistic cannibalism
Creutzfeldt-Jakob disease	CJD	Unknown mechanism
Sporadic CJD	sCJD	Unknown mechanism; possibly somatic mutation or spontaneous conversion of PrP^c to PrP^{Sc}
Variant CJD	vCJD	Infection presumably from consumption of BSE-contaminated cattle products and secondary bloodborne transmission
Familial CJD	fCJD	Germline mutations in PrP gene
Iatrogenic CJD	iCJD	Infection from contaminated corneal and dural grafts, pituitary hormone, or neurosurgical equipment
Gerstmann-Sträussler-Scheinker syndrome	GSS	Germline mutations in PrP gene
Fatal familial insomnia	FFI	Germline mutations in PrP gene

Table 8. The Animal Prion Diseases

Disease	Abbreviation	Natural Host	Mechanism of Pathogenesis
Scrapie		Sheep, goats, mouflon	Infection in genetically susceptible sheep
Bovine spongiform encephalopathy	BSE	Cattle	Infection with prion-contaminated feedstuffs
Chronic wasting disease	CWD	Mule, deer, white-tailed deer, Rocky Mountain elk	Unknown mechanism; possibly from direct animal contact or indirectly from contaminated feed and water sources
Exotic ungulate encephalopathy	EUE	Nyala, greater kudu and oryx	Infection with BSE-contaminated feedstuffs
Feline spongiform encephalopathy	FSE	Domestic and wild cats in captivity	Infection with BSE-contaminated feedstuffs
Transmissible mink encephalopathy	TME	Mink (farm raised	Infection with prion-contaminated feedstuffs

Laboratory Safety and Containment Recommendations

In the laboratory setting prions from human tissue and human prions propagated in animals should be manipulated at BSL-2. BSE prions can likewise be manipulated at BSL-2. Due to the high probability that BSE prions have been transmitted to humans, certain circumstances may require the use of BSL-3 facilities and practices. All other animal prions are manipulated at BSL-2. However, when a prion from one species is inoculated into another the resultant infected animal should be treated according to the guidelines applying to the source of the inoculum. Contact APHIS National Center for Import and Export at (301) 734-5960 for specific guidance.

Although the exact mechanism of spread of scrapie among sheep and goats developing natural scrapie is unknown, there is considerable evidence that one of the primary sources is oral inoculation with placental membranes from infected ewes. There has been no evidence for transmission of scrapie to humans, even though the disease was recognized in sheep for over 200 years. The diseases TME, BSE, FSE, and EUE are all thought to occur after the consumption of prion-infected foods.[1,2] The exact mechanism of CWD spread among mule deer, white-tailed deer and Rocky Mountain elk is unknown. There is strong evidence that CWD is laterally transmitted and environmental contamination may play an important role in local maintenance of the disease.[2]

In the care of patients diagnosed with human prion disease, Standard Precautions are adequate. However, the human prion diseases in this setting are not communicable or contagious.[3] There is no evidence of contact or aerosol transmission of prions from one human to another. However, they are infectious under some circumstances, such as ritualistic cannibalism in New Guinea causing kuru, the administration of prion-contaminated growth hormone causing iatrogenic CJD, and the transplantation of prion-contaminated dura mater and corneal grafts. It is highly suspected that variant CJD can also be transmitted by blood transfusion.[4] However, there is no evidence for bloodborne transmission of non-variant forms of CJD. Familial CJD, GSS, and FFI are all dominantly inherited prion diseases; many different mutations of the PrP gene have been shown to be genetically linked to the development of inherited prion disease. Prions from many cases of inherited prion disease have been transmitted to apes, monkeys, and mice, especially those carrying human PrP transgenes.

Special Issues

Inactivation of Prions Prions are characterized by resistance to conventional inactivation procedures including irradiation, boiling, dry heat, and chemicals (formalin, betapropiolactone, alcohols). While prion infectivity in purified samples is diminished by prolonged digestion with proteases, results from boiling in sodium dodecyl sulfate and urea are variable. Likewise, denaturing organic solvents such as phenol or chaotropic reagents such as guanidine isothiocyanate have also resulted in greatly reduced but not complete inactivation. The use of conventional autoclaves as the sole treatment has not resulted in complete inactivation of prions.[5] Formalin-fixed and paraffin-embedded tissues, especially of the brain, remain infectious. Some investigators recommend that formalin-fixed tissues from suspected cases of prion disease be immersed for 30 min in 96% formic acid or phenol before histopathologic processing (Table 9), but such treatment may severely distort the microscopic neuropathology.

The safest and most unambiguous method for ensuring that there is no risk of residual infectivity on contaminated instruments and other materials is to discard and destroy them by incineration.[6] Current recommendations for inactivation of prions on instruments and other materials are based on the use of sodium hypochlorite, NaOH, Environ LpH and the moist heat of autoclaving with combinations of heat and chemical being most effective (Table 9).[5,6]

Surgical Procedures Precautions for surgical procedures on patients diagnosed with prion disease are outlined in an infection control guideline for transmissible spongiform encephalopathies developed by a consultation convened by the WHO in 1999.[6] Sterilization of reusable surgical instruments and decontamination of surfaces should be performed in accordance with recommendations described by the CDC (*www.cdc.gov*) and the WHO infection control guidelines.[6] Table 9 summarizes the key recommendations for decontamination of reusable instruments and surfaces. Contaminated disposable instruments or materials should be incinerated at 1000° C or greater.[7]

Autopsies Routine autopsies and the processing of small amounts of formalin-fixed tissues containing human prions can safely be done using Standard Precautions.[8] The absence of any known effective treatment for prion disease demands caution. The highest concentrations of prions are in the central nervous system and its coverings. Based on animal studies, it is likely that prions are also found in spleen, thymus, lymph nodes, and intestine. The main precaution to be taken by laboratorians working with prion-infected or contaminated material is to avoid accidental puncture of the skin.[3] Persons handling contaminated specimens should wear cut-resistant gloves if possible. If accidental contamination of unbroken skin occurs, the area should be washed with detergent and abundant quantities of warm water (avoid scrubbing); brief exposure (1 minute to 1N NaOH or a 1:10 dilution of bleach) can be considered for maximum safety.[6] Additional guidance related to occupational injury are provided in the WHO infection control guidelines.[6] Unfixed samples of brain, spinal cord, and other tissues containing human prions should be processed with extreme care in a BSL-2 facility utilizing BSL-3 practices.

Bovine Spongiform Encephalopathy Although the eventual total number of variant CJD cases resulting from BSE transmission to humans is unknown, a review of the epidemiological data from the United Kingdom indicates that BSE transmission to humans is not efficient.[9] The most prudent approach is to study BSE prions at a minimum in a BSL-2 facility utilizing BSL-3 practices. When performing necropsies on large animals where there is an opportunity that the worker may be accidentally splashed or have contact with high-risk materials (e.g., spinal column, brain) personnel should wear full body coverage personal protective equipment (e.g., gloves, rear closing gown and face shield). Disposable plasticware, which can be discarded as a dry regulated medical waste, is highly recommended. Because the paraformaldehyde vaporization procedure does not diminish prion titers, BSCs must be decontaminated with 1N NaOH and rinsed with water. HEPA filters should be bagged out and incinerated. Although there is no evidence to suggest that aerosol transmission occurs in the natural disease, it is prudent to avoid the generation of aerosols or droplets during the manipulation of tissues or fluids and during the necropsy of experimental animals. It is further strongly recommended that impervious gloves be worn for activities that provide the opportunity for skin contact with infectious tissues and fluids.

Animal carcasses and other tissue waste can be disposed by incineration with a minimum secondary temperature of 1000°C (1832°F).[6] Pathological incinerators should maintain a primary chamber temperature in compliance with design and applicable state regulations, and employ good combustion practices. Medical waste incinerators should comply with applicable state and federal regulations.

The alkaline hydrolysis process, using a pressurized vessel that exposes the carcass or tissues to 1 N NaOH or KOH heated to 150°C, can be used as an alternative to incineration for the disposal of carcasses and tissue.[5,10] The process has been shown to completely inactive TSEs (301v agent used) when used for the recommended period.

Table 9. Tissue Preparation for Human CJD and Related Diseases

1. Histology technicians wear gloves, apron, laboratory coat, and face protection.

2. Adequate fixation of small tissue samples (e.g., biopsies) from a patient with suspected prion disease can be followed by post-fixation in 96% absolute formic acid for 30 minutes, followed by 45 hours in fresh 10% formalin.

3. Liquid waste is collected in a 4L waste bottle initially containing 600 ml 6N NaOH.

4. Gloves, embedding molds, and all handling materials are disposed s regulated medical waste.

5. Tissue cassettes are processed manually to prevent contamination of tissue processors.

6. Tissues are embedded in a disposable embedding mold. If used, forceps are decontaminated as in Table 10.

7. In preparing sections, gloves are worn, section waste is collected and disposed in a regulated medical waste receptacle. The knife stage is wiped with 2N NaOH, and the knife used is discarded immediately in a "regulated medical waste sharps" receptacle. Slides are labeled with "CJD Precautions." The sectioned block is sealed with paraffin.

8. Routine staining:
   a. slides are processed by hand;
   b. reagents are prepared in 100 ml disposable specimen cups;
   c. after placing the cover slip on, slides are decontaminated by soaking them for 1 hour in 2N NaOH;
   d. slides are labeled as "Infectious-CJD."

9. Other suggestions:
   a. disposable specimen cups or slide mailers may be used for reagents;
   b. slides for immunocytochemistry may be processed in disposable Petri dishes;
   c. equipment is decontaminated as described above or disposed as regulated medical waste.

Handling and processing of tissues from patients with suspected prion disease The special characteristics of work with prions require particular attention to the facilities, equipment, policies, and procedures involved.[10] The related considerations outlined in Table 9 should be incorporated into the laboratory's risk management for this work.

Table 10. Prion Inactivation Methods for Reusable Instruments and Surfaces

1. Immerse in 1 N NaOH, heat in a gravity displacement autoclave at 121°C for 30 minutes. Clean and sterilize by conventional means.

2. Immerse in 1 N NaOH or sodium hypochlorite (20,000 ppm) for 1 hours. Transfer into water and autoclave (gravity displacement) at 121°C for 1 hour. Clean and sterilize by conventional means.

3. Immerse in 1N NaOH or sodium hypochlorite (20,000) for 1 hour. Rinse instruments with water, transfer to open pan and autoclave at 121°C (gravity displacement) or 134°C (porous load) for 1 hour. Clean and sterilize by conventional means.

4. Surfaces or heat-sensitive instruments can be treated with 2N NaOH or sodium hypochlorite (20,000 ppm) for 1 hour. Ensure surfaces remain wet for entire period, then rinse well with water. Before chemical treatment, it is strongly recommended that gross contamination of surfaces be reduced because the presence of excess organic material will reduce the strength of either NaOH or sodium hypochlorite solutions.

5. Environ LpH (EPA Reg. No. 1043-118) may be used on washable, hard, non-porous surfaces (such as floors, tables, equipment, and counters), items (such as non-disposable instruments, sharps, and sharp containers), and/or laboratory waste solutions (such as formalin or other liquids). This product is currently being used under FIFRA Section 18 exemptions in a number of states. Users should consult with the state environmental protection office prior to use.

(Adapted from *www.cdc.gov* [11,12])

Working Solutions 1 N NaOH equals 40 grams of NaOH per liter of water. Solution should be prepared daily. A stock solution of 10 N NaOH can be prepared and fresh 1:10 dilutions (1 part 10 N NaOH plus 9 parts water) used daily.

20,000 ppm sodium hypochlorite equals a 2% solution. Most commercial household bleach contains 5.25% sodium hypochlorite, therefore, make a 1:2.5 dilution (1 part 5.25% bleach plus 1.5 parts water) to produce a 20,000 ppm solution. This ratio can also be stated as two parts 5.25% bleach to three parts water. Working solutions should be prepared daily.

CAUTION: Above solutions are corrosive and require suitable personal protective equipment and proper secondary containment. These strong corrosive solutions require careful disposal in accordance with local regulations.

Precautions in using NaOH or sodium hypochlorite solutions in autoclaves: NaOH spills or gas may damage the autoclave if proper containers are not used. The use of containers with a rim and lid designed for condensation to collect and drip back into the pan is recommended. Persons who use this procedure should be cautious in handling hot NaOH solution (post-autoclave) and in avoiding potential exposure to gaseous NaOH; exercise caution during all sterilization steps; and allow the autoclave, instruments, and solutions to cool down before removal. Immersion in sodium hypochlorite bleach can cause severe damage to some instruments.

References

1. Prusiner SB. Prion diseases and the BSE crisis. Science. 1997;278:245-51.
2. Williams ES, Miller MW. Transmissible spongiform encephalopathies in non-domestic animals: origin, transmission and risk factors. Rev Sci Tech Off Int Epiz. 2003;22:145-56.
3. Ridley RM, Baker HF. Occupational risk of Creutzfeldt-Jakob disease. Lancet. 1993;341:641-2.
4. Llewelyn CA, Hewitt PE, Knight RS, et al. Possible transmission of variant Creutzfeldt-Jakob disease by blood transfusion. Lancet. 2004;7;363:417-21.
5. Taylor DM, Woodgate SL. Rendering practices and inactivation of transmissible spongiform encephalopathy agents. Rev Sci Tech Off Int Epiz. 2003;22:297-310.
6. World Health Organization. [*http://www.who.int/en/*]. Geneva (Switzerland): The Organization; [updated 2006 Sept 21; cited 2006 Sept 21]. 2000. WHO Infection Control Guidelines for Transmissible Spongiform Encephalopathies. Report of a WHO Consultation, Geneva, Switzerland, 23-26 March 1999. Available from: *http://www.who.int/csr/resources/publications/bse/WHO_CDS_CSR_APH_2000_3/en/*
7. Brown P, Rau EH, Johnson BK, Bacote AE, Gibbs CJ, Jr, Gajdusek DC. New studies on the heat resistance of hamster-adapted scrapie agent: threshold survival after ashing at 600 degrees C suggests an inorganic template of replication. Proc Natl Acad Sci USA. 2000;97:3418–3421. doi: 10.1073/pnas.050566797.
8. Ironside JW and JE Bell. The 'high-risk' neuropathological autopsy in AIDS and Creutzfeldt-Jakob disease: principles and practice. Neuropathol Appl Neurobiol. 1996;22:388-393.
9. Hilton DA. Pathogenesis and prevalence of variant Creutzfeldt-Jakob disease. Pathol J. 2006;208:134-41.
10. Richmond JY, Hill RH, Weyant RS, et al. What's hot in animal biosafety? ILAR J. 2003;44:20-7.
11. Ernst DR, Race RE. Comparative analysis of scrapie agent inactivation methods. J. Virol. Methods. 1994; 41:193-202.
12. Race RE, Raymond GJ. Inactivation of transmissible spongiform encephalopathy (prion) agents by Environ LpH. J Virol. 2004;78:2164-5.

Appendix A – Primary Containment for Biohazards: Selection, Installation and Use of Biological Safety Cabinets

Section I—Introduction

This document presents information on the design, selection, function and use of Biological Safety Cabinets (BSCs), which are the primary means of containment developed for working safely with infectious microorganisms. Brief descriptions of the facility and engineering concepts for the conduct of microbiological research are also provided. BSCs are only one part of an overall biosafety program, which requires consistent use of good microbiological practices, use of primary containment equipment and proper containment facility design. Detailed descriptions of acceptable work practices, procedures and facilities, known as Biosafety Levels 1 through 4, are presented in the CDC/NIH publication Biosafety in Microbiological and Biomedical Laboratories (BMBL).[1]

BSCs are designed to provide personnel, environmental and product protection when appropriate practices and procedures are followed. Three kinds of biological safety cabinets, designated as Class I, II and III, have been developed to meet varying research and clinical needs.

Most BSCs use high efficiency particulate air (HEPA) filters in the exhaust and supply systems. The exception is a Class I BSC, which does not have HEPA filtered supply air. These filters and their use in BSCs are briefly described in Section II. Section III presents a general description of the special features of BSCs that provide varying degrees of personnel, environmental, and product protection.

Laboratory hazards and risk assessment are discussed in Section IV. Section V presents work practices, procedures and practical tips to maximize information regarding the protection afforded by the most commonly used BSCs. Facility and engineering requirements needed for the operation of each type of BSC are presented in Section VI. Section VII reviews requirements for routine annual certification of cabinet operation and integrity.

These sections are not meant to be definitive or all encompassing. Rather, an overview is provided to clarify the expectations, functions and performance of these critical primary barriers. This document has been written for the biosafety officer, laboratorian, engineer or manager who desires a better understanding of each type of cabinet; factors considered for the selection of a BSC to meet specific operational needs; and the services required to maintain the operational integrity of the cabinet.

Proper maintenance of cabinets used for work at all biosafety levels cannot be over emphasized. Biosafety Officers (BSOs) should understand that an active cabinet is a primary containment device. A BSC must be routinely inspected and tested by training personnel, following strict protocols, to verify that it is working properly. This process is referred to as certification of the cabinet and should be performed annually.

Section II—The High Efficiency Particulate Air (HEPA) Filter and the Development of Biological Containment Devices

From the earliest laboratory-acquired typhoid infections to the hazards posed by bioterrorism, antibiotic-resistant bacteria and rapidly mutating viruses, threats to worker safety have stimulated the development and refinement of workstations in which infectious microorganisms could be safely handled. The needs to work with tissue cultures, maintain sterility of cell lines, and minimize cross-contamination have contributed to concerns regarding product integrity.

The use of proper procedures and equipment (as described in BMBL)[1] cannot be overemphasized in providing primary personnel and environmental protection. For example, high-speed blenders designed to reduce aerosol generation, needle-locking syringes, micro burners and safety centrifuge cups or sealed rotors are among the engineered devices that protect laboratory workers from biological hazards. An important piece of safety equipment is the biological safety cabinet in which manipulations of infectious microorganisms are performed.

Background

Early prototype clean air cubicles were designed to protect the materials being manipulated from environmental or worker-generated contamination rather than to protect the worker from the risks associated with the manipulation of potentially hazardous materials. Filtered air was blown across the work surface directly at the worker. Therefore, these cubicles could not be used for handling infectious agents because the worker was in a contaminated air stream.

To protect the worker during manipulations of infectious agents, a small workstation was needed that could be installed in existing laboratories with minimum modification to the room. The earliest designs for primary containment devices were essentially non-ventilated "boxes" built of wood and later of stainless steel, within which simple operations such as weighing materials could be accomplished.[2]

Early versions of ventilated cabinets did not have adequate or controlled directional air movement. They were characterized by mass airflow into the cabinets albeit with widely varying air volumes across openings. Mass airflow into cabinet drew "contaminated" air away from the laboratory worker. This was the forerunner of the Class I BSC. However, since the air was unfiltered, the cabinet was contaminated with environmental microorganisms and other undesirable particulate matter.

Control of airborne particulate materials became possible with the development of filters, which efficiently removed microscopic contaminants from the air. The HEPA filter was developed to create dust-free work environments (e.g., "clean rooms" and "clean benches") in the 1940s.[2]

HEPA filters remove the most penetrating particle size (MPPS) of 0.3 µm with an efficiency of at least 99.97%. Particles both larger and smaller than the MPPS are removed with greater efficiency. Bacteria, spores and viruses are removed from the air by these filters. HEPA filter efficiency and the mechanics of particle collection by these filters have been studied and well-documented [3,4] therefore only a brief description is included here.

The typical HEPA filter medium is a single sheet of borosilicate fibers treated with a wet-strength water-repellant binder. The filter medium is pleated to increase the overall surface area inside the filter frames and the pleats are often divided by corrugated aluminum separators (Figure 1). The separators prevent the pleats from collapsing in the air stream and provide a path for airflow. Alternate designs providing substitutions for the aluminum separators may also be used. The filter is glued into a wood, metal or plastic frame. Careless handling of the filter (e.g., improper storage or dropping) can damage the medium at the glue joint and cause tears or shifting of the filter resulting in leaks in the medium. This is the primary reason why filter integrity must be tested when a BSC is installed initially and each time it is moved or relocated. (See Section VII.)

Various types of containment and clean air devices incorporate the use of HEPA filters in the exhaust and/or supply air system to remove airborne particulate material. Depending on the configuration of these filters and the direction of the airflow, varying degrees of personnel, environmental and product protection can be achieved.[5] Section V describes the proper practices and procedures necessary to maximize the protection afforded by the device.

Section III—Biological Safety Cabinets

The similarities and differences in protection offered by the various classes of BSCs are reflected in Table 1. Please also refer to Table 2 and Section IV for further considerations pertinent to BSC selection and risk assessment.

The Class I BSC

The Class I BSC provides personnel and environmental protection, but no product protection. It is similar in terms of air movement to a chemical fume hood, but has a HEPA filter in the exhaust system to protect the environment (Figure 2). In the Class I BSC, unfiltered room air is drawn in through the work opening and across the work surface. Personnel protection is provided by this inward airflow as long as a minimum velocity of 75 linear feet per minute (lfm) is maintained[6] through the front opening. Because product protection is provided by the Class II BSCs, general usage of the Class I BSC has declined. However, in many cases, Class I BSCs are used specifically to enclose equipment (e.g., centrifuges, harvesting equipment or small fermenters), or procedures with potential to generate aerosols (e.g., cage dumping, culture aeration or tissue homogenation).

The classical Class I BSC is hard-ducted (i.e., direct connection) to the building exhaust system and the building exhaust fan provides the negative pressure necessary to draw room air into the cabinet. Cabinet air is drawn through a HEPA filter as it enters the cabinet exhaust plenum. A second HEPA filter may be installed at the terminal end of the building exhaust system prior to the exhaust fan.

Some Class I BSCs are equipped with an integral exhaust fan. The cabinet exhaust fan must be interlocked with the building exhaust fan. In the event that the building exhaust fan fails, the cabinet exhaust fan must turn off so that the building exhaust ducts are not pressurized. If the ducts are pressurized and the HEPA filter has developed a leak, contaminated air could be discharged into other parts of the building or the environment. The use of two filters in the cabinet increases the static pressure on the fan.

A panel with openings to allow access for the hands and arms to the work surface can be added to the Class I cabinet. The restricted opening results in increased inward air velocity, increasing worker protection. For added safety, arm-length gloves can be attached to the panel. Makeup air is then drawn through an auxiliary air supply opening (which may contain a filter) and/or around a loose-fitting front panel.

Some Class I models used for animal cage changing are designed to allow recirculation of air into the room after HEPA filtration and may require more frequent filter replacement due to filter loading and odor from organic materials captured on the filter. This type of Class I BSC should be certified annually for sufficient airflow and filter integrity.

The Class II BSC

As biomedical researchers began to use sterile animal tissue and cell culture systems, particularly for the propagation of viruses, cabinets were needed that also provided product protection. In the early 1960s, the "laminar flow" principle evolved. Unidirectional air moving at a fixed velocity along parallel lines was demonstrated to reduce turbulence resulting in predictable particle behavior. Biocontainment technology also incorporated this laminar flow principle with the use of the HEPA filter to aid in the capture and removal of airborne contaminants from the air stream.[7] This combination of technologies serves to help protect the laboratory worker from potentially infectious aerosols[4] generated within the cabinet and provides necessary product protection, as well. Class II BSCs are partial barrier systems that rely on the directional movement of air to provide containment. As the air curtain is disrupted (e.g., movement of materials in and out of a cabinet, rapid or sweeping movement of the arms) the potential for contaminant release into the laboratory work environment is increased, as is the risk of product contamination.

The Class II (Types A1, A2, B1 and B2)[8] BSCs provide personnel, environmental and product protection. Airflow is drawn into the front grille of the cabinet, providing personnel protection. In addition, the downward flow of HEPA-filtered air provides product protection by minimizing the chance of cross-contamination across the work surface of the cabinet. Because cabinet exhaust air is passed through a certified HEPA filter, it is particulate-free (environmental protection), and may be recirculated to the laboratory (Type A1 and A2 BSCs) or discharged from the building via a canopy or "thimble" connected to the building exhaust. Exhaust air from Types B1 and B2 BSCs must be discharged directly to the outdoors via a hard connection.

HEPA filters are effective at trapping particulates and thus infectious agents but do not capture volatile chemicals or gases. Only Type A2-exhausted or Types B1and B2 BSCs exhausting to the outside should be used when working with volatile, toxic chemicals, but amounts must be limited (Table 2).

All Class II cabinets are designed for work involving microorganisms assigned to biosafety levels 1, 2, 3 and 4.[1] Class II BSCs provide the microbe-free work environment necessary for cell culture propagation and also may be used for the formulation of nonvolatile antineoplastic or chemotherapeutic drugs.[9] Class II BSCs may be used with organisms requiring BSL-4 containment in a BSL-4 suit laboratory by a worker wearing a positive pressure protective suit.

1. *The Class II, Type A1 BSC:* An internal fan (Figure 3) draws sufficient room air through the front grille to maintain a minimum calculated or measured average inflow velocity of at least 75 lfm at the face opening of the cabinet. The supply air flows through a HEPA filter and provides particulate-free air to the work surface. Airflow provided in this manner reduces turbulence in the work zone and minimizes the potential for cross-contamination.

    The downward moving air "splits" as it approaches the work surface; the fan[6] draws part of the air to the front grille and the remainder to the rear grille. Although there are variations among different cabinets, this split generally occurs about halfway between the front and rear grilles and two to six inches above the work surface.

    The air is drawn through the front and rear grilles by a fan pushed into the space between the supply and exhaust filters. Due to the relative size of these two filters, approximately 30% of the air passes through the exhaust HEPA filter and 70% recirculates through the supply HEPA filter back into the work zone of the cabinet. Most Class II, Type A1 and A2 cabinets have dampers to modulate this division of airflow.

    A Class II Type A1 BSC is not to be used for work involving volatile toxic chemicals. The buildup of chemical vapors in the cabinet (by recirculated air) and in the laboratory (from exhaust air) could create health and safety hazards (See Section IV).

It is possible to exhaust the air from a Type A1 or A2 cabinet outside of the building. However, it must be done in a manner that does not alter the balance of the cabinet exhaust system, thereby disturbing the internal cabinet airflow. The proper method of connecting a Type A1 or A2 cabinet to the building exhaust system is through use of a canopy hood,[8,10] which provides a small opening or air gap (usually 1 inch) around the cabinet exhaust filter housing (Figure 4). The airflow of the building exhaust must be sufficient to maintain the flow of room air into the gap between the canopy unit and the filter housing. The canopy must be removable or be designed to allow for operational testing of the cabinet. (See Section VI.) Class II Type A1 or A2 cabinets should never be hard-ducted to the building exhaust system.[8] Fluctuations in air volume and pressure that are common to all building exhaust systems sometimes make it difficult to match the airflow requirements of the cabinet.

2. *The Class II, Type B1 BSC:* Some biomedical research requires the use of small quantities of hazardous chemicals, such as organic solvents or carcinogens. Carcinogens used in cell culture or microbial systems require both biological and chemical containment.[11]

    The Class II, Type B cabinet originated with the National Cancer Institute (NCI)-designed Type 212 (later called Type B) BSC (Figure 5A), and was designed for manipulations of minute quantities of hazardous chemicals with *in vitro* biological systems. The NSF International NSF/ANSI Standard 49—2007 definition of Type B1 cabinets[8] includes this classic NCI design Type B, and cabinets without supply HEPA filters located immediately below the work surface (Figure 5B), and/or those with exhaust/recirculation down flow splits other than exactly 70/30%.

    The cabinet supply blowers draw room air (plus a portion of the cabinet's recirculated air) through the front grille and through the supply HEPA filters located immediately below the work surface. This particulate-free air flows upward through a plenum at each side of the cabinet and then downward to the work area through a backpressure plate. In some cabinets, there is an additional supply HEPA filter to remove particulates that may be generated by the blower-motor system.

    Room air is drawn through the face opening of the cabinet at a minimum measured inflow velocity of 100 lfm. As with the Type A1 and A2 cabinets, there is a split in the down-flowing air stream just above the work surface. In the Type B1 cabinet, approximately 70 percent of the down flow air exits through the rear grille, passes through the exhaust HEPA filter, and is discharged from the building. The remaining 30 percent of the down flow air is drawn through the front grille. Since the air that flows to the rear grille is discharged into the exhaust system, activities that may generate hazardous chemical vapors or particulates should be conducted toward the rear of the cabinetwork area.[13]

Type B1 cabinets must be hard-ducted, preferably to a dedicated, independent exhaust system. As indicated earlier, fans for laboratory exhaust systems should be located at the terminal end of the ductwork to avoid pressuring the exhaust ducts. A failure in the building exhaust system may not be apparent to the user, as the supply blowers in the cabinet will continue to operate. A pressure-independent monitor and alarm should be installed to provide warning and shut off the BSC supply fan, should failure in exhaust airflow occur. Since this feature is not supplied by all cabinet manufacturers, it is prudent to install a sensor such as a flow monitor and alarm in the exhaust system as necessary. To maintain critical operations, laboratories using Type B1 BSCs should connect the exhaust blower to the emergency power supply.

3. *The Class II, Type B2 BSC:* This BSC is a total-exhaust cabinet; no air is recirculated within it (Figure 6). This cabinet provides simultaneous primary biological and chemical (small quantity) containment. Consideration must be given to the chemicals used in BSCs as some chemicals can destroy the filter medium, housings and/or gaskets causing loss of containment. The supply blower draws either room or outside air in at the top of the cabinet, passes it through a HEPA filter and down into the work area of the cabinet. The building exhaust system draws air through both the rear and front grills, capturing the supply air plus the additional amount of room air needed to produce a minimum calculated or measured inflow face velocity of 100 lfm. All air entering this cabinet is exhausted, and passes through a HEPA filter (and perhaps some other air-cleaning device such as a carbon filter if required for the work being performed) prior to discharge to the outside. This cabinet exhausts as much as 1200 cubic feet per minute of conditioned room air making this cabinet expensive to operate. The higher static air pressure required to operate this cabinet also results in additional costs associated with heavier gauge ductwork and higher capacity exhaust fan. Therefore, the need for the Class II, Type B2 should be justified by the research to be conducted.

    Should the building exhaust system fail, the cabinet will be pressurized, resulting in a flow of air from the work area back into the laboratory. Cabinets built since the early 1980's usually have an interlock system, installed by the manufacturer, to prevent the supply blower from operating whenever the exhaust flow is insufficient; systems can be retrofitted if necessary. Exhaust air movement should be monitored by a pressure-independent device, such as a flow monitor.

4. *The Class II, Type A2 BSC (Formerly called A/B3):* Only when this BSC (Figure 7) is ducted to the outdoors does it meet the requirements of the

former Class II Type B3.[8] The Type A2 cabinet has a minimum calculated or measured inflow velocity of 100 lfm. All positive pressure contaminated plenums within the cabinet are surrounded by a negative air pressure plenum thus ensuring that any leakage from a contaminated plenum will be drawn into the cabinet and not released to the environment. Minute quantities of volatile toxic chemicals or radionuclides can be used in a Type A2 cabinet only if it exhausts to the outside via a properly functioning canopy connection.[8]

5. *Special Applications:* Class II BSCs can be modified to accommodate special tasks. For example, the front sash can be modified by the manufacturer to accommodate the eyepieces of a microscope. The work surface can be designed to accept a carboy, a centrifuge or other equipment that may require containment. A rigid plate with openings for the arms can be added if needed. Good cabinet design, microbiological aerosol tracer testing of the modification and appropriate certification (see Section VII) are required to ensure that the basic systems operate properly after modification. Maximum containment potential is achieved only through strict adherence to proper practices and procedures (see Section V).

The Class III BSC

The Class III BSC (Figure 8) was designed for work with highly infectious microbiological agents and for the conduct of hazardous operations and provides maximum protection for the environment and the worker. It is a gas-tight (no leak greater than 1x10-7 cc/sec with 1% test gas at 3 inches pressure Water Gauge[14]) enclosure with a non-opening view window. Access for passage of materials into the cabinet is through a dunk tank, that is accessible through the cabinet floor, or double-door pass-through box (e.g., an autoclave) that can be decontaminated between uses. Reversing that process allows materials to be removed from the Class III BSC safely. Both supply and exhaust air are HEPA filtered on a Class III cabinet. Exhaust air must pass through two HEPA filters, or a HEPA filter and an air incinerator, before discharge directly to the outdoors. Class III cabinets are not exhausted through the general laboratory exhaust system. Airflow is maintained by an exhaust system exterior to the cabinet, which keeps the cabinet under negative pressure (minimum of 0.5 inches of water gauge.)

Long, heavy-duty rubber gloves are attached in a gas-tight manner to ports in the cabinet to allow direct manipulation of the materials isolated inside. Although these gloves restrict movement, they prevent the user's direct contact with the hazardous materials. The trade-off is clearly on the side of maximizing personal safety. Depending on the design of the cabinet, the supply HEPA filter provides particulate-free, albeit somewhat turbulent, airflow within the work environment. Laminar airflow is not a characteristic of a Class III cabinet.

Several Class III BSCs can be joined together in a "line" to provide a larger work area. Such cabinet lines are custom-built; the equipment installed in the cabinet line (e.g., refrigerators, small elevators, shelves to hold small animal cage racks, microscopes, centrifuges, incubators) is generally custom-built as well.

Horizontal Laminar Flow "Clean Bench"

Horizontal laminar flow "clean benches" (Figure 9A) are not BSCs. These pieces of equipment discharge HEPA-filtered air from the back of the cabinet across the work surface and toward the user. These devices only provide product protection. They can be used for certain clean activities, such as the dust-free assembly of sterile equipment or electronic devices. Clean benches should never be used when handling cell culture materials, drug formulations, potentially infectious materials, or any other potentially hazardous materials. The worker will be exposed to the materials being manipulated on the clean bench potentially resulting in hypersensitivity, toxicity or infection depending on the materials being handled. Horizontal airflow "clean benches" must never be used as a substitute for a biological safety cabinet. Users must be aware of the differences between these two devices.

Vertical Flow "Clean Bench"

Vertical flow clean benches (Figure 9B) also are not BSCs. They may be useful, for example, in hospital pharmacies when a clean area is needed for preparation of intravenous solutions. While these units generally have a sash, the air is usually discharged into the room under the sash, resulting in the same potential problems presented by the horizontal laminar flow clean benches. These benches should never be used for the manipulation of potentially infectious or toxic materials or for preparation of antineoplastic agents.

Section IV—Other Laboratory Hazards and Risk Assessment

Primary containment is an important strategy in minimizing exposure to the many chemical, radiological and biological hazards encountered in the laboratory. An overview is provided, in Table 2, of the various classes of BSCs, the level of containment afforded by each and the appropriate risk assessment considerations. Microbiological risk assessment is addressed in depth in BMBL.[1]

Working with Chemicals in BSCs

Work with infectious microorganisms often requires the use of various chemical agents, and many commonly used chemicals vaporize easily. Therefore, evaluation of the inherent hazards of the chemicals must be part of the risk assessment when selecting a BSC. Flammable chemicals should not be used in Class II, Type A1 or A2 cabinets since vapor buildup inside the cabinet presents a fire hazard. In order to determine the greatest chemical concentration, which might be entrained in the air stream following an accident or spill, it is necessary to evaluate the quantities to be used. Mathematical models are available to assist in these determinations.[13] For more information regarding the risks associated with exposure to chemicals, the reader should consult the Threshold Limit Values (TLVs) for various chemical substances established by the American Conference of Governmental Industrial Hygienists.[15]

The electrical systems of Class II BSCs are not spark-proof. Therefore, a chemical concentration approaching the lower explosive limits of the compound must be prohibited. Furthermore, since non-exhausted Class II, Type A1 and A2 cabinets return chemical vapors to the cabinetwork space and the room, they may expose the operator and other room occupants to toxic chemical vapors.

A chemical fume hood should be used for procedures using volatile chemicals instead of a BSC. Chemical fume hoods are connected to an independent exhaust system and operate with single-pass air discharged, directly or through a manifold, outside the building. They may also be used when manipulating chemical carcinogens.[11] When manipulating small quantities of volatile toxic chemicals required for use in microbiological studies, Class I and Class II (Type B2) BSCs, exhausted to the outdoors, can be used. The Class II, Type B1 and A2 canopy-exhausted cabinets may be used with minute or tracer quantities of nonvolatile toxic chemicals.[8]

Many liquid chemicals, including nonvolatile antineoplastic agents, chemotherapeutic drugs and low-level radionuclides, can be safely handled inside Class II, Type A cabinets.[9] Class II BSCs should not be used for labeling of biohazardous materials with radioactive iodine. Hard-ducted, ventilated containment devices incorporating both HEPA and charcoal filters in the exhaust systems are necessary for the conduct of this type of work (Figure 10).

Many virology and cell culture laboratories use diluted preparations of chemical carcinogens[11,16] and other toxic substances. Prior to maintenance, careful evaluation must be made of potential problems associated with decontaminating the cabinet and the exhaust system. Air treatment systems, such as a charcoal filter in a bag-in/bag-out housing,[17] (Figure 13) may be required so that discharged air meets applicable emission regulations.

National Sanitation Foundation (NSF)/ANSI Standard 49—2007[8] requires biologically-contaminated ducts and plenums of Class II, Type A2 and B cabinets be maintained under negative air pressure, or surrounded by negative pressure ducts and plenums.

Radiological Hazards in the BSC

As indicated above, volatile radionuclides such as I^{125} should not be used within Class II BSCs. When using nonvolatile radionuclides inside a BSC, the same hazards exist as if working with radioactive materials on the bench top. Work that has the potential for splatter or creation of aerosols can be done within the BSC.

Radiologic monitoring must be performed. A straight, vertical (not sloping) beta shield may be used inside the BSC to provide worker protection. A sloping shield can disrupt the air curtain and increase the possibility of contaminated air being released from the cabinet. A radiation safety professional should be contacted for specific guidance.

Risk Assessment

The potential for adverse events must be evaluated to eliminate or reduce to the greatest extent possible worker exposure to infectious organisms and to prevent release to the environment. Agent summary statements detailed in BMBL[1] provide data for microorganisms known to have caused laboratory-associated infections that may be used in protocol-driven risk assessment. Through the process of risk assessment, the laboratory environment and the work to be conducted are evaluated to identify hazards and develop interventions to ameliorate risks.

A properly certified and operational BSC is an effective engineering control (see Section VI) that must be used in concert with the appropriate practices, procedures and other administrative controls to further reduce the risk of exposure to potentially infectious microorganisms. Suggested work practices and procedures for minimizing risks when working in a BSC are detailed in the next section.

Section V — BSC Use by the Investigator: Work Practices and Procedures

Preparing for Work Within a Class II BSC

Preparing a written checklist of materials necessary for a particular activity and placing necessary materials in the BSC before beginning work serves to minimize the number and extent of air curtain disruptions compromising the fragile air barrier of the cabinet. The rapid movement of a worker's arms in a sweeping motion into and out of the cabinet will disrupt the air curtain and compromise the partial containment barrier provided by the BSC. Moving arms in and out slowly, perpendicular to the face opening of the cabinet will reduce this risk. Other personnel activities in the room (e.g., rapid movements near the face of the cabinet, walking traffic, room fans, open/closing room doors) may also disrupt the cabinet air barrier.[6]

Laboratory coats should be worn buttoned over street clothing; latex, vinyl, nitrile or other suitable gloves are worn to provide hand protection. Increasing levels of PPE may be warranted as determined by an individual risk assessment. For example, a solid front, back-closing laboratory gown provides better protection of personal clothing than a traditional laboratory coat and is a recommended practice at BSL-3.

Before beginning work, the investigator should adjust the stool height so that his/her face is above the front opening. Manipulation of materials should be delayed for approximately one minute after placing the hands/arms inside the cabinet. This allows the cabinet to stabilize, to "air sweep" the hands and arms, and to allow time for turbulence reduction. When the user's arms rest flatly across the front grille, occluding the grille opening, room air laden with particles may flow directly into the work area, rather than being drawn down through the front grille. Raising the arms slightly will alleviate this problem. The front grille must not be blocked with toweling, research notes, discarded plastic wrappers, pipetting devices, etc. All operations should be performed on the work surface at least four inches in from the front grille. If there is a drain valve under the work surface, it should be closed prior to beginning work in the BSC.

Materials or equipment placed inside the cabinet may cause disruption of the airflow, resulting in turbulence, possible cross-contamination and/or breach of containment. Extra supplies (e.g., additional gloves, culture plates or flasks, culture media) should be stored outside the cabinet. Only the materials and equipment required for the immediate work should be placed in the BSC.

BSCs are designed for 24-hour per day operation and some investigators find that continuous operation helps to control the laboratory's level of dust and other airborne particulates. Although energy conservation may suggest BSC operation only when needed, especially if the cabinet is not used routinely, room air balance is an overriding consideration. Air discharged through ducted BSCs must be considered in the overall air balance of the laboratory.

If the cabinet has been shut down, the blowers should be operated at least four minutes before beginning work to allow the cabinet to "purge." This purge will remove any suspended particulates in the cabinet. The work surface, the interior walls (except the supply filter diffuser), and the interior surface of the window should be wiped with 70% ethanol (EtOH), a 1:100 dilution of household bleach (i.e., 0.05% sodium hypochlorite), or other disinfectant as determined by the investigator to meet the requirements of the particular activity. When bleach is used, a second wiping with sterile water is needed to remove the residual chlorine, which may eventually corrode stainless steel surfaces. Wiping with non-sterile water may recontaminate cabinet surfaces, a critical issue when sterility is essential (e.g., maintenance of cell cultures).

Similarly, the surfaces of all materials and containers placed into the cabinet should be wiped with 70% EtOH to reduce the introduction of contaminants to the cabinet environment. This simple step will reduce introduction of mold spores and thereby minimize contamination of cultures. Further reduction of microbial load on materials to be placed or used in BSCs may be achieved by periodic decontamination of incubators and refrigerators.

Material Placement Inside the BSC

Plastic-backed absorbent toweling can be placed on the work surface but not on the front or rear grille openings. The use of toweling facilitates routine cleanup and reduces splatter and aerosol generation[19] during an overt spill. It can be folded and placed in a biohazard bag or other appropriate receptacle when work is completed.

All materials should be placed as far back in the cabinet as practical, toward the rear edge of the work surface and away from the front grille of the cabinet (Figure 11). Similarly, aerosol-generating equipment (e.g., vortex mixers, tabletop centrifuges) should be placed toward the rear of the cabinet to take advantage of the air split described in Section III. Bulky items such as biohazard bags, discard pipette trays and vacuum collection flasks should be placed to one side of the interior of the cabinet. If placing those items in the cabinet requires opening the sash, make sure that the sash is returned to its original position before work is initiated. The correct sash position (usually 8″ or 10″ above the base of the opening) should be indicated on the front of the cabinet. On most BSCs, an audible alarm will sound if the sash is in the wrong position while the fan is operating.

Certain common practices interfere with the operation of the BSC. The biohazard collection bag should not be taped to the outside of the cabinet. Upright pipette collection containers should not be used in BSCs nor placed on the floor outside the cabinet. The frequent inward/outward movement needed to place objects in these containers is disruptive to the integrity of the cabinet air barrier and can compromise both personnel and product protection. Only horizontal pipette discard trays containing an appropriate chemical disinfectant should be used within the cabinet. Furthermore, potentially contaminated materials should not be brought out of the cabinet until they have been surface decontaminated. Alternatively, contaminated materials can be placed into a closable container for transfer to an incubator, autoclave or another part of the laboratory.

Operations Within a Class II BSC

Laboratory Hazards

Many procedures conducted in BSCs may create splatter or aerosols. Good microbiological techniques should always be used when working in a BSC. For example, techniques used to reduce splatter and aerosol generation will also minimize the potential for personnel exposure to infectious materials manipulated within the cabinet. Class II cabinets are designed so that horizontally nebulized spores introduced into the cabinet will be captured by the downward flowing cabinet air within fourteen inches[8] of travel. Therefore, as a general rule of thumb, keeping clean materials at least one foot away from aerosol-generating activities will minimize the potential for cross-contamination.

The workflow should be from "clean to dirty" (Figure 11). Materials and supplies should be placed in the cabinet in such a way as to limit the movement of "dirty" items over "clean" ones.

Several measures can be taken to reduce the chance for cross-contamination of materials when working in a BSC. Opened tubes or bottles should not be held in a vertical position. Investigators working with Petri dishes and tissue culture plates should hold the lid above the open sterile surface to minimize direct impaction of downward air. Bottle or tube caps should not be placed on the toweling. Items should be recapped or covered as soon as possible.

Open flames are not required in the near microbe-free environment of a biological safety cabinet. On an open bench, flaming the neck of a culture vessel will create an upward air current that prevents microorganisms from falling into the tube or flask. An open flame in a BSC, however, creates turbulence that disrupts the pattern of HEPA-filtered air being supplied to the work surface. When deemed absolutely necessary, touch-plate micro burners equipped with a pilot light to provide a flame on demand may be used. Internal cabinet air disturbance and heat buildup will be minimized. The burner must be turned off when work is completed. Small electric "furnaces" are available for decontaminating bacteriological loops and needles and are preferable to an open flame inside the BSC. Disposable or recyclable sterile loops should be used whenever possible.

Aspirator bottles or suction flasks should be connected to an overflow collection flask containing appropriate disinfectant, and to an in-line HEPA or equivalent filter (Figure 12). This combination will provide protection to the central building vacuum system or vacuum pump, as well as to the personnel who service this equipment. Inactivation of aspirated materials can be accomplished by placing sufficient chemical decontamination solution into the flask to inactivate the microorganisms as they are collected. Once inactivation occurs, liquid materials can be disposed of as noninfectious waste.

Investigators must determine the appropriate method of decontaminating materials that will be removed from the BSC at the conclusion of the work. When chemical means are appropriate, suitable liquid disinfectant should be placed into the discard pan before work begins. Items should be introduced into the pan with minimum splatter and allowed appropriate contact time as per manufacturer's instructions. Alternatively, liquids can be autoclaved prior to disposal. Contaminated items should be placed into a biohazard bag, discard tray, or other suitable container prior to removal from the BSC.

When a steam autoclave is used, contaminated materials should be placed into a biohazard bag or discard pan containing enough water to ensure steam generation during the autoclave cycle. The bag should be taped shut or the discard pan should be covered in the BSC prior to transfer to the autoclave. The bag should be transported and autoclaved in a leak proof tray or pan. It is a prudent practice to decontaminate the exterior surface of bags and pans just prior to removal from the cabinet.

Decontamination

Cabinet Surface Decontamination

With the cabinet blower running, all containers and equipment should be surface decontaminated and removed from the cabinet when work is completed. At the end of the workday, the final surface decontamination of the cabinet should include a wipe-down of the work surface, the cabinet's sides and back and the interior of the glass. If necessary, the cabinet should also be monitored for radioactivity and decontaminated when necessary. Investigators should remove their gloves and gowns in a manner to prevent contamination of unprotected skin and aerosol generation and wash their hands as the final step in safe microbiological practices. The cabinet blower may be turned off after these operations are completed, or left on.

Small spills within the operating BSC can be handled immediately by removing the contaminated absorbent paper toweling and placing it into the biohazard bag or receptacle. Any splatter onto items within the cabinet, as well as the cabinet interior, should be immediately cleaned up with a towel dampened with an appropriate decontaminating solution. Gloves should be changed after the work surface is decontaminated and before placing clean absorbent toweling in the cabinet. Hands should be washed whenever gloves are changed or removed.

Spills large enough to result in liquids flowing through the front or rear grilles require decontamination that is more extensive. All items within the cabinet should be surface decontaminated and removed. After ensuring that the drain valve is closed, decontaminating solution can be poured onto the work surface and through the grille(s) into the drain pan.

Twenty to 30 minutes is generally considered an appropriate contact time for decontamination, but this varies with the disinfectant and the microbiological agent. Manufacturer's directions should be followed. The spilled fluid and disinfectant solution on the work surface should be absorbed with paper towels and discarded into a biohazard bag. The drain pan should be emptied into a collection vessel containing disinfectant. A hose barb and flexible tube should be attached to the drain valve and be of sufficient length to allow the open end to be submerged in the disinfectant within the collection vessel. This procedure serves to minimize aerosol generation. The drain pan should be flushed with water and the drain tube removed.

Should the spilled liquid contain radioactive material, a similar procedure can be followed. Radiation safety personnel should be contacted for specific instructions.

Periodic removal of the cabinetwork surface and/or grilles after the completion of drain pan decontamination may be justified because of dirty drain pan surfaces and grilles, which ultimately could occlude the drain valve or block airflow. However, extreme caution should be observed on wiping these surfaces to avoid injury from broken glass that may be present and sharp metal edges. Always use disposable paper toweling and avoid applying harsh force. Wipe dirty surfaces gently. Never leave toweling on the drain pan because the paper could block the drain valve or the air passages in the cabinet.

Gas Decontamination

BSCs that have been used for work involving infectious materials must be decontaminated before HEPA filters are changed or internal repair work is done.[20-23] Before a BSC is relocated, a risk assessment considering the agents manipulated within the BSC must be performed to determine the need and method for decontamination. The most common decontamination method uses formaldehyde gas, although more recently, hydrogen peroxide vapor[21] and chlorine dioxide gas have been used successfully.

Section VI—Facility and Engineering Requirements

Secondary Barriers

Whereas BSCs are considered to be the primary containment barrier for manipulation of infectious materials, the laboratory room itself is considered to be the secondary containment barrier.[24] Inward directional airflow is established[25] by exhausting a greater volume of air than is supplied to a given laboratory and by drawing makeup air from the adjacent space. This is optional at BSL-2 but must be maintained at BSL-3 and BSL-4.[1,26] The air balance for the entire facility should be established and maintained to ensure that airflow is from areas of least to greater potential contamination.

Building Exhaust

At BSL-3 and BSL-4, exhaust laboratory air must be directly exhausted to the outside since it is considered potentially contaminated. This concept is referred to as a dedicated, single-pass exhaust system. The exhausted room air can be HEPA-filtered when a high level of aerosol containment is needed, which is always true at BSL-4 and may be optional at BSL-3. When the building exhaust system is used to vent a ducted BSC, the system must have sufficient capacity to maintain the exhaust flow if changes in the static pressure within the system should occur. The connection to a BSC must be constant air volume (CAV).

The HVAC exhaust system must be sized to handle both the room exhaust and the exhaust requirements of all containment devices that may be present. Adequate supply air must be provided to ensure appropriate function of the exhaust system. Right angle bends changing duct diameters and transitional connections within the systems will add to the demand on the exhaust fan. The building exhaust air should be discharged away from supply air intakes, to prevent re-entrainment of laboratory exhaust air into the building air supply system. Refer to recognized design guides for locating the exhaust terminus relative to nearby air intakes.[27]

Utility Services

Utility services needed within a BSC must be planned carefully. Protection of vacuum systems must be addressed (Figure 12). Electrical outlets inside the cabinet must be protected by ground fault circuit interrupters and should be supplied by an independent circuit. When propane or natural gas is provided, a clearly marked emergency gas shut-off valve outside the cabinet must be installed for fire safety. All non-electrical utility services should have exposed, accessible shut-off valves. The use of compressed air within a BSC must be carefully considered and controlled to prevent aerosol production and reduce the potential for vessel pressurization.

Ultraviolet Lamps

Ultraviolet (UV) lamps are not recommended in BSCs[8] nor are they necessary. If installed, UV lamps must be cleaned weekly to remove any dust and dirt that may block the germicidal effectiveness of the ultraviolet light. The lamps should be checked weekly with a UV meter to ensure that the appropriate intensity of UV light is being emitted. UV lamps must be turned off when the room is occupied to protect eyes and skin from UV exposure, which can burn the cornea and cause skin cancer. If the cabinet has a sliding sash, close the sash when operating the UV lamp.

BSC Placement

BSCs were developed (see Section I) as workstations to provide personnel, environmental and product protection during the manipulation of infectious microorganisms. Certain considerations must be met to ensure maximum effectiveness of these primary barriers. Whenever possible, adequate clearance should be provided behind and on each side of the cabinet to allow easy access for maintenance and to ensure that the cabinet air re-circulated to the laboratory is not hindered. A 12 to 14 inch clearance above the cabinet may be required to provide for accurate air velocity measurement across the exhaust filter surface[28,29] and for exhaust filter changes. When the BSC is hard-ducted or connected by a canopy unit to the ventilation system, adequate space must be provided so that the configuration of the ductwork will not interfere with airflow. The canopy unit must provide adequate access to the exhaust HEPA filter for testing.

The ideal location for the biological safety cabinet is remote from the entry (i.e., the rear of the laboratory away from traffic), since people walking parallel to the face of a BSC can disrupt the air curtain.[16,20,30] The air curtain created at the front of the cabinet is quite fragile, amounting to a nominal inward and downward velocity of 1 mph. Open windows, air supply registers, portable fans or laboratory equipment that creates air movement (e.g., centrifuges, vacuum pumps) should not be located near the BSC. Similarly, chemical fume hoods must not be located close to BSCs.

HEPA Filters

HEPA filters, whether part of a building exhaust system or part of a cabinet, will require replacement when they become loaded to the extent that sufficient airflow can no longer be maintained. In most instances, filters must be decontaminated before removal. To contain the formaldehyde gas typically used for microbiological decontamination, exhaust systems containing HEPA filters require airtight dampers to be installed on both the inlet and discharge side of the filter housing. This ensures containment of the gas inside the filter housing during decontamination. Access panel ports in the filter housing also allow for performance testing of the HEPA filter. (See Section VII.)

A bag-in/bag-out filter assembly[3,17] (Figure 13) can be used in situations where HEPA filtration is necessary for operations involving biohazardous materials and hazardous or toxic chemicals. The bag-in/bag-out system is used when it is not possible to gas or vapor decontaminate the HEPA filters, or when hazardous chemicals or radionuclides have been used in the BSC, and provides protection against exposure for the maintenance personnel and the environment. Note, however, that this requirement must be identified at the time of purchase and installation; a bag-in/bag-out assembly cannot be added to a cabinet after-the-fact without an extensive engineering evaluation.

Section VII — Certification of BSCs

Development of Containment Standards

The evolution of containment equipment for varied research and diagnostic applications created the need for consistency in construction and performance. Federal Standard 209[a,32,33] was developed to establish classes of air cleanliness and methods for monitoring clean workstations and clean rooms where HEPA filters are used to control airborne particulates.

The first "standard" to be developed specifically for BSCs[12] served as a Federal procurement specification for the NIH Class II, Type 1 (now called Type A1) BSC, which had a fixed or hinged front window or a vertical sliding sash, vertical downward airflow and HEPA-filtered supply and exhaust air. This specification described design criteria and defined prototype tests for microbiological aerosol challenge, velocity profiles, and leak testing of the HEPA filters. A similar procurement specification was generated[31] when the Class II, Type 2 (now called Type B1) BSC was developed.

---

a Federal Standard No. 209E9 has been replaced by ISO 14644. This standard does not apply to BSCs and should not be considered a basis for their performance or integrity certification. However, the methodology of ISO 14644 can be used to quantify the particle count within the work area of a BSC. ISO 14644 defines how to classify a clean room/clean zone. Performance tests and procedures needed to achieve a specified cleanliness classification are outlined by the Institute of Environmental Sciences and Technology's IEST-RP-CC-006.

National Sanitation Foundation (NSF) Standard #49 for Class II BSCs was first published in 1976, providing the first independent standard for design, manufacture and testing of BSCs. This standard "replaced" the NIH specifications, which were being used by other institutions and organizations purchasing BSCs. NSF/ANSI Standard 49—2007[b,8] incorporates current specifications regarding design, materials, construction, and testing. This Standard for BSCs establishes performance criteria and provides the minimum testing requirements that are accepted in the United States. Cabinets, which meet the Standard and are certified by NSF bear an NSF Mark.

NSF/ANSI Standard 49—2007 pertains to all models of Class II cabinets (Type A1, A2, B1, B2) and provides a series of specifications regarding:

- Design/construction

- Performance

- Installation recommendations

- Recommended microbiological decontamination procedure

- References and specifications pertinent to Class II Biosafety Cabinetry

Annex F of NSF/ANSI Standard 49—2007, which covers field-testing of BSCs, is now a normative part of the Standard. This Standard is reviewed periodically by a committee of experts to ensure that it remains consistent with developing technologies.

The operational integrity of a BSC must be validated before it is placed into service and after it has been repaired or relocated. Relocation may break the HEPA filter seals or otherwise damage the filters or the cabinet. Each BSC should be tested and certified at least annually to ensure continued, proper operation.

On-site field-testing (NSF/ANSI Standard 49—2007 Annex F plus Addendum #1) must be performed by experienced, qualified personnel. Some basic information is included in the Standard to assist in understanding the frequency and kinds of tests to be performed. In 1993, NSF began a program for accreditation of certifiers based on written and practical examinations. Education and training programs for persons seeking accreditation as qualified to perform all field certification tests are offered by a variety of organizations. Selecting competent individuals to perform testing and certification is important. It is suggested that the institutional BSO be consulted when identifying companies qualified to conduct the necessary field performance tests.

---

b The standard can be ordered from the NSF for a nominal fee at NSF International, 789 North Dixboro Road, P.O. Box 130140, Ann Arbor, Michigan, 48113-0140; Telephone: 734-769-8010; Fax: 734-769-0190; e-mail: info@nsf.org; Telex 753215 NSF INTL.

It is strongly recommended that, whenever possible, accredited field certifiers are used to test and certify BSCs. If in-house personnel are performing the certifications, then these individuals should become accredited.

The annual tests applicable to each of the three classes of BSCs are listed in Table 3. Table 4 indicates where to find information regarding the conduct of selected tests. BSCs consistently perform well when proper annual certification procedures are followed; cabinet or filter failures tend to occur infrequently.

Performance Testing BSCs in the Field

Class II BSCs are the primary containment devices that protect the worker, product and environment from exposure to microbiological agents. BSC operation, as specified by NSF/ANSI Standard 49—2007, Annex F plus Addendum #1 needs to be verified at the time of installation and, as a minimum, annually thereafter. The purpose and acceptance level of the operational tests (Table 3) ensure the balance of inflow and exhaust air, the distribution of air onto the work surface, and the integrity of the cabinet and the filters. Other tests check electrical and physical features of the BSC.

A. *Down flow Velocity Profile Test:* This test is performed to measure the velocity of air moving through the cabinet workspace, and is to be performed on all Class II BSCs.

B. *Inflow Velocity Test:* This test is performed to determine the calculated or directly measured velocity through the work access opening, to verify the nominal set point average inflow velocity and to calculate the exhaust airflow volume rate.

C. *Airflow Smoke Patterns Test:* This test is performed to determine if: 1) the airflow along the entire perimeter of the work access opening is inward; 2) if airflow within the work area is downward with no dead spots or refluxing; 3) if ambient air passes onto or over the work surface; and 4) if there is no escape to the outside of the cabinet at the sides and top of the window. The smoke test is an indicator of airflow direction, not velocity.

D. *HEPA Filter Leak Test:* This test is performed to determine the integrity of supply and exhaust HEPA filters, filter housing and filter mounting frames while the cabinet is operated at the nominal set point velocities. An aerosol in the form of generated particulates of dioctylphthalate (DOP) or an accepted alternative (e.g., poly alpha olefin (PAO), di(2-ethylhexyl) sebecate, polyethylene glycol and medical grade light mineral oil) is required for leak-testing HEPA filters and their seals. The aerosol is generated on the intake side of the filter and particles passing through the filter or around the seal are measured with a photometer on the discharge side. This test is suitable for ascertaining the integrity of all HEPA filters.

E   *Cabinet Integrity Test (A1 Cabinets only):* This pressure holding test is performed to determine if exterior surfaces of all plenums, welds, gaskets and plenum penetrations or seals are free of leaks. In the field, it need only be performed on Type A1 cabinets at the time of initial installation when the BSC is in a free-standing position (all four sides are easily accessible) in the room in which it will be used, after a cabinet has been relocated to a new location, and again after removal of access panels to plenums for repairs or a filter change. This test may also be performed on fully installed cabinets. Cabinet integrity can also be checked using the bubble test; liquid soap can be spread along welds, gaskets and penetrations to visualize air leaks that may occur.

F. *Electrical Leakage and Ground Circuit Resistance and Polarity Tests:* Electrical testing has been taken out of NSF/ANSI 49 Standard—2007 for new cabinets certified under the this Standard. This responsibility has been turned over to UL. All new cabinets must meet UL 61010A-1 in order to be certified by NSF. These safety tests are performed to determine if a potential shock hazard exists by measuring the electrical leakage, polarity, ground fault interrupter function and ground circuit resistance to the cabinet connection. An electrical technician other than the field certification personnel may perform the tests at the same time the other field certification tests are conducted. The polarity of electrical outlets is checked (Table 3, E). The ground fault circuit interrupter should trip when approximately five milliamperes (mA) is applied.

G. *Lighting Intensity Test:* This test is performed to measure the light intensity on the work surface of the cabinet as an aid in minimizing cabinet operator fatigue.

H. *Vibration Test:* This test is performed to determine the amount of vibration in an operating cabinet as a guide to satisfactory mechanical performance, as an aid in minimizing cabinet operator fatigue and to prevent damage to delicate tissue culture specimens.

I.   *Noise Level Test:* This test is performed to measure the noise levels produced by the cabinets, as a guide to satisfactory mechanical performance and an aid in minimizing cabinet operator fatigue.

J. *UV Lamp Test:* A few BSCs have UV lamps. When used, they must be tested periodically to ensure that their energy output is sufficient to kill microorganisms. The surface on the bulb should be cleaned with 70% ethanol prior to performing this test. Five minutes after the lamp has been turned on, the sensor of the UV meter is placed in the center of the work surface. The radiation output should not be less than 40 microwatts per square centimeter at a wavelength of 254 nanometers (nm).

Finally, accurate test results can only be assured when the testing equipment is properly maintained and calibrated. It is appropriate to request the calibration information for the test equipment being used by the certifier.

Table 1. Selection of a Safety Cabinet through Risk Assessment

Biological Risk Assessed	Protection Provided			BSC Class
	Personnel	Product	Environmental	
BSL 1 – 3	Yes	No	Yes	I
BSL 1 – 3	Yes	Yes	Yes	II (A1, A2, B1, B2)
BSL – 4	Yes	Yes	Yes	III; II—When used in suit room with suit

Table 2. Comparison of Biosafety Cabinet Characteristics

BSC Class	Face Velocity	Airflow Pattern	Applications	
			Nonvolatile Toxic Chemicals and Radionuclides	Volatile Toxic Chemicals and Radionuclides
I	75	In at front through HEPA to the outside or into the room through HEPA (Figure 2)	Yes	When exhausted outdoors[1,2]
II, A1	75	70% recirculated to the cabinet work area through HEPA; 30% balance can be exhausted through HEPA back into the room or to outside through a canopy unit (Figure 3)	Yes (minute amounts)	No
II, B1	100	30% recirculated, 70% exhausted. Exhaust cabinet air must pass through a dedicated duct to the outside through a HEPA filter (Figures 5A, 5B)	Yes	Yes (minute amounts)[1,2]
I, B2	100	No recirculation; total exhaust to the outside through a HEPA filter (Figure 6)	Yes	Yes (small amounts)[1,2]
II, A2	100	Similar to II, A1, but has 100 lfm intake air velocity and plenums are under negative pressure to room; exhaust air can be ducted to the outside through a canopy unit (Figure 7)	Yes	When exhausted outdoors (FORMALLY "B3") (minute amounts)[1,2]

BSC Class	Face Velocity	Airflow Pattern	Applications	
			Nonvolatile Toxic Chemicals and Radionuclides	Volatile Toxic Chemicals and Radionuclides
III	N/A	Supply air is HELP filtered. Exhaust air passes through two HEPA filters in series and is exhausted to the outside via a hard connection (Figure 8)	Yes	Yes (small amounts)[1,2]

[1] Installation requires a special duct to the outside, an in-line charcoal filter, and a spark proof (explosion proof) motor and other electrical components in the cabinet. Discharge of a Class I or Class II, Type A2 cabinet into a room should not occur if volatile chemicals are used.
[2] In no instance should the chemical concentration approach the lower explosion limits of the compounds.

Table 3. Field Performance Tests Applied to the Three Classes of Biological Safety Cabinets

Test Performed for	Biosafety Cabinet		
	Class I	Class II	Class III
Primary Containment			
Cabinet Integrity	N/A	A (A1 Only)	A
HEPA Filter Leak	Required	Required	Required
Down flow Velocity	N/A	Required	N/A
Face Velocity	Required	Required	N/A
Negative Pressure / Ventilation Rate	B	N/A	Required
Airflow Smoke Patterns	Required	Required	E, F
Alarms and Interlocks	C, D	C, D	Required
Electrical Safety			
Electrical Leakage, etc.	E, D	E, D	E, D
Ground Fault Interrupter	D	D	D

	Biosafety Cabinet		
Test Performed for	**Class I**	**Class II**	**Class III**
Other			
Lighting Intensity	E	E	E
UV Intensity	C, E	C, E	C, E
Noise Level	E	E	E
Vibration	E	E	E

Required Required during certification.
A   Required for proper certification if the cabinet is new, has been moved or panels have been removed for maintenance.
B   If used with gloves.
C   If present.
D   Encouraged for electrical safety.
E   Optional, at the discretion of the user.
F   Used to determine air distribution within cabinet for clean to dirty procedures.
N/A   Not applicable.

Table 4. Reference for Applicable Containment Test

	Biosafety Cabinet Type		
Test	**Class I**	**Class II**	**Class III**
HEPA Filter Leak	(F.5)[1]	(F.5)	(F.5)
Airflow Smoke Pattern	No smoke shall reflux out of BSC once drawn in	(F.4)	N/A
Cabinet Integrity	N/A	(F.6)	[p.138 – 141][2]
Face Velocity Open Front	[75-125 lfm]	75 lfm—type A1; 100 lfm type A2, B1 & B2: (F.3)	N/A
Face VelocityGloves Ports / No Gloves	150 lfm	N/A	N/A
Water Gauge Pressure Glove Ports and Gloves	N/A	N/A	(-0.5 "w.c.") [p. 145]
Down flow Velocity	N/A	(F.2)	N/A

[1] Parenthetical references are to the NSF/ANSI Standard 49—2007; letters and numerals indicate specific sections and subsections.
[2] Bracketed reference [ ] is to the Laboratory Safety Monograph; page numbers are indicated.

Figure 1. HEPA filters are typically constructed of paper-thin sheets of borosilicate medium, pleated to increase surface area, and affixed to a frame. Aluminum separators are often added for stability.

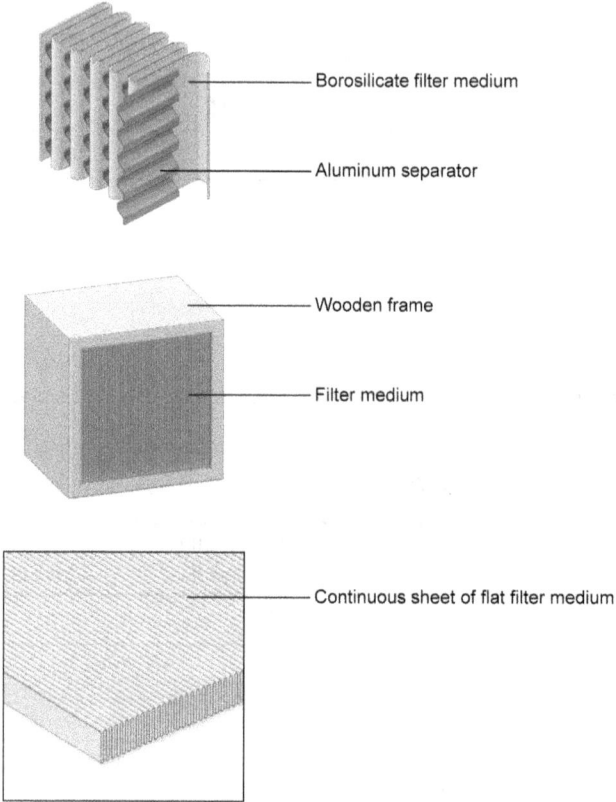

Borosilicate filter medium

Aluminum separator

Wooden frame

Filter medium

Continuous sheet of flat filter medium

Figure 2. *The Class I BSC* (A) front opening; (B) sash; (C) exhaust HEPA filter; (D) exhaust plenum. *Note:* The cabinet needs to be hard connected to the building exhaust system if toxic vapors are to be used.

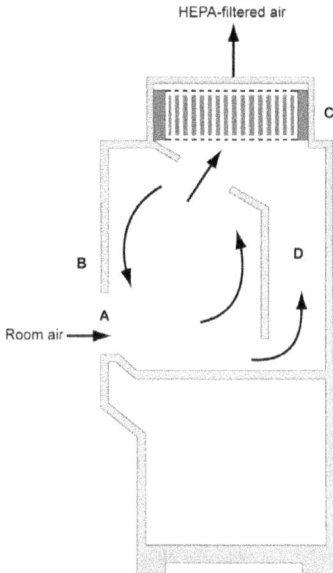

Figure 3. *The Class II, Type A1 BSC* (A) front opening; (B) sash; (C) exhaust HEPA filter; (D) supply HEPA filter; (E) common plenum; (F) blower.

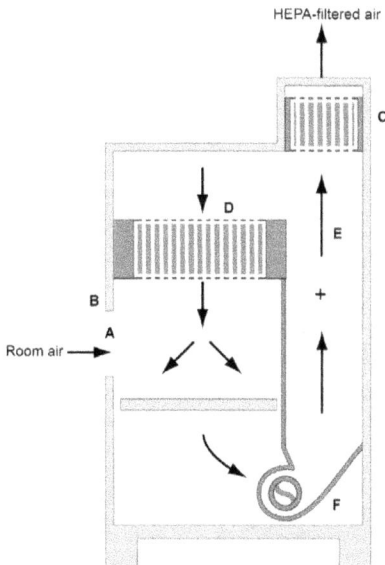

Figure 4. *Canopy (thimble) unit for ducting a Class II, Type A BSC* (A) balancing damper; (B) flexible connector to exhaust system; (C) cabinet exhaust HEPA filter housing; (D) canopy unit; (E) BSC. *Note:* There is a 1" gap between the canopy unit (D) and the exhaust filter housing (C), through which room air is exhausted.

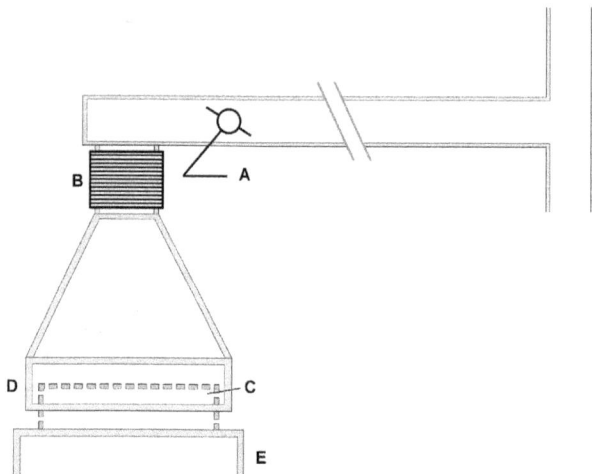

Figure 5A. *The Class II, Type B1 BSC (classic design)* (A) front opening; (B) sash; (C) exhaust HEPA filter; (D) supply HEPA filter; (E) negative pressure dedicated exhaust plenum; (F) blower; (G) additional HEPA filter for supply air. *Note:* The cabinet exhaust needs to be hard connected to the building exhaust system.

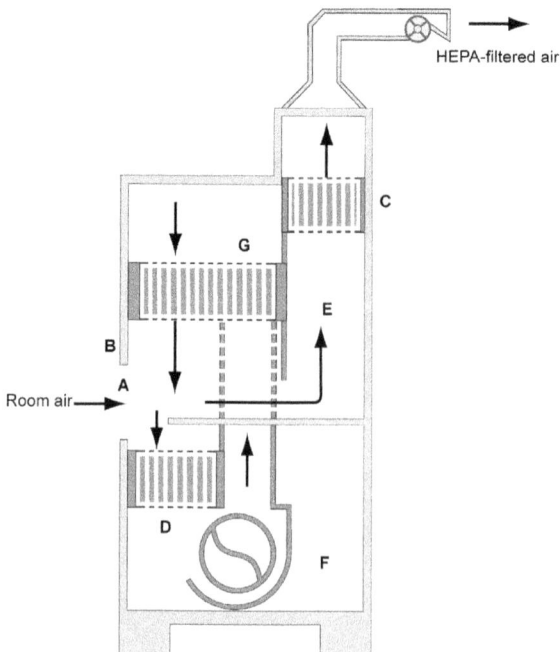

Figure 5B. *The Class II, Type B1 BSC (bench top design)* (A) front opening; (B) sash; (C) exhaust HEPA filter; (D) supply plenum; (E) supply HEPA filter; (F) blower; (G) negative pressure exhaust plenum. *Note:* The cabinet exhaust needs to be hard connected to the building exhaust system.

Connection to the building exhaust system is required.

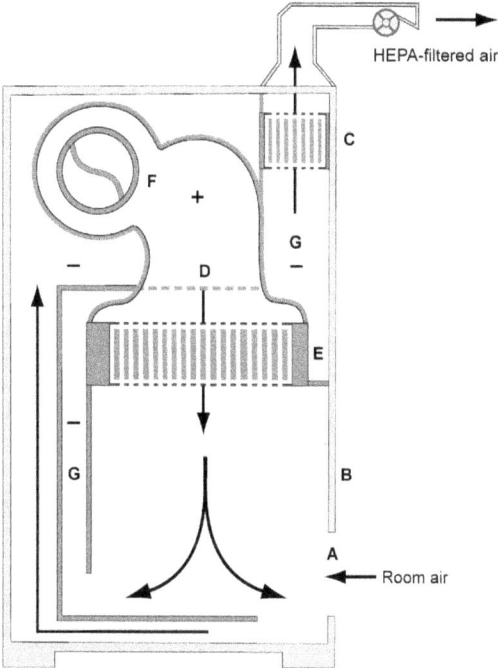

Figure 6. *The Class II, Type B2 BSC* (A) front opening; (B) sash; (C) exhaust HEPA filter; (D) supply HEPA filter; (E) negative pressure exhaust plenum. *Note:* The carbon filter in the exhaust system is not shown. The cabinet needs to be hard connected to the building exhaust system.

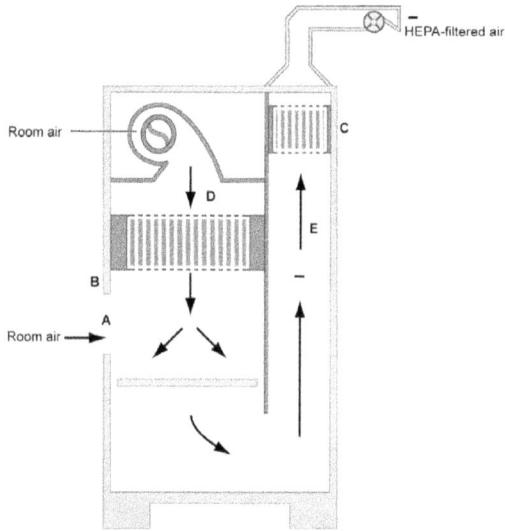

Figure 7. *The tabletop model of a Class II, Type A2 BSC* (A) front opening; (B) sash; (C) exhaust HEPA filter; (D) supply HEPA filter; (E) positive pressure common plenum; (F) negative pressure plenum. The Class II Type A2 BSC **is not equivalent to what was formerly called a** *Class II Type B3* **unless it is connected to the laboratory exhaust system.** *Note:* The A2 BSC should be canopy connected to the exhaust system.

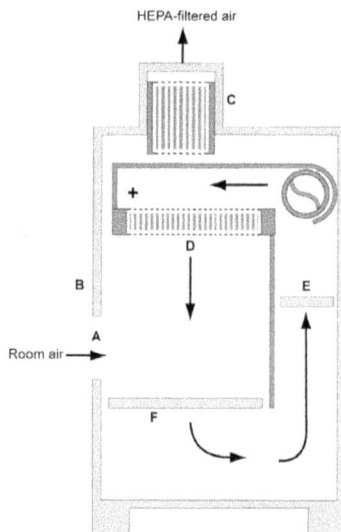

Figure 8. *The Class III BSC* (A) glove ports with O-ring for attaching arm-length gloves to cabinet; (B) sash; (C) exhaust HEPA filter; (D) supply HEPA filter; (E) double-ended autoclave or pass-through box. *Note:* A chemical dunk tank may be installed which would be located beneath the work surface of the BSC with access from above. The cabinet exhaust needs to be hard connected to an exhaust system where the fan is generally separate from the exhaust fans of the facility ventilation system. The exhaust air must be double HEPA-filtered or HEPA-filtered and incinerated.

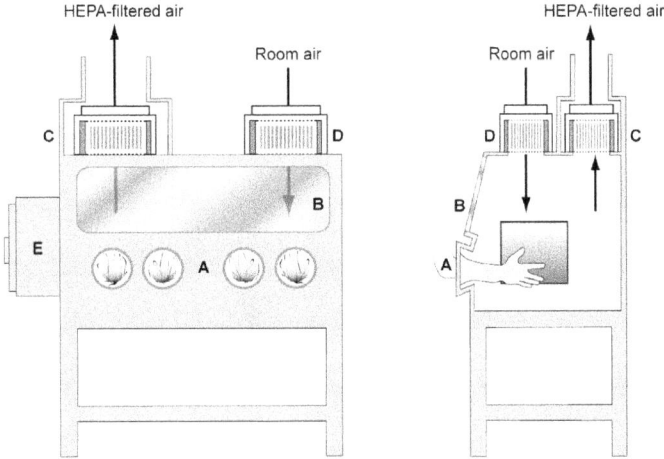

Figure 9A. *The horizontal laminar flow "clean bench"* (A) front opening; (B) supply grille; (C) supply HEPA filter; (D) supply plenum; (E) blower.

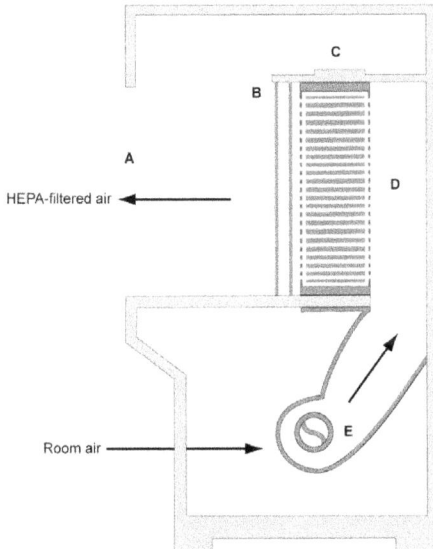

Figure 9B. *The vertical laminar flow "clean bench"* (A) front opening; (B) sash; (C) supply HEPA filter; (D) blower. *Note:* Some vertical flow clean benches have recirculated air through front and/or rear perforated grilles.

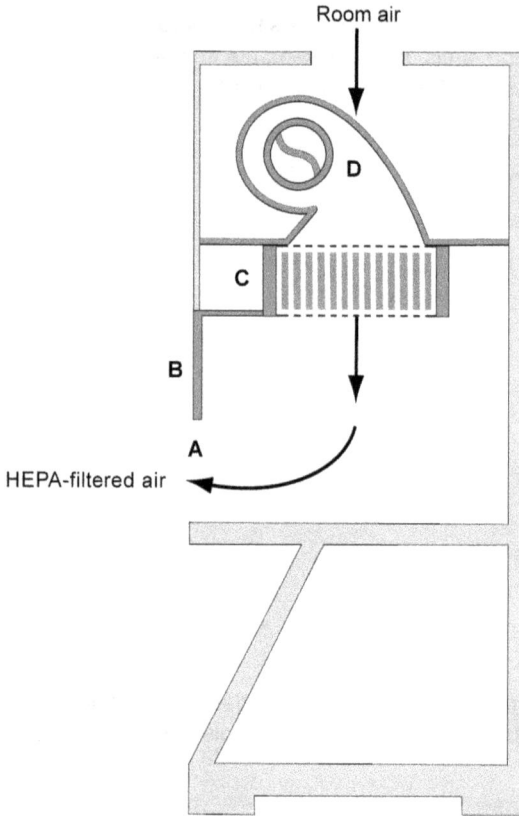

Figure 10. A modified containment cabinet or Class I BSC can be used for labeling infectious microorganisms with I^{125}. (A) arm holes; (B) Lexan^R hinged doors; (C) exhaust charcoal filter; (D) exhaust HEPA filter; (E) filter housing with required connection to building exhaust (see also Figure 13).

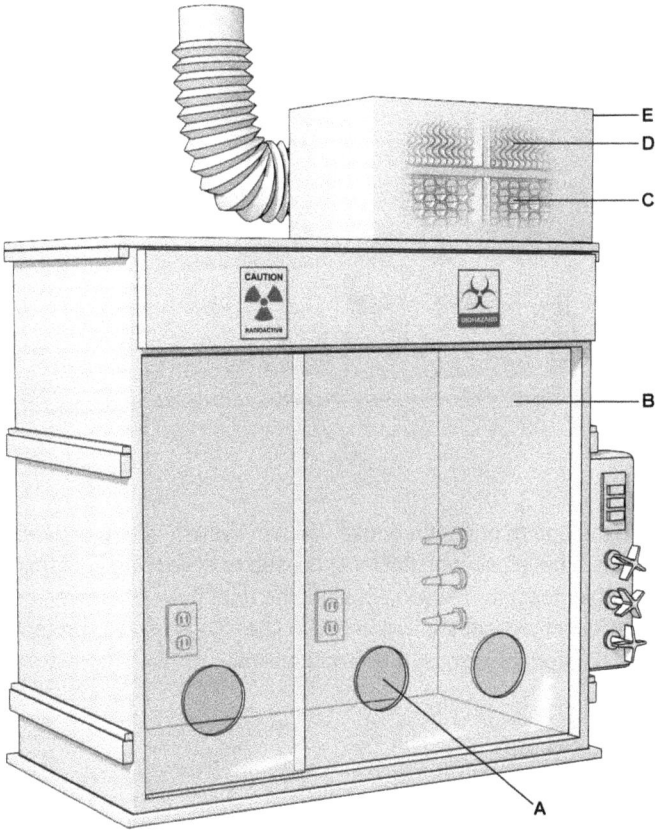

Figure 11. A typical layout for working "clean to dirty" within a Class II BSC. Clean cultures (left) can be inoculated (center); contaminated pipettes can be discarded in the shallow pan and other contaminated materials can be placed in the biohazard bag (right). This arrangement is reversed for left-handed persons.

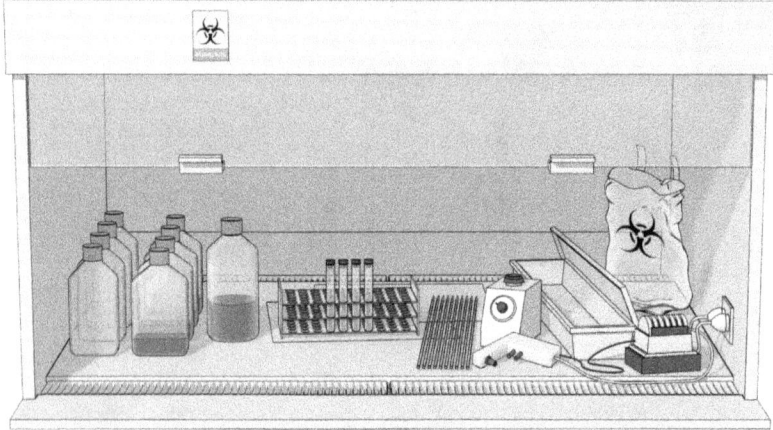

Figure 12. One method to protect a house vacuum system during aspiration of infectious fluids. The left suction flask (A) is used to collect the contaminated fluids into a suitable decontamination solution; the right flask (B) serves as a fluid overflow collection vessel. An in-line HEPA filter (C) is used to protect the vacuum system (D) from aerosolized microorganisms.

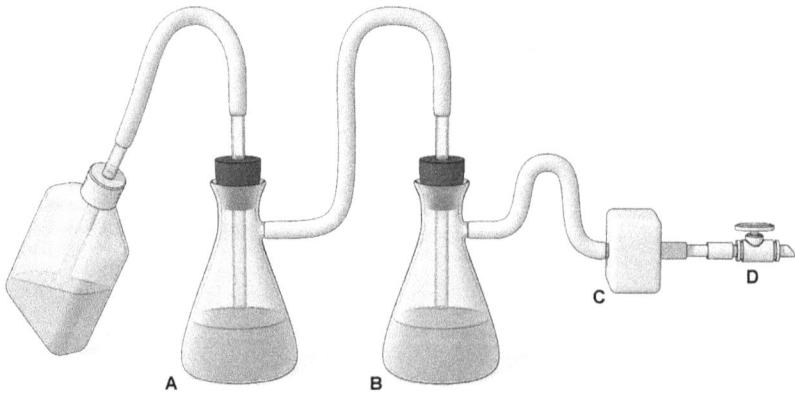

Figure 13. A bag-in-bag-out filter enclosure allows for the removal of the contaminated filter without worker exposure. (A) filters; (B) bags; (C) safety straps; (D) cinching straps; (E) shock cord located in the mouth of the PVC bag restricts the bag around the second rib of the housing lip.

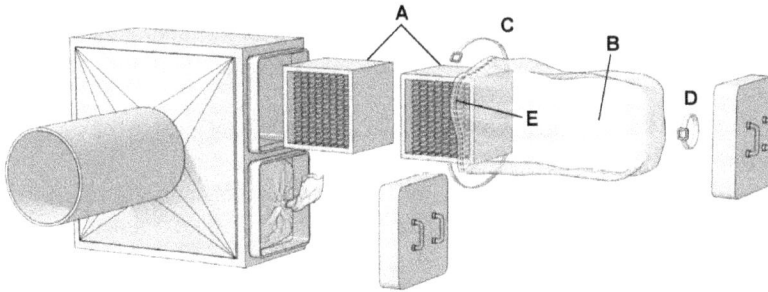

Acknowledgements

We gratefully acknowledge the Baker Company; Filtration Group, Inc.; Flanders Filters; and Forma Scientific, Inc., for use of some drawings and figures reproduced herein.

Referrences

1. Centers for Disease Control and Prevention; The National Institutes of Health. Biosafety in microbiological and biomedical laboratories. 4th ed. Washington, DC. 1999.
2. Kruse RH, Puckett WH, Richardson JH. Biological safety cabinetry. Clin Microbiol Rev. 1991;4:207-541.
3. First MW. Filters, high capacity filters and high efficiency filters: review and production. In-Place Filter Testing Workshop; 1971; Boston, Massachusetts. 1971.
4. The National Institutes of Health (NIH); Rockville Bio-Engineering Services, Dow Chemical USA. A Workshop for Certification of Biological Safety Cabinets. Bethesda: The Institute, Office of Biohazards and Environmental Control; 1974.
5. Richmond, JY. Safe practices and procedures for working with human specimens in biomedical research laboratories. J Clin Immunoassay. 1988;13:115-9.
6. Barbeito MS, Taylor LA. Containment of microbial aerosols in a microbiological safety cabinet. Appl Microbiol. 1968;16:1255-29.
7. Whitfield WJ. A new approach to clean room design. Final report. Albuquerque (NM): Sandia Corporation; 1962. Report No.: SC-4673 (RR).
8. NSF International (NSF); American National Standards Institute (ANSI). NSF/ANSI Standard 49-2007. Class II (laminar flow) biosafety cabinetry. Ann Arbor (MI); 2004.

9. Occupational Safety and Health Administration, Office of Occupational Medicine. Guidelines for cytotoxic (antineoplastic) drugs. Appendix A. Standard 01-23-001. 1986.
10. Jones RL Jr., Tepper B, Greenier TG, et al. Effects of Thimble Connections of Biological Safety Cabinets. Abstracts of 32nd Biological Safety Conference; 1989; New Orleans, LA: 1989.
11. National Institutes of Health (NIH). NIH guidelines for the laboratory use of chemical carcinogens. Washington, DC: U. S. Government Printing Office; 1981.
12. Class II. Type 1 safety cabinet specification the National Institute of Health-030112. The Institute; 1973, May.
13. Stuart DG, First MW, Jones RL Jr., et al. Comparison of chemical vapor handling by three types of class II biological safety cabinets. Particulate and Microbial Control. 1983;2:18-24.
14. Stuart D, Kiley M, Ghidoni D, et al. The class III biological safety cabinet. In: Richmond JY, editor. Anthology of biosafety vol. VII. biosafety level 3. Mundelein, IL: American Biological Safety Association; 2004.
15. American Conference of Government Industrial Hygienists (ACGIH). Threshold limit values for chemical substances and physical agents and biological exposure indices. Cincinnati, OH: ACGIH; 2006.
16. National Institutes of Health NIH, National Cancer Institute (NCI), Office of Research Safety and the Special Committee of Safety and Health Experts. Laboratory safety monograph: a supplement to the NIH guidelines for recombinant DNA research. Bethesda (MD): The Institute; 1979.
17. Barbeito M, West, D, editors. Laboratory ventilation for hazard control. Proceedings of a Cancer Research Safety Symposium; l976 Oct 21-22; Frederick, MD: Litton Bionetics, Inc.; 1976.
18. National Cancer Institute Safety Standards for Research Involving Chemical Carcinogens. Bethesda: The National Institutes of Health (US), National Cancer Institute; 1975.
19. Anderson RE, Stein L, Moss ML, et al. Potential infectious hazards of bacteriological techniques. J Bacteriol. 1952;64:473-81.
20. Baker Company. Factors to consider before selecting a laminar flow biological safety cabinet. Sanford (ME): Baker Company; 1993.
21. Jones R, Drake J, Eagleson D. 1993. US hydrogen peroxide vapor to decontaminate biological cabinets. Acumen. 1993;1:1.
22. Jones R, Stuart D, Large S, et al. 1993. Cycle parameters for decontaminating a biological safety cabinet using H2O2 vapor. Acumen, 1993;1:2.
23. Jones R, Stuart D, PhD, Large S, et al. Decontamination of a HEPA filter using hydrogen peroxide vapor. Acumen. 1993;1:3.
24. Fox D, editor. Proceedings of the National Cancer Institute symposium on design of biomedical research facilities. Monograph series. Volume 4; l979 Oct 18-19; Frederick, MD: Litton Bionetics, Inc.; 1979.

25. American Society of Heating, Refrigeration and Air Conditioning Engineers, Inc. (ASHRAE). Laboratories. In: ASHRAE Handbook, 1982 Applications. Atlanta (GA): ASHRAE; 1982.

26. US Department of Agriculture (USDA), Agricultural Research Services (ARS). ARS construction project design standard manual, 242.1. Beltsville (MD): USDA; 1991.

27. American Conference of Governmental Industrial Hygienists (ACGIH). Industrial ventilation, a manual of recommended practice. Cincinnati (OH): ACGIH; 1998.

28. Jones RL, Jr., Stuart DG, Eagleson D, et al. The effects of changing intake and supply air flow on biological safety performance. Appl Occup Environ Hyg. 1990;5:370.

29. Jones RL, Jr., Stuart DG, Eagleson, D, et al. Effects of ceiling height on determining calculated intake air velocities for biological safety cabinets. Appl Occup Environ Hyg. 1991;6:683.

30. Rake BW. Influence of cross drafts on the performance of a biological safety cabinet. Appl Env Microbiol. 1978;36:278-83.

31. Specifications for general purpose clean air biological safety cabinet. Bethesda: The National Institutes of Health (US), National Cancer Institute; 1973.

32. Airborne Particulate Cleanliness Classes in Clean rooms and Clean Zones, Federal Standard No. 209E. (1992).

33. Clean Room and Work Station Requirements, Controlled Environment, Federal Standard No. 209B. (1973).

Appendix B—Decontamination and Disinfection

This section describes basic strategies for decontaminating surfaces, items, and areas in laboratories to eliminate the possibility of transmission of infectious agents to laboratory workers, the general public, and the environment. Factors necessary for environmentally mediated infection transmission are reviewed as well as methods for sterilization and disinfection and the levels of antimicrobial activity associated with liquid chemical germicides. General approaches are emphasized, not detailed protocols and methods. The principles of sterilization and disinfection are stated and compared.

Environmentally Mediated Infection Transmission

Environmentally associated laboratory infections can be transmitted directly or indirectly from environmental sources (e.g., air, contaminated fomites and laboratory instruments, and aerosols) to laboratory staff. Fortunately, LAI are rare events[1] because there are a number of requirements necessary for environmental transmission to occur.[2] Commonly referred to as the "chain of infection" they include: presence of a pathogen of sufficient virulence, relatively high concentration of the pathogen (i.e., infectious dose), and a mechanism of transmission of the pathogen from environment to the host, a correct portal of entry to a susceptible host.

To accomplish successful transmission from an environmental source, all of these requirements for the "chain of infection" must be present. The absence of any one element will prevent transmission. Additionally, the pathogen in question must overcome environmental stresses to retain viability, virulence, and the capability to initiate infection in the host. In the laboratory setting, high concentrations of pathogens can be common. Reduction of environmental microbial contamination by conventional cleaning procedures is often enough to prevent environmentally mediated transmission. However, it is the general practice in laboratories to use sterilization methods to remove the potential for infection transmission.

Principles of Sterilization and Disinfection

In order to implement a laboratory biosafety program it is important to understand the principles of decontamination, cleaning, sterilization, and disinfection. We review here the definitions of sterilization, disinfection, antisepsis, decontamination, and sanitization to avoid misuse and confusion. The definitions and implied capabilities of each inactivation procedure are discussed with an emphasis on achievement and in some cases, monitoring of each state.

Sterilization

Any item, device, or solution is considered to be sterile when it is completely free of all living microorganisms and viruses. The definition is categorical and absolute (i.e., an item is either sterile or it is not). A sterilization *procedure* is one that kills all microorganisms, including high numbers of bacterial endospores. Sterilization can be accomplished by heat, ethylene oxide gas, hydrogen peroxide gas, plasma, ozone, and radiation (in industry). From an operational standpoint, a sterilization procedure cannot be categorically defined. Rather, the procedure is defined as a process, after which the probability of a microorganism surviving on an item subjected to treatment is less than one in one million (10-6). This is referred to as the "sterility assurance level." [3,4]

Disinfection

Disinfection is generally a less lethal process than sterilization. It eliminates nearly all recognized pathogenic microorganisms but not necessarily all microbial forms (e.g., bacterial spores) on inanimate objects. Disinfection does not ensure an "overkill" and therefore lacks the margin of safety achieved by sterilization procedures. The effectiveness of a disinfection procedure is controlled significantly by a number of factors, each one of which may have a pronounced effect on the end result. Among these are:

- the nature and number of contaminating microorganisms (especially the presence of bacterial spores);

- the amount of organic matter present (e.g., soil, feces, and blood);

- the type and condition of instruments, devices, and materials to be disinfected;

- the temperature.

Disinfection is a procedure that reduces the level of microbial contamination, but there is a broad range of activity that extends from sterility at one extreme to a minimal reduction in the number of microbial contaminants at the other. By definition, chemical disinfection and in particular, high-level disinfection differs from chemical sterilization by its lack of sporicidal power. This is an over simplification of the actual situation because a few chemical germicides used as disinfectants do, in fact, kill large numbers of spores even though high concentrations and several hours of exposure may be required. Non-sporicidal disinfectants may differ in their capacity to accomplish disinfection or decontamination. Some germicides rapidly kill only the ordinary vegetative forms of bacteria such as staphylococci and streptococci, some forms of fungi, and lipid-containing viruses, whereas others are effective against such relatively resistant organisms as *Mycobacterium tuberculosis* var. *bovis*, non-lipid viruses, and most forms of fungi.

Spaulding Classification

In 1972, Dr. Earl Spaulding[5] proposed a system for classifying liquid chemical germicides and inanimate surfaces that has been used subsequently by CDC, FDA, and opinion leaders in the United States. This system, as it applies to device surfaces, is divided into three general categories based on the theoretical risk of infection if the surfaces are contaminated at time of use. From the laboratory perspective, these categories are:

- critical—instruments or devices that are exposed to normally sterile areas of the body require sterilization;

- semi-critical—instruments or devices that touch mucous membranes may be either sterilized or disinfected;

- non-critical—instruments or devices that touch skin or come into contact with persons only indirectly can be either cleaned and then disinfected with an intermediate-level disinfectant, sanitized with a low-level disinfectant, or simply cleaned with soap and water.

In 1991, microbiologists at CDC proposed an additional category, environmental surfaces (e.g., floors, walls, and other "housekeeping surfaces") that do not make direct contact with a person's skin.[6] Spaulding also classified chemical germicides by activity level:

High-level Disinfection

This procedure kills vegetative microorganisms and inactivates viruses, but not necessarily high numbers of bacterial spores. Such disinfectants are capable of sterilization when the contact time is relatively long (e.g., 6 to 10 hours). As high-level disinfectants, they are used for relatively short periods of time (e.g., 10 to 30 minutes). These chemical germicides are potent sporicides and, in the United States, are classified by the FDA as sterilant/disinfectants. They are formulated for use on medical devices, but not on environmental surfaces such as laboratory benches or floors.[7]

Intermediate-level Disinfection

This procedure kills vegetative microorganisms, including *Mycobacterium tuberculosis*, all fungi, and inactivates most viruses. Chemical germicides used in this procedure often correspond to Environmental Protection Agency (EPA)-approved "hospital disinfectants" that are also "tuberculocidal." They are used commonly in laboratories for disinfection of laboratory benches and as part of detergent germicides used for housekeeping purposes.

Low-level Disinfection

This procedure kills most vegetative bacteria except *M. tuberculosis*, some fungi, and inactivates some viruses. The EPA approves chemical germicides used in this procedure in the US as "hospital disinfectants" or "sanitizers."

Decontamination in the Microbiology Laboratory

Decontamination in the microbiology laboratory must be carried out with great care. In this arena, decontamination may entail disinfection of work surfaces, decontamination of equipment so it is safe to handle, or may require sterilization. Regardless of the method, the purpose of decontamination is to protect the laboratory worker, the environment, and anyone who enters the laboratory or handles laboratory products away from the laboratory. Reduction of cross-contamination in the laboratory is an added benefit.

Decontamination and Cleaning

Decontamination renders an area, device, item, or material safe to handle (i.e., safe in the context of being reasonably free from a risk of disease transmission). The primary objective is to reduce the level of microbial contamination so that infection transmission is eliminated. The decontamination process may be ordinary soap and water cleaning of an instrument, device, or area. In laboratory settings, decontamination of items, spent laboratory materials, and regulated laboratory wastes is often accomplished by a sterilization procedure such as steam autoclaving, perhaps the most cost-effective way of decontaminating a device or an item.

The presence of any organic matter necessitates longer contact time with a decontamination method if the item or area is not pre-cleaned. For example, a steam cycle used to sterilize pre-cleaned items is 20 minutes at 121°C. When steam sterilization is used to decontaminate items that have a high bio-burden and there is no pre-cleaning (i.e., infectious waste) the cycle is longer. Decontamination in laboratory settings often requires longer exposure times because pathogenic microorganisms may be protected from contact with the decontaminating agents.

Table 1. Descending Order of Resistance to Germicidal Chemicals

Bacterial Spores
Bacillus subtilis, Clostridium sporogenes
▼

Mycobacteria
Mycobacterium tuberculosis var. *bovis*, Nontuberculous mycobacteria
▼

Nonlipid or Small Viruses
Poliovirus, Coxsackievirus, Rhinovirus
▼

Fungi
Trichophyton spp., *Cryptococcus* spp., *Candida* spp.
▼

Vegetative Bacteria
Pseudomonas aeruginosa, Staphylococcus aureus, Salmonella choleraesuis,
Enterococci
▼

Lipid or Medium-size Viruses
Herpes simplex virus, CMV, Respiratory syncytial virus, HBV, HCV, HIV, Hantavirus,
Ebola virus

Note: There are exceptions to this list. *Pseudomonas* spp are sensitive to high-level disinfectants, but if they grow in water and form biofilms on surfaces, the protected cells can approach the resistance of bacterial spores to the same disinfectant. The same is true for the resistance to glutaraldehyde by some nontuberculous mycobacteria, some fungal ascospores of *Microascus cinereus* and *Cheatomium globosum,* and the pink pigmented *Methylobacteria*. Prions are also resistant to most liquid chemical germicides and are discussed in the last part of this section.

Chemical germicides used for decontamination range in activity from high-level disinfectants (i.e., high concentrations of sodium hypochlorite [chlorine bleach]), which might be used to decontaminate spills of cultured or concentrated infectious agents in research or clinical laboratories, to low-level disinfectants or sanitizers for general housekeeping purposes or spot decontamination of environmental surfaces in healthcare settings. Resistance of selected organisms to decontamination is presented in descending order in Table 1. If dangerous and highly infectious agents are present in a laboratory, the methods for decontamination of spills, laboratory equipment, BSC, or infectious waste are very significant and may include prolonged autoclave cycles, incineration or gaseous treatment of surfaces.

Decontamination of Large Spaces

Space decontamination is a specialized activity and should be performed by specialists with proper training and protective equipment.[8] Decontamination requirements for BSL-3 and BSL-4 laboratory space have an impact on the design of these facilities. The interior surfaces of BSL-3 laboratories must be water resistant in order for them to be easily cleaned and decontaminated. Penetrations in these surfaces should be sealed or capable of being sealed for decontamination purposes. Thus, in the BSL-3 laboratory, surface decontamination, not fumigation, is the primary means of decontaminating space. Care should be taken that penetrations in the walls, floors and ceilings are kept to a minimum and are "sight sealed." Verification of the seals is usually not required for most BSL-3 laboratories. The BSL-4 laboratory design requires interior surfaces that are water resistant AND sealed to facilitate fumigation. These seals must be tested and verified to ensure containment in order to permit both liquid disinfection and fumigation. Periodic fumigation is required in the BSL-4 suit laboratory to allow routine maintenance and certification of equipment. Procedures for decontamination of large spaces such as incubators or rooms are varied and influenced significantly by the type of etiologic agent involved, the characteristics of the structure containing the space, and the materials present in the space. The primary methods for space decontamination follow.

Formaldehyde—Paraformaldehyde

Formaldehyde gas at a concentration of 0.3 grams/cubic foot for four hours is often used for space decontamination. Gaseous formaldehyde can be generated by heating flake paraformaldehyde (0.3 grams per cubic foot) in a frying pan, thereby converting it to formaldehyde gas. The humidity must be controlled and the system works optimally at 80% relative humidity. This method is effective in killing microorganisms but toxicity issues are present.[1,9] Additional information on environmental and safety issues related to paraformaldehyde is available from the EPA Web site: *www.epa.gov/pesticides*.

Hydrogen Peroxide Vapor

Hydrogen peroxide can be vaporized and used for the decontamination of glove boxes as well as small room areas. Vapor phase hydrogen peroxide has been shown to be an effective sporicide at concentrations ranging from 0.5 mg/L to <10 mg/L. The optimal concentration of this agent is about 2.4 mg/L with a contact time of at least one hour. This system can be used to decontaminate glove boxes, walk in incubators and small rooms. An advantage of this system is that the end products (i.e., water) are not toxic. Low relative humidity can be used.[10-14]

Chlorine Dioxide Gas

Chlorine dioxide gas sterilization can be used for decontamination of laboratory rooms, equipment, glove boxes, and incubators. The concentration of gas at the site of decontamination should be approximately 10 mg/L with contact time of one to two hours.

Chlorine dioxide possesses the bactericidal, virucidal and sporicidal properties of chlorine, but unlike chlorine, does not lead to the formation of trihalomethanes or combine with ammonia to form chlorinated organic products (chloramines). The gas cannot be compressed and stored in high-pressure cylinders, but is generated upon demand using a column-based solid phase generation system. Gas is diluted to the use concentration, usually between 10 and 30 mg/L. Within reasonable limits, a chlorine dioxide gas generation system is unaffected by the size or location of the ultimate destination for the gas. Relative humidity does need to be controlled and high humidities are optimal. Although most often used in closed sterilizers, the destination enclosure for the chlorine dioxide gas does not, in fact, need to be such a chamber. Because chlorine dioxide gas exits the generator at a modest positive pressure and flow rate, the enclosure also need not be evacuated and could be a sterility-testing isolator, a glove box or sealed BSC, or even a small room that could be sealed to prevent gas egress.[15] Chlorine dioxide gas is rapidly broken down by light; care must be taken to eliminate light sources in spaces to be decontaminated.

Decontamination of Surfaces

Liquid chemical germicides formulated as disinfectants may be used for decontamination of large areas. The usual procedure is to flood the area with a disinfectant for periods up to several hours. This approach is messy and with some of the disinfectants used represents a toxic hazard to laboratory staff. For example, most of the "high-level" disinfectants on the United States market are formulated to use on instruments and medical devices and not on environmental surfaces. Intermediate and low-level disinfectants are formulated to use on fomites and environmental surfaces but lack the potency of a high-level disinfectant. For the most part intermediate and low level disinfectants can be safely used and, as with all disinfectants, the manufacturer's instructions should be closely followed.[7] Disinfectants that have been used for decontamination include sodium hypochlorite solutions at concentrations of 500 to 6000 parts per million (ppm), oxidative disinfectants such as hydrogen peroxide and peracetic acid, phenols, and iodophors.

Concentrations and exposure times vary depending on the formulation and the manufacturer's instructions for use.[6,16] See Table 2 for a list of chemical germicides and their activity levels. A spill control plan must be available in the laboratory. This plan should include the rationale for selection of the disinfecting agent, the approach to its application, contact time and other parameters. Agents requiring BSL-3 and BSL-4 containment pose a high risk to workers and possibly to the environment and should be managed by well-informed professional staff trained and equipped to work with concentrated material.

Table 2. Activity Levels of Selected Liquid Germicides[a]

Procedure / Product	Aqueous Concentration	Activity Level
Sterilization		
glutaraldehyde	variable	
hydrogen peroxide	6 – 30%	
formaldehyde	6 – 8%	
chlorine dioxide	variable	
peracetic acid		
Disinfection		
glutaraldehyde	variable	high to intermediate
ortho-phthalaldehyde	0.5%	high
hydrogen peroxide	3 – 6%	high to intermediate
formaldehyde [b]	1 – 8%	high to low
chlorine dioxide	variable	high
peracetic acid	variable	high
chlorine compounds [c]	500 to 5000 ml/L free/available chlorine	Intermediate
alcohols (ethyl, isopropyl) [d]	70%	Intermediate
phenolic compounds	0.5 to 3%	intermediate to low
iodophor compounds [e]	30 – 50 mg/L free iodine up to 10,000 mg/L available iodine 0.1 – 0.2%	intermediate to low
quaternary ammonium compounds		low

[a] This list of chemical germicides centers on generic formulations. A large number of commercial products based on these generic components can be considered for use. Users should ensure that commercial formulations are registered with EPA or by the FDA.

[b] Because of the ongoing controversy of the role of formaldehyde as a potential occupational carcinogen, the use of formaldehyde is limited to certain specific circumstances under carefully controlled conditions (e.g., for the disinfection of certain hemodialysis equipment). There are no FDA cleared liquid chemical sterilant/disinfectants that contain formaldehyde.

[c] Generic disinfectants containing chlorine are available in liquid or solid form (e.g., sodium or calcium hypochlorite). Although the indicated concentrations are rapid acting and broad-spectrum (tuberculocidal, bactericidal, fungicidal, and virucidal), no proprietary hypochlorite formulations are formally registered with EPA or cleared by FDA. Common household bleach is an excellent and inexpensive source of sodium hypochlorite. Concentrations between 500 and 1000 mg/L chlorine are appropriate for the vast majority of uses requiring an intermediate level of germicidal activity; higher concentrations are extremely corrosive as well as irritating to personnel, and their use should be limited to situations where there is an excessive amount of organic material or unusually high concentrations of microorganisms (e.g., spills of cultured material in the laboratory).

^d The effectiveness of alcohols as intermediate level germicides is limited because they evaporate rapidly, resulting in short contact times, and also lack the ability to penetrate residual organic material. They are rapidly tuberculocidal, bactericidal and fungicidal, but may vary in spectrum of virucidal activity (see text). Items to be disinfected with alcohols should be carefully pre-cleaned then totally submerged for an appropriate exposure time (e.g., 10 minutes).

^e Only those iodophors registered with EPA as hard-surface disinfectants should be used, closely following the manufacturer's instructions regarding proper dilution and product stability. Antiseptic iodophors are not suitable to disinfect devices, environmental surfaces, or medical instruments.

Special Infectious Agent Issues

Transmissible Spongiform Encephalopathy Agents (Prions)

The major exception to the rule in the previous discussion of microbial inactivation and decontamination is the causative agent of CJD or other prion agents responsible for transmissible spongiform encephalopathies of the central nervous system in humans or animals. Studies show that prions are resistant to conventional uses of heat and/or chemical germicides for the sterilization of instruments and devices. (See Section IX.)

References

1. Vesley D, Lauer J, Hawley R. Decontamination, sterilization, disinfection, and antisepsis. In: Fleming DO, Hunt DL, editors. Laboratory safety: principles and practices. 3rd ed. Washington, DC: ASM Press; 2001. p. 383-402.
2. Rhame FS. The inanimate environment. In: Bennett JV, Brachmann PS, editors. Hospital infections. 4th ed. Philadelphia: Lippincott-Raven; 1998. p. 299-324.
3. Favero M. Developing indicators for sterilization. In: Rutala W, editor. Disinfection, sterilization and antisepsis in health care. Washington, DC: Association for Professionals in Infection Control and Epidemiology, Inc.; 1998. p. 119-132.
4. Favero M. Sterility assurance: concepts for patient safety. In: Rutala W, editor. Disinfection, sterilization and antisepsis: principles and practices in healthcare facilities. Washington, DC: Association for Professionals in Infection Control and Epidemiology, Inc.; 2001. p. 110-9.
5. Spaulding EH. Chemical disinfection and antisepsis in the hospital. J Hosp Res. 1972:9;5-31.
6. Favero M, Bond W. Chemical disinfection of medical surgical material. In: Block S, editor. Disinfection, sterilization, and preservation. 5th edition. Philadelphia: Lippencott, Williams and Wilkens; 2001. p. 881-917.
7. Centers for Disease Control and Prevention [www.cdc.gov]. Atlanta: The Centers for Disease Control and Prevention; [updated 2006 Sept 21]. Guidelines for Environmental Infection Control in Health-Care Facilities, 2003; [about 2 screens] Available from: http://www.cdc.gov/hicpac/pubs.html
8. Tearle P. Decontamination by fumigation. Commun Dis Public Health. 2003;6:166-8.

9. Fink R, Liberman DF, Murphy K, et al. Biological safety cabinets, decontamination or sterilization with paraformaldehyde. Am Ind Hyg Assoc J. 1988;49:277-9.

10. Klapes NA, Vesley D. Vapor-phase hydrogen peroxide as a surface decontaminant and sterilant. Appl Environ Microbiol. 1990;56:503-6.

11. Graham GS, Rickloff JR. Development of VHP sterilization technology. J Healthc Mater Manage. 1992;54:56-8.

12. Johnson JW, Arnold JF, Nail SL, et al. Vaporized hydrogen peroxide sterilization of freeze dryers. J Parenter Sci Technol. 1992;46:215-25.

13. Heckert RA, Best M, Jordan LT, et al. Efficacy of vaporized hydrogen peroxide against exotic animal viruses. Appl Environ Microbiol. 1997;63:3916-8.

14. Krause J, McDonnell G, Riedesel H. Biodecontamination of animal rooms and heat-sensitive equipment with vaporized hydrogen peroxide. Contemp Top Lab Anim Sci. 2001;40:18-21.

15, Knapp JE, Battisti DL. Chlorine Dioxide. In: Block S, editor. Disinfection, sterilization, and preservation. 5th ed. Philadelphia: Lippencott, Williams and Wilkens; 2001. p. 215-27.

16. Weber DJ, Rutala WA. Occupational risks associated with the use of selected disinfectants and sterilants. In: Rutala WA, editor. Disinfection, sterilization and antisepsis in health care. Champlain (NY): Polyscience Publications: 1998 p. 211-26.

Appendix C—Transportation of Infectious Substances

An infectious substance is a material known to contain or reasonably expected to contain a pathogen. A pathogen is a microorganism (including bacteria, viruses, rickettsiae, parasites, fungi) or other agent, such as a proteinaceous infectious particle (prion), that can cause disease in humans or animals. Infectious substances may exist as purified and concentrated cultures, but may also be present in a variety of materials, such as body fluids or tissues. Transportation of infectious substances and materials that are known or suspected to contain them are regulated as hazardous materials by the United State Department of Transportation (DOT), foreign governments, and the International Civil Aviation Organization, and their transportation is subject to regulatory controls. For transport purposes, the term "infectious substance" is understood to include the term "etiologic agent."

Transportation Regulations

International and domestic transport regulations for infectious substances are designed to prevent the release of these materials in transit to protect the public, workers, property, and the environment from the harmful effects that may occur from exposure to these materials. Protection is achieved through rigorous packaging requirements and hazard communication. Packages must be designed to withstand rough handling and other forces experienced in transportation, such as changes in air pressure and temperature, vibration, stacking, and moisture. Hazard communication includes shipping papers, labels, markings on the outside of packagings, and other information necessary to enable transport workers and emergency response personnel to correctly identify the material and respond efficiently in an emergency situation. In addition, shippers and carriers must be trained on these regulations so they can properly prepare shipments and recognize and respond to the risks posed by these materials.

Select agents include infectious substances that have been identified by the CDC and the USDA as having the potential to pose a severe threat to public health and safety. Persons who offer for transportation or transport select agents in commerce in the United States must develop and implement security plans for such transportation. A security plan must include an assessment of the possible transportation security risks for materials covered by the security plan and specific measures to reduce or eliminate the assessed risks. At a minimum, a security plan must include measures to address those risks associated with personnel security, en route security, and unauthorized access.

Regulations

Department of Transportation. 49 CFR Part 171-180, Hazardous Materials Regulations. Applies to the shipment of infectious substances in commercial transportation within the United States. Information on these regulations is available at: *http://www.phmsa.dot.gov/hazmat.*

United States Postal Service (USPS). 39 CFR Part 20, International Postal Service (International Mail Manual), and Part 111, General Information on Postal Service (Domestic Mail Manual). Regulations on transporting infectious substances through the USPS are codified in Section 601.10.17 of the Domestic Mail Manual and Section 135 of the International Mail Manual. A copy of the Domestic and International Mail Manuals may be obtained from the U.S. Government Printing Office by calling Monday through Friday, 7:30 a.m. – 9:00 p.m. EST: (202) 512-1800; toll free (866) 512-1800; or at the USPS Web site: *http://bookstore.gpo.gov/*.

Occupational Health and Safety Administration (OSHA). 29 CFR Part 1910.1030, Occupational Exposure to Bloodborne Pathogens. These regulations provide minimal packaging and labeling for blood and body fluids when transported within a laboratory or outside of it. Information may be obtained from your local OSHA office or at the OSHA Web site: *http://www.osha.gov*.

Technical Instructions for the Safe Transport of Dangerous Goods by Air (Technical Instructions). International Civil Aviation Organization (ICAO). Applies to the shipment of infectious substances by air and is recognized in the United States and by most countries worldwide. A copy of these regulations may be obtained from the ICAO Document Sales Unit at (514) 954-8022, fax: (514) 954-6769; e-mail: sales_unit@icao.int; or from the ICAO Web site: *http://www.icao.int*.

Dangerous Goods Regulations. International Air Transport Association (IATA). These regulations are issued by an airline association, are based on the ICAO Technical Instructions, and are followed by most airline carriers. A copy of these regulations is available at: *http://www.iata.org/index.htm* or *http://www.who.int/en/* ; or by contacting the IATA Customer Care office at: telephone: +1 (514) 390 6726; fax: +1 (514) 874 9659; for Canada and USA (800) 716-6326 (toll free); Europe, Africa and Middle East +41 (22) 770 2751; fax: +41 (22) 770 2674; TTY: YMQTPXB, or e-mail: *custserv@iata.org*.

Transfers

Regulations governing the transfer of biological agents are designed to ensure that possession of these agents is in the best interest of the public and the nation. These regulations require documentation of personnel, facilities, justification of need and pre-approval of the transfer by a federal authority. The following regulations apply to this category:

Importation of Etiologic Agents of Human Disease. 42 CFR Part 71 Foreign Quarantine. Part 71.54 Etiological Agents, Hosts and Vectors. This regulation requires an import permit from the CDC for importation of etiologic agents, hosts or vectors of human disease. The regulation, application form, and additional guidance is available at the CDC Web site: *http://www.cdc.gov/od/eaipp*.

Completed application forms may be submitted to the CDC Etiologic Agent Import Permit Program by fax: (404) 718-2093, or by mail:

Centers for Disease Control and Prevention
Etiologic Agent Import Permit Program
1600 Clifton Road, N.E., Mailstop A-46
Atlanta, GA 30333

Importation of select agents or toxins into the U.S. also requires the intended recipient to be registered with the Select Agent Program and submit an APHIS/ CDC Form 2 to obtain approval to import the select agent or toxin prior to each importation event (see 42 CFR 73 and/or 9 CFR 121). More information regarding select agents and toxins is available at: *www.selectagents.gov*.

Importation of Etiologic Agents of Livestock, Poultry and Other Animal Diseases and Other Materials Derived from Livestock, Poultry or Other Animal. 9 CFR Parts 122. Organisms and Vectors. The USDA, APHIS, Veterinary Services (VS) requires that a permit be issued prior to the importation or domestic transfer (interstate movement) of etiologic disease agents of livestock, poultry, other animals. Information may be obtained at (301) 734-5960, or from the USDA Web site: *http://www. aphis.usda.gov/animal_health*. Completed permit applications may be submitted electronically at: *http://www.aphis.usda.gov/permits/learn_epermits.shtml*; or by fax to (301) 734-3652; or by mail to:

USDA APHIS VS
National Center for Import and Export
4700 River Road
Unit 2, Mailstop 22, Cubicle 1A07
Riverdale, MD 20737

Importation of select agents into the United States also requires the intended recipient to be registered with the Select Agent Program and submit an APHIS/ CDC Form 2 to obtain approval to import the select agent or toxin prior to each importation event (see 42 CFR 73 and/or 9 CFR 121). More information regarding select agents and toxins is available at: *http://www.aphis.usda.gov/ programs/ag_selectagent/index.shtml*.

Importation of Plant Pests 7 CFR Part 330. Federal Plant Pest Regulations; General; Plant Pests; Soil; Stone and Quarry Products; Garbage. This regulation requires a permit for movement into or through the United States, or interstate any plant pest or a regulated product, article, or means of conveyance in accordance with this part. Information can be obtained by calling (877) 770-5990 or at the USDA Web site: *http://www.aphis.usda.gov/permits*.

Export of Etiologic Agents of Humans, Animals, Plants and Related Materials; Department of Commerce (DoC); 5 CFR Parts 730 to 799. This regulation requires that exporters of a wide variety of etiologic agents of human, plant and animal diseases, including genetic material, and products which might be used for culture of large amounts of agents, will require an export license. Information may be obtained by calling the DoC Bureau of Export Administration at (202) 482-4811, or at the DoC Web site: *http://www.ntis.gov/products/export-regs.aspx*; or at *http://www.access.gpo.gov/bis/index.html*; and *http://www.bis.doc.gov*.

Transfer of CDC Select Agents and Toxins. 42 CFR Part 73 Possession, Use, and Transfer of Select Agents and Toxins. The CDC regulates the possession, use, and transfer of select agents and toxins that have the potential to pose a severe threat to public health and safety. The CDC Select Agent Program registers all laboratories and other entities in the United States that possess, use, or transfer a select agent or toxin. Entities transferring or receiving select agents and toxins must be registered with the Select Agent Program and submit an APHIS/CDC Form 2 (see 42 CFR 73 and/or 9 CFR 121) to obtain approval prior to transfer of a select agent or toxin. The regulations, Select Agent Program forms, and additional guidance is available at the CDC Web site: *www.selectagents.gov*.

Transfer of USDA Select Agents and Toxins. 9 CFR Part 121 Possession, Use, and Transfer of Select Agents and Toxins. The USDA, APHIS, VS regulates the possession, use, and transfer of select agents and toxins that have the potential to pose a severe threat to animal health or animal products. The VS Select Agent Program oversees these activities and registers all laboratories and other entities in the U.S. that possess, use, or transfer a VS select agent or toxin. Entities transferring or receiving select agents and toxins must be registered with either the CDC or APHIS Select Agent Program, and submit an APHIS/CDC Form 2 (see 42 CFR 73 and/or 9 CFR 121) to obtain approval prior to transfer of a select agent or toxin. The regulations, Select Agent Program forms, and additional guidance is available at the APHIS Web site: *http://www.aphis.usda.gov/ programs/ag_selectagent/index.shtml*.

Transfer of USDA Plant Pests

The movement of Plant Pests is regulated under two distinct and separate regulations: (1) 7 CFR Part 331. Agricultural Bioterrorism Protection Act of 2002; Possession, Use, and Transfer of Biological Agents and Toxins; and (2) 7 CFR Part 330 Federal Plant Pest Regulations; General; Plant Pests; Soil; Stone and Quarry Products; Garbage. The regulation found at 7 CFR Part 331 requires an approved Transfer Form (APHIS/CDC Form 2) prior to importation, interstate, or intrastate movement of a Select Agent Plant Pest. In addition, under 7 CFR Part 330, the movement of a Plant Pest also requires a permit for movement into or through the United States, or interstate any plant pest or a regulated product, article, or means of conveyance in accordance with this part. Information can be obtained by calling (301) 734-5960 or at the USDA Web site: *http://www.aphis. usda.gov/programs/ag_selectagent/index.shtml*.

General DOT Packaging Requirements for Transport of Infectious Substances by Aircraft

The DOT packagings for transporting infectious substances by aircraft are required by domestic and international aircraft carriers, and are the basis for infectious substance packagings for motor vehicle, railcar, and vessel transport. The following is a summary of each packaging type and related transportation requirements.

Category A Infectious Substance (UN 2814 and UN 2900): Figure 1. A Category A material is an infectious substance that is transported in a form that is capable of causing permanent disability or life-threatening or fatal disease to otherwise healthy humans or animals when exposure to it occurs. An exposure occurs when an infectious substance is released outside of its protective packaging, resulting in physical contact with humans or animals. Category A infectious substances are assigned to identification number "UN 2814" for substances that cause disease in humans or in both humans and animals, or "UN 2900" for substances that cause disease in animals only.

Figure 1 shows an example of the UN standard triple packaging system for materials known or suspected of being a Category A infectious substance. The package consists of a watertight primary receptacle or receptacles; a watertight secondary packaging; for liquid materials, the secondary packaging must contain absorbent material in sufficient quantities to absorb the entire contents of all primary receptacles; and a rigid outer packaging of adequate strength for its capacity, mass, and intended use. Each surface of the external dimension of the packaging must be 100 mm (3.9 inches) or more. The completed package must pass specific performance tests, including a drop test and a water-spray test, and must be capable of withstanding, without leakage, an internal pressure producing a pressure differential of not less than 95 kPa (0.95 bar, 14 psi). The completed package must also be capable of withstanding, without leakage, temperatures in the range of -40°C to +55°C (-40°F to 131°F). The completed package must be marked "Infectious substances, affecting humans, UN 2814" or "Infectious substances, affecting animals, UN 2900" and labeled with a Division 6.2 (infectious substance) label. In addition, the package must be accompanied by appropriate shipping documentation, including a shipping paper and emergency response information.

Figure 1. *A Category A UN Standard Triple Packaging*

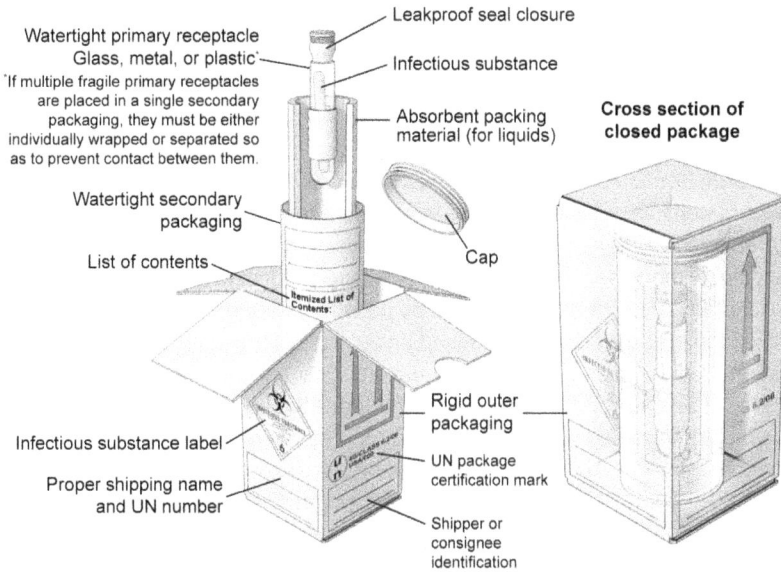

Leakproof seal closure

Watertight primary receptacle
Glass, metal, or plastic

Infectious substance

'If multiple fragile primary receptacles
are placed in a single secondary
packaging, they must be either
individually wrapped or separated so
as to prevent contact between them.

Absorbent packing
material (for liquids)

**Cross section of
closed package**

Watertight secondary
packaging

Cap

List of contents

Itemized List of
Contents:

Rigid outer
packaging

Infectious substance label

UN package
certification mark

Proper shipping name
and UN number

Shipper or
consignee
identification

Biological specimen, Category B (UN 3373): Figure 2. (previously known as Clinical specimen and Diagnostic Specimen). A Category B infectious substance is one that does not meet the criteria for inclusion in Category A. A Category B infectious substance does not cause permanent disability or life-threatening or fatal disease to humans or animals when exposure to it occurs. The proper shipping name for a Category B infectious substance, "Biological specimen, Category B," is assigned to identification number "UN 3373." The proper shipping names "Diagnostic specimen" and "Clinical specimen" may no longer be used (as of January 1, 2007).

Figure 2 shows an example of the triple packaging system for materials known or suspected of containing a Category B infectious substance. A Category B infectious substance must be placed in a packaging consisting of a leak proof primary receptacle, leak proof secondary packaging, and rigid outer packaging. At least one surface of the outer packaging must have a minimum dimension of 100 mm by 100 mm (3.9 inches). The packaging must be of good quality and strong enough to withstand the shocks and loadings normally encountered during transportation. For liquid materials, the secondary packaging must contain absorbent material in sufficient quantities to absorb the entire contents of all primary receptacles. The primary or secondary packaging must be capable of withstanding, without leakage, an internal pressure producing a pressure differential of 95 kPa. The package must be constructed and closed to prevent any loss of contents that might be caused under normal transportation conditions by vibration or changes in temperature, humidity, or pressure. The completed package must be capable of passing a 1.2-meter (3.9 feet) drop test. The package must be marked with a diamond-shaped marking containing the identification number "UN 3373" and with the proper shipping name "Biological substance, Category B." In addition, the name, address, and telephone number of a person knowledgeable about the material must be provided on a written document, such as an air waybill, or on the package itself.

Figure 2. *A Category B Non-specification Triple Packaging*

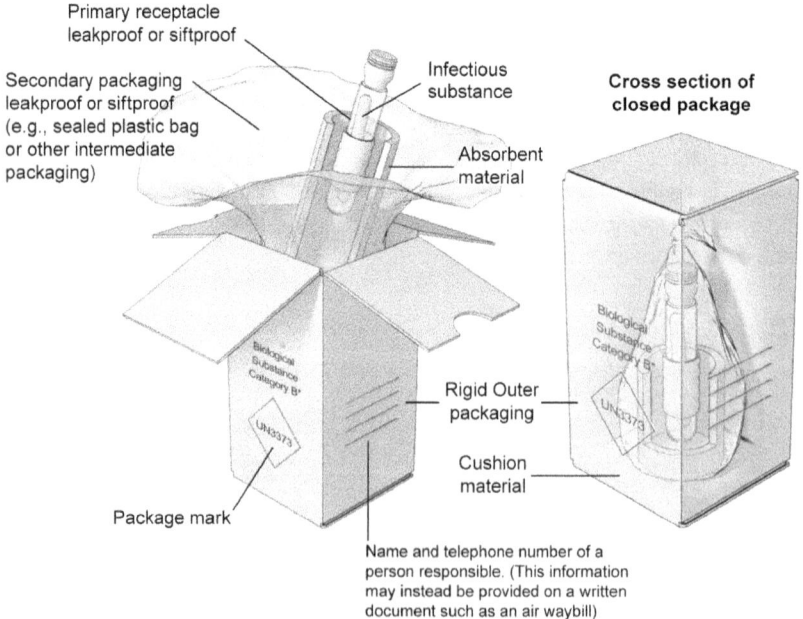

Appendix D—Agriculture Pathogen Biosafety

The contents of this Appendix were provided by USDA. All questions regarding its contents should be forwarded to the USDA.

Contents

I. *Introduction*

Risk assessment and management guidelines for agriculture differ from human public health standards. Risk management for agriculture research is based on the potential economic impact of animal and plant morbidity, and mortality, and the trade implications of disease. Agricultural guidelines take this difference into account. Worker protection is important but great emphasis is placed on reducing the risk of agent escape into the environment. This Appendix describes the facility parameters and work practices of what has come to be known as BSL-3-Ag. BSL-3-Ag is unique to agriculture because of the necessity to protect the environment from an economic, high risk pathogen in a situation where studies are conducted employing large agricultural animals or other similar situations in which the *facility barriers now serve as primary containment*. Also described are some of the enhancements beyond BSL-3 that may be required by USDA/APHIS when working in the laboratory or vivarium with veterinary agents of concern. This Appendix provides guidance and is not regulatory nor is it meant to describe policy. Conditions for approval to work with specific agricultural agents are provided at the time USDA/APHIS permits a location to work with an agent.

II. *BSL-3-Ag for Work with Loose-housed Animals*

In agriculture, special biocontainment features are required for certain types of research involving high consequence livestock pathogens in animal species or other research where the room provides the primary containment. To support such research, USDA has developed a special facility designed, constructed and operated at a unique animal containment level called BSL-3-Ag. Using the containment features of the standard ABSL-3 facility as a starting point, BSL-3-Ag facilities are specifically designed to protect the environment by including almost all of the features ordinarily used for BSL-4 facilities as enhancements. All BSL-3-Ag containment spaces must be designed, constructed and certified as primary containment barriers.

The BSL-3-Ag facility can be a separate building, but more often, it is an isolated zone within a facility operating at a lower biosafety level, usually at BSL-3. This isolated zone has strictly controlled access with special physical security measures and functions on the "box within a box" principle. All BSL-3-Ag facilities that cannot readily house animals in primary containment devices require the features for an ABSL-3 facility with the following enhancements typical of BSL-4 facilities:

1. Personnel change and shower rooms that provide for the separation of laboratory clothing from animal facility clothing and that control access to the containment spaces. The facility is arranged so that personnel ingress and egress are only through a series of rooms consisting of: a ventilated vestibule with compressible gaskets on the two doors, a "clean" change room outside containment, a shower room at the non-containment/containment boundary, and a "dirty" change room within containment. Complete animal facility clothing (including undergarments, pants and shirts or jump suits, and shoes and gloves) is typically provided in the "dirty" change room, and put on by personnel before entering the research areas. In some facilities, complete animal facility clothing and personal protective equipment are provided in the "clean" change room, where they can be stored and stowed for use without entry into containment. When leaving a BSL-3-Ag animal space that acts as the primary barrier and that contains large volumes of aerosols containing highly infectious agents (an animal room, necropsy room, carcass disposal area, contaminated corridor, etc.), personnel usually would be required to remove "dirty" lab clothing, take a shower, and put on "clean" lab clothing immediately after leaving this high risk animal space and before going to any other part of the BSL-Ag facility. When leaving the facility, these personnel would take another shower at the access control shower and put on their street clothing. Soiled clothing worn in a BSL-3-Ag space is autoclaved before being laundered. Personnel moving from one space within containment to another will follow the practices and procedures described in the biosafety manual specifically developed for the particular facility and adopted by the laboratory director.

2. Access doors are self closing and lockable. Emergency exit doors are provided, but are locked on the outside against unauthorized use. The architect or engineer shall consider the practicality of providing vestibules at emergency exits.

3. Supplies, materials and equipment enter the BSL-3-Ag space only through an airlock, fumigation chamber, an interlocked and double-door autoclave or shower.

4. Double-door autoclaves engineered with bioseals are provided to decontaminate laboratory waste passing out of the containment area. The double doors of the autoclaves must be interlocked so that the outer door can be opened only after the completion of the sterilizing cycle, and to prevent the simultaneous opening of both doors. All double door autoclaves are situated through an exterior wall of the containment area, with the autoclave unit forming an airtight seal with the barrier wall and the bulk of the autoclave situated outside the containment space so that autoclave maintenance can be performed conveniently. A gas sterilizer, a pass-through liquid dunk tank, or a cold gas decontamination chamber must be provided for the safe removal of materials and equipment that are steam or heat sensitive. Disposable materials must be decontaminated through autoclaving or other validated decontamination method followed by incineration.

5. Dedicated, single pass, directional, and pressure gradient ventilation systems must be used. All BSL-3-Ag facilities have independent air supply and exhaust systems that are operated to provide directional airflow and a negative air pressure within the containment space. The directional airflow within the containment spaces moves from areas of least hazard potential toward areas of greatest hazard potential. A visible means of displaying pressure differentials is provided. The pressure differential display/gauge can be seen inside and outside of the containment space, and an alarm sounds when the preset pressure differential is not maintained. The air supply and exhaust systems must be interlocked to prevent reversal of the directional airflow and positive pressurization of containment spaces in the event of an exhaust system failure.

6. Supply and exhaust air to and from the containment space is HEPA filtered. Exhaust air is discharged in such a manner that it cannot be drawn into outside air intake systems. The HEPA filters are outside of containment but are located as near as possible to the containment space to minimize the length of potentially contaminated air ducts. The HEPA filter housings are fabricated to permit scan testing of the filters in place after installation, and to permit filter decontamination before removal. Backup HEPA filter units are strongly recommended to allow filter changes without disrupting research. The most severe requirements for these modern, high level biocontainment facilities include HEPA filters arranged both in series and in parallel on the exhaust side, and parallel HEPA filters on the supply side of the HVAC systems serving "high risk" areas where large amounts of aerosols containing BSL-3-Ag agents could be expected (e.g., animal rooms, contaminated corridors, necropsy areas, carcass disposal facilities). For these high-risk areas, redundant supply and exhaust fans are recommended. The supply and exhaust air systems should be equipped with pre-filters (80-90% efficient) to prolong

the life of the HEPA filters. Air handling systems must provide 100% outside conditioned air to the containment spaces.

7. Liquid effluents from BSL-3-Ag areas must be collected and decontaminated in a central liquid waste sterilization system before disposal into the sanitary sewers. Typically, a heat decontamination system is utilized in these facilities and equipment must be provided to process, heat and hold the contaminated liquid effluents to temperatures, pressures and times sufficient to inactivate all biohazardous materials that reasonably can be expected to be studied at the facility in the future. The system may need to operate at a wide range of temperatures and holding times to process effluents economically and efficiently. Double containment piping systems with leak alarms and annular space decontaminating capability should be considered for these wastes. Effluents from laboratory sinks, cabinets, floors and autoclave chambers are sterilized by heat treatment. Under certain conditions, liquid wastes from shower rooms and toilets may be decontaminated by chemical treatment systems. Facilities must be constructed with appropriate basements or piping tunnels to allow for inspection of plumbing systems.

8. Each BSL-3-Ag containment space shall have its interior surfaces (walls, floors, and ceilings) and penetrations sealed to create a functional area capable of being certified as airtight. It is recommended that a pressure decay test be used (new construction only). Information on how to conduct a pressure decay test may be found within Appendix 9B of the ARS Facilities Design Manual (Policy and Procedure 242.1M-ARS; *http://www.afm. ars.usda.gov/*). This requirement includes all interior surfaces of all animal BSL-3-Ag spaces, not just the surfaces making up the external containment boundary. All walls are constructed slab to slab, and all penetrations, of whatever type, are sealed airtight to prevent escape of contained agents and to allow gaseous fumigation for biological decontamination. This requirement prevents cross contamination between individual BSL-3-Ag spaces and allows gaseous fumigation in one space without affecting other spaces. Exterior windows and vision panels, if required, are breakage-resistant and sealed. Greenhouses constructed to meet the BSL-3-Ag containment level will undergo the following tests, or the latest subsequent standards: (a) an air infiltration test conducted according to ASTM E 283-91; (b) a static pressure water resistance test conducted according to ASTM E 331-93; and (c) a dynamic pressure water resistance test conducted according to AAMA 501.1-94.

9. All ductwork serving BSL-3-Ag spaces shall be airtight (pressure tested-consult your facility engineer for testing and certification details).

10. The hinges and latch/knob areas of all passage doors shall be sealed to airtight requirements (pressure decay testing).

11. All airlock doors shall have air inflated or compressible gaskets. The compressed air lines to the air inflated gaskets shall be provided with HEPA filters and check valves.

12. Restraining devices shall be provided in large animal rooms.

13. Necropsy rooms shall be sized and equipped to accommodate large farm animals.

14. Pathological incinerators, or other approved means, must be provided for the safe disposal of the large carcasses of infected animals. Redundancy and the use of multiple technologies need to be considered and evaluated.

15. HEPA filters must be installed on all atmospheric vents serving plumbing traps, as near as possible to the point of use, or to the service cock, of central or local vacuum systems, and on the return lines of compressed air systems. All HEPA filters are installed to allow in-place decontamination and replacement. All traps are filled with liquid disinfectant.

16. If BSCs are installed, they should be located such that their operation is not adversely affected by air circulation and foot traffic. Class II BSCs use HEPA filters to treat their supply and exhaust air. Selection of the appropriate type of Class II BSCs will be dependent upon the proposed procedures and type of reagents utilized. BSC selection should be made with input from a knowledgeable safety professional well versed on the operational limitations of class II biohazard cabinetry. Supply air to a Class III cabinet is HEPA filtered, and the exhaust air must be double filtered (through a cabinet HEPA and then through a HEPA in a dedicated building exhaust system) before being discharged to the atmosphere.

III. BSL-3 and ABSL-3 Plus Potential Facility Enhancements for Agriculture Agent Permitting

The descriptions and requirements listed above for BSL-3-Ag studies are based on the use of high-risk organisms in animal systems or other types of agriculture research where the facility barriers, usually considered secondary barriers, now act as primary barriers. Certain agents that typically require a BSL-3-Ag facility for research that utilizes large agricultural animals may be studied in small animals in an enhanced BSL-3 laboratory or enhanced ABSL-3 when the research is done within primary containment devices. In these situations, the facility no longer serves as the primary barrier as with the large animal rooms. Therefore, when manipulating high consequence livestock pathogens in the laboratory or small animal facility, facility design and work procedures must meet the requirements of BSL-3 or ABSL-3 with additional enhancements unique to agriculture. Agriculture enhancements are agent, site and protocol dependent. The facility may have personnel enter and exit through a clothing change and shower room, have a double-door autoclave and/or fumigation chamber, HEPA filter supply and exhaust air, and a validated or approved system in place to decontaminate research materials and waste. Surfaces must be smooth to support wipe-down decontamination and penetrations should be sealed and the room capable of sealing in case gaseous decontamination is required. Because all work with infectious material is conducted within primary containment, there is no requirement for pressure decay testing the room itself.

The need for any potential agriculture enhancements is dependant upon a risk assessment. Therefore, after an assessment and in consultation with USDA/APHIS, the required agriculture enhancement(s) may include:

1. Personnel change and shower rooms that provide for the separation of street clothing from laboratory clothing and that control access to the containment spaces. The facility is arranged so that personnel ingress and egress are only through a series of rooms (usually one series for men and one for women) consisting of: a ventilated vestibule with a "clean" change room outside containment, a shower room at the non-containment/containment boundary, and a "dirty" change room within containment. Complete laboratory clothing (including undergarments, pants and shirts or jump suits, and shoes and gloves) is provided in the "dirty" change room, and put on by personnel before entering the research areas. In some facilities, complete laboratory clothing and personal protective equipment are provided in the "clean" change room, where they can be stored and stowed for use without entry into containment. When leaving a BSL-3 enhanced space, personnel usually would be required to remove their "dirty" laboratory clothing, take a shower, and put on "clean" laboratory clothing immediately after leaving the BSL-3 enhanced space and before going to any other part of the facility. Soiled clothing worn in a BSL-3 enhanced space should be autoclaved before being laundered outside of the containment space. Personnel moving from one space within containment to another will follow the practices and procedures described in the biosafety manual specifically developed for the particular facility and adopted by the laboratory director.

2. Access doors to these facilities are self closing and lockable. Emergency exit doors are provided but are locked on the outside against unauthorized use. The architect or engineer shall consider the practicality of providing vestibules at emergency exits.

3. Supplies, materials and equipment enter the BSL-3 enhanced space only through the double-door ventilated vestibule, fumigation chamber or an interlocked and double-door autoclave.

4. Double-door autoclaves engineered with bioseals are provided to decontaminate laboratory waste passing out of the containment area. The double doors of the autoclaves must be interlocked so that the outer door can be opened only after the completion of the sterilizing cycle, and to prevent the simultaneous opening of both doors. All double door autoclaves are situated through an exterior wall of the containment area, with the autoclave unit forming an airtight seal with the barrier wall and the bulk of the autoclave situated outside the containment space so that autoclave maintenance can be performed conveniently. A gas sterilizer, a pass-through liquid dunk tank, or a cold gas decontamination chamber must be provided for the safe removal of materials and equipment that are steam or heat sensitive. All other materials must be autoclaved or otherwise decontaminated by a method validated to inactivate the agent before being removed from the BSL-3 enhanced space. Wastes and other materials being removed from the BSL-3 enhanced space must be disposed of through incineration or other approved process.

5. Dedicated, single pass, directional, and pressure gradient ventilation systems must be used. All BSL-3 enhanced facilities have independent air supply and exhaust systems operated to provide directional airflow and a negative air pressure within the containment space. The directional airflow within the containment spaces moves from areas of least hazard potential toward areas of greatest hazard potential. A visible means of displaying pressure differentials is provided. The pressure differential display/gauge can be seen inside and outside of the containment space, and an alarm sounds when the preset pressure differential is not maintained. Supply and exhaust air to and from the containment space is HEPA filtered, with special electrical interlocks to prevent positive pressurization during electrical or mechanical breakdowns.

6. The exhaust air is discharged in such a manner that it cannot be drawn into outside air intake systems. HEPA filters located outside of the containment barrier are located as near as possible to the containment space to minimize the length of potentially contaminated air ducts. The HEPA filter housings are fabricated to permit scan testing of the filters in place after installation, and to permit filter decontamination before removal. Backup parallel HEPA filter units are strongly recommended to allow filter changes without disrupting research. Air handling systems must provide 100% outside conditioned air to the containment spaces.

7. Contaminated liquid wastes from BSL-3 enhanced areas must be collected and decontaminated by a method validated to inactivate the agent being used before disposal into the sanitary sewers. Treatment requirement will be determined by a site-specific, agent-specific risk assessment. Floor drains are discouraged in ABSL-3 and BSL-3

agriculture enhanced laboratories lacking a liquid waste central sterilization system. If floor drains are present, they should be capped and sealed. Facilities should be constructed with appropriate basements or piping tunnels to allow for inspection of plumbing systems, if a central liquid waste sterilization system is used.

8. Each BSL-3 enhanced containment space shall have its interior surfaces (walls, floors, and ceilings) and penetrations sealed to create a functional area capable of being decontaminated using a gaseous or vapor phase method. All walls are contiguous with the floor and ceiling, and all penetrations, of whatever type, are sealed. Construction materials should be appropriate for the intended end use. Exterior windows and vision panels, if required, are breakage-resistant and sealed.

9. All exhaust ductwork prior to the HEPA exhaust filter serving BSL-3 enhanced spaces shall be subjected to pressure decay testing before acceptance of the facility for use. Consult your facility engineer for testing and commissioning details.

IV. *Pathogens of Veterinary Significance*

Some pathogens of livestock, poultry and fish may require special laboratory design, operation, and containment features. This may be BSL-3, BSL-3 plus enhancements or BSL-4 and for animals ABSL-2, ABSL-3 or BSL-3-Ag. The importation, possession, or use of the following agents is prohibited or restricted by law or by USDA regulations or administrative policies.

This Appendix does not cover manipulation of diagnostic samples; however, if a foreign animal disease agent is suspected, samples should be immediately forwarded to a USDA diagnostic laboratory (The National Veterinary Services Laboratories, Ames, IA or the Foreign Animal Disease Diagnostic Laboratory, Plum Island, NY). A list of agents and their requirements follows.

African horse sickness virus [a, b]	Louping ill virus [a]
African swine fever virus [a, b, c]	Lumpy skin disease virus [a, b, c]
Akabane virus [b]	Malignant catarrhal fever virus (exotic strains or alcelaphine herpesvirus type 1) [b]
Avian influenza virus (highly pathogenic) [a, b, c]	Menangle virus [b]
Bacillus anthracis [a, b]	*Mycobacterium bovis*
Besnoitia besnoiti	*Mycoplasma agalactiae*
Bluetongue virus (exotic) [a, b]	*Mycoplasma mycoides* subsp. *mycoides* (small colony type) [a, b, c]

Borna disease virus	*Mycoplasma capricolum* [a, b, c]
Bovine infectious petechial fever agent	Nairobi sheep disease virus (Ganjam virus)
Bovine spongiform encephalopathy prion [b]	Newcastle disease virus (velogenic strains) [a, b, c]
Brucella abortus [a, b]	Nipah virus [a, b, d]
Brucella melitensis [a, b]	Peste des petits ruminants virus (plague of small ruminants) [a, b, c]
Brucella suis [a, b]	Rift Valley fever virus [a, b, c]
Burkholderia mallei/Pseudomonas mallei (Glanders) [a, b]	Rinderpest virus [a, b, c]
Burkholderia pseudomallei [a, b]	Sheep pox virus [a, b]
Camelpox virus [b]	Spring Viremia of Carp virus
Classical swine fever virus [a, b, c]	Swine vesicular disease virus [b]
Coccidioides immitis [b]	Teschen disease virus [a]
Cochliomyia hominivorax (Screwworm)	*Theileria annulata*
Coxiella burnetti (Q fever) [b]	*Theileria lawrencei*
Ephemeral fever virus	*Theileria bovis*
Ehrlichia (Cowdria) ruminantium (heartwater) [b]	*Theileria hirci*
Eastern equine encephalitis virus [a, b]	*Trypanosoma brucei*
Foot and mouth disease virus [a, b, c]	*Trypanosoma congolense*
Francisella tularensis [b]	*Trypanosoma equiperdum* (dourine)
Goat pox [a, b]	*Trypanosoma evansi*
Hemorrhagic disease of rabbits virus	*Trypanosoma vivax*
Hendra virus [a, b, d]	Venezuelan equine encephalomyelitis virus [a, b]
Histoplasma (Zymonema) farciminosum	Vesicular exanthema virus
Infectious salmon anemia virus	Vesicular stomatitis virus (exotic) [a, b]
Japanese encephalitis virus [a, b]	Wesselsbron disease virus

Notes:

[a] Export license required by Department of Commerce (See: *http://www.bis.doc.gov/index.htm*).

[b] Agents regulated as Select Agents under the Bioterrorism Act of 2002. Possession of these agents requires registration with either the CDC or APHIS and a permit issued for interstate movement or importation by APHIS-VS. Most require BSL-3/ABSL-3 or higher containment (enhancements as described in this Appendix or on a case-by-case basis as determined by APHIS-VS).

[c] Requires BSL-3-Ag containment for all work with the agent in loose-housed animals.

[d] Requires BSL-4 containment for all work with the agent.

A USDA/APHIS import or interstate movement permit is required to obtain any infectious agent of animals or plants that is regulated by USDA/APHIS. An import permit is also required to import any livestock or poultry product such as blood, serum, or other tissues.

V. Summaries of Selected Agriculture Agents

African Swine Fever Virus (ASFV)

ASF is a tick-borne and contagious, febrile, systemic viral disease of swine.[1,2,3] The ASF virus (ASFV) is a large (about 200 nm) lipoprotein-enveloped, icosahedral, double-stranded DNA virus in the family *Asfarviridae*, genus *Asfivirus*. This virus is quite stable and will survive over a wide range of pH. The virus will survive for 15 weeks in putrefied blood, three hours at 50°C, 70 days in blood on wooden boards, 11 days in feces held at room temperature, 18 months in pig blood held at 4°C, 150 days in boned meat held at 39°F, and 140 days in salted dried hams. Initially, domestic and wild pigs (Africa: warthog, bush pig, and giant forest hog; Europe: feral pig) were thought to be the only hosts of ASFV. Subsequently, researchers showed that ASFV replicates in *Ornithodoros* ticks and that there is transstadial, transovarial, and sexual transmission. ASF in wild pigs in Africa is now believed to cycle between soft ticks living in warthog burrows and newborn warthogs. *Ornithodoros* ticks collected from Haiti, the Dominican Republic, and southern California have been shown to be capable vectors of ASFV, but in contrast to the African ticks, many of the ticks from California died after being infected with ASFV. Because ASFV-infected ticks can infect pigs, ASFV is the only DNA virus that can qualify as an arbovirus.

Even though the soft tick has been shown to be a vector (and in Africa probably the reservoir of ASFV), the primary method of spread from country to country has been through the feeding of uncooked garbage containing ASFV-infected pork scraps to pigs.

Aerosol transmission is not important in the spread of ASF. Because ASFV does not replicate in epithelial cells, the amount of virus shed by an ASF-infected pig is much less than the amount of virus shed by a hog-cholera-infected pig. The blood of a recently infected pig contains a very high ASFV titer.

Laboratory Safety and Containment Recommendations

Humans are not susceptible to ASFV infection. The greatest risk of working with the virus is the escape of the organism into a susceptible pig population, which would necessitate USDA emergency procedures to contain and eradicate the disease.

ASF is considered a foreign animal disease in the United States. Due to the highly contagious nature of the agent and the severe economic consequences of disease in the United States, this organism should only be handled *in vitro* in a BSL-3 laboratory with enhancements as required by the USDA and *in vivo* in a

USDA-approved BSL-3-Ag facility for loosely housed animals. Special consideration should be given to infected vector control.

Special Issues

The importation, possession, or use of this agent is prohibited or restricted by law or by USDA regulations or administrative policies. A USDA/APHIS import or interstate movement permit is required to obtain this agent or any livestock or poultry product, such as blood, serum, or other tissues containing the agent.

African Horse Sickness Virus (AHSV)

AHSV is a member of genus *Orbivirus* in the family *Reoviridae*. Nine serotypes, numbers 1 – 9, are recognized. AHSV grows readily in embryonated chicken eggs, suckling mice, and a variety of standard cell cultures. AHSV infects and causes viremia in equids. Most horses die from the disease, about half of donkeys and most mules survive, but zebras show no disease. Viremias may last up to one month despite the rapid development of neutralizing antibodies. AHSV may cause disease in dogs, but these are not thought to be important in the natural history of the disease.[4,5]

AHSV has been recognized in central Africa and periodically spreads to naive populations in South and North Africa, the Iberian Peninsula, the Middle East, Pakistan, Afghanistan, and India. AHSV is vectored by *Culicoides* species and perhaps by mosquitoes, biting flies, and ticks limiting viral spread to climates and seasons favorable to the vectors. At least one North American *Culicoides* species transmits AHSV. AHSV may infect carnivores that consume infected animals but these are not thought to be relevant to natural transmission to equids.

Occupational Infections

Encephalitis and uveochorioretinitis were observed in four laboratory workers accidentally exposed to freeze-dried modified live vaccine preparations. Although AHSV could not be conclusively linked to disease, all four had neutralizing antibodies. Encephalitis was documented in experimentally infected monkeys.

Laboratory Safety and Containment Recommendations

Virus may be present in virtually any sample taken from an infected animal, but the highest concentrations are found in spleen, lung, and lymph nodes. The only documented risk to laboratory workers involves aerosol exposure to large amounts of vaccine virus. AHSV is unusually stable in blood or serum stored at 4°C.

AHS is considered a foreign animal disease in the United States. Due to the severe economic consequences of disease presence in the United States, this organism should only be handled *in vitro* in a BSL-3 laboratory with enhancements as required by the USDA and *in vivo* in a USDA-approved ABSL-3 animal facility with enhancements. Blood, serum, or tissues taken from equids in areas where AHSV exists are potential means of transmitting the agent long distances. Special consideration should be given to infected vector containment.

Special Issues

The importation, possession, or use of this agent is prohibited or restricted by law or by USDA regulations or administrative policies. A USDA/APHIS import or interstate movement permit is required to obtain this agent or any livestock or poultry product, such as blood, serum, or other tissues containing the agent.

Akabane Virus (AKAV)

AKAV is a member of the genus *Orthobunyavirus* in the Simbu serogroup of the family *Bunyaviridae*. The Simbu serogroup also includes Aino, Peaton, and Tinaroo viruses that can cause similar disease. Experimental infection of pregnant hamsters leads to death of the fetus. This virus grows and causes disease in chick embryos. Isolated in suckling mice and hamster lung cell cultures, AKAV is an important cause of disease in ruminants. The virus does not cause overt disease in adults but infects the placenta and fetal tissues in cattle, sheep, and goats to cause abortions, stillbirths, and congenital malformations. The broad range of clinical signs in the fetus is related primarily to central nervous system damage that occurs during the first trimester of pregnancy.[6,7]

AKAV is not known to infect or cause disease in humans; concern focuses only on effects to agriculture and wildlife. Common names of disease include congenital arthrogryposis-hydranencephaly syndrome, Akabane disease, acorn calves, silly calves, curly lamb disease, curly calf disease, and dummy calf disease. The host range of naturally occurring Akabane disease appears limited to cattle, sheep, swine, and goats but other animals including swine and numerous wildlife species become infected. AKAV is an Old World virus, being found in Africa, Asia, and Australia. Disease is unusual in areas where the virus is common because animals generally become immune before pregnancy. AKAV spreads naturally only in gnat and mosquito insect vectors that become infected after feeding on viremic animals.

Laboratory Safety and Containment Recommendations

AKAV may be present in blood, sera, and tissues from infected animals, as well as vectors from endemic regions. Parenteral injection of these materials into naive animals and vector-borne spread to other animals represents a significant risk to agricultural interests.

Akabane disease is considered a foreign animal disease in the United States. Due to the severe economic consequences of disease presence in the United States, this organism should only be handled *in vitro* in a BSL-3 laboratory with enhancements as required by the USDA and *in vivo* in a USDA-approved ABSL-3 animal facility with enhancements. Special consideration should be given to infected vector containment.

Specials Issues

Although it is virtually certain AKAV will grow and cause disease in New World livestock, it is not known if it will cause viremias in New World wildlife high enough to infect vectors, if it can be vectored by New World insects, or if it will cause disease in New World wildlife. Because fetal disease may not become evident until months after virus transmission, an introduction into a new ecosystem may not be recognized before the virus has become firmly entrenched.

The importation, possession, or use of this agent is prohibited or restricted by law or by USDA regulations or administrative policies. A USDA/APHIS import or interstate movement permit is required to obtain this agent or any livestock or poultry product, such as blood, serum, or other tissues containing the agent.

Bluetongue Virus (BTV)

BTV is a member of the family *Reoviridae*, genus *Orbivirus*. There are 24 recognized serotypes numbered 1 through 24. BTV is notable for causing disease in sheep and cattle and is very similar to other orbiviruses that cause disease in deer (epizootic hemorrhagic disease of deer virus) and horses (AHSV), and a few that cause disease in man (Colorado tick fever virus and others). These viruses have dsRNA genomes distributed amongst 10 segments, enabling efficient reassortment. Growth on a wide variety of cultured cells is usually cytocidal. Growth in animals results in viremia within three to four days that endures as long as 50 days despite the presence of high levels of neutralizing antibodies.[8,9]

BTV infects all ruminants, but bluetongue disease is unusual except in sheep and is unpredictable even in sheep. Disease is evidenced by fever, hyperemia, swelling, and rarely erosions and ulceration of the buccal and nasal mucosa. Hyperemia of the coronary bands of the hooves may cause lameness. In the worst cases, the disease progresses through weakness, depression, rapid weight loss, prostration, and death. Maternal transmission to the fetus may cause abortion or fetal abnormalities in the first trimester. Bluetongue disease also occurs in cattle but is rarely diagnosed. BTV may infect fetal calves and result in abortion or fetal brain damage. The full host range of BTV is still unknown but includes wild ruminants, neonatal mice, dogs, and chicken embryos.

BTV infection occurs in tropical, subtropical, and temperate climates where the *Culicoides* vectors exist. Global warming may be expanding the geographic range of *Culicoides*, and therefore BTV, into higher latitudes. Most countries have a unique assortment of the 24 serotypes. For example, BTV serotypes 2, 10, 11, 13, and 17 are currently active in the United States, but serotypes 1, 3, 4, 6, 8, 12, and 17 were present in the Caribbean basin when last surveyed. Concern over the spread of individual serotypes by trade in animals and animal products has engendered costly worldwide trade barriers.

The primary natural mode of transmission is by *Culicoides* midges. Only a few of more than 1,000 species of *Culicoides* transmit BTV. A strong correlation between the vector species and the associated BTV suggests these viruses may have adapted to their local vector. Thus, BTV does not exist in areas such as the Northeast United States where the local *Culicoides* fails to transmit BTV. Virus is present in semen at peak of viremia, but this is not considered a major route of transmission. Because of the prolonged viremia, iatrogenic transmission is possible. Only modified-live (attenuated) virus vaccines are available and a vaccine for only one serotype is currently available in the United States.

Laboratory Safety and Containment Recommendations

BTV is not known to cause disease in humans under any conditions. BTV commonly enters the laboratory in blood samples. The virus is stable at -70°C and in blood or washed blood cells held at 4°C. Sera prepared from viremic animals may represent some risk if introduced parenterally into naive animals. Blood, sera, and bovine semen can carry BTV across disease control boundaries.

The most significant threat from BTV occurs when virus is inoculated parenterally into naive animals. If appropriate *Culicoides* are present, virus can be transmitted to other hosts. Therefore, BTV-infected animals must be controlled for the two-month period of viremia and protected against *Culicoides* by physical means and/ or performing experiments at least two months before local *Culicoides* emerge. Thus, BTV exotic to the United States should only be handled *in vitro* in a BSL-3 laboratory with enhancements as required by the USDA and *in vivo* in a USDA-approved ABSL-3 with enhancements. Special consideration should be given to infected vector containment. Special containment is only needed when working with serotypes of BTV that are exotic to the country or locality. BTV on laboratory surfaces is susceptible to 95% ethanol and 0.5% sodium hypochlorite solution.

Special Issues

The importation, possession, or use of this agent is prohibited or restricted by law or by USDA regulations or administrative policies. A USDA/APHIS import or interstate movement permit is required to obtain this agent or any livestock or poultry product, such as blood, serum, or other tissues containing the agent.

Classical Swine Fever Virus (Hog Cholera)

Classical swine fever is a highly contagious viral disease of swine that occurs worldwide in an acute, a subacute, a chronic, or a persistent form.[10-12] In the acute form, the disease is characterized by high fever, severe depression, multiple superficial and internal hemorrhages, and high morbidity and mortality. In the chronic form, the signs of depression, anorexia, and fever are less severe than in the acute form, and recovery is occasionally seen in mature animals. Transplacental infection with viral strains of low virulence often results in persistently infected piglets, which constitute a major cause of virus dissemination to noninfected farms. Although minor antigenic variants of classical swine fever virus (CSFV) have been reported, there is only one serotype. Hog cholera virus is a lipid-enveloped pathogen belonging to the family *Flaviviridae*, genus *Pestivirus*. The organism has a close antigenic relationship with the bovine viral diarrhea virus (BVDV) and the border disease virus (BDV). In a protein-rich environment, hog cholera virus is very stable and can survive for months in refrigerated meat and for years in frozen meat. The virus is sensitive to drying (desiccation) and is rapidly inactivated by a pH of less than 3 and greater than 11.

The pig is the only natural reservoir of CSFV. Blood, tissues, secretions and excretions from an infected animal contain virus. Transmission occurs mostly by the oral route, though infection can occur through the conjunctiva, mucous membrane, skin abrasion, insemination, and percutaneous blood transfer (e.g., common needle, contaminated instruments). Airborne transmission is not thought to be important in the epizootiology of classical swine fever. Introduction of infected pigs is the principal source of infection in classical swine fever-free herds. Farming activities such as auction sales, livestock shows, visits by feed dealers, and rendering trucks also are potential sources of contagion. Feeding of raw or insufficiently cooked garbage is a potent source of hog cholera virus. During the warm season, insect vectors common to the farm environment may spread hog cholera virus mechanically. There is no evidence, however, that hog cholera virus replicates in invertebrate vectors.

Laboratory Safety and Containment Recommendations

Humans being are not susceptible to infection by CSFV. The greatest risk of working with these viruses is the escape of the organism into susceptible domestic or feral pig populations, which would necessitate USDA emergency procedures to contain and eradicate the diseases.

The virus is considered cause of a foreign animal disease in the United States. Due to the highly contagious nature of the agent and the severe economic consequences of disease presence in the United States, this organism should only be handled *in vitro* in a BSL-3 laboratory with enhancements as required by the USDA and *in vivo* in a USDA-approved BSL-3-Ag facility for loosely housed animals. Laboratory workers should have no contact with susceptible hosts for five days after working with the agent.

Specials Issues

The importation, possession, or use of this agent is prohibited or restricted by law or by USDA regulations or administrative policies. A USDA/APHIS import or interstate movement permit is required to obtain this agent or any livestock or poultry product, such as blood, serum, or other tissues containing the agent.

Contagious Bovine Pleuropneumonia Agent (CBPP)

CBPP is a highly infectious acute, subacute, or chronic disease, primarily of cattle, affecting the lungs and occasionally the joints, caused by *Mycoplasma mycoides mycoides*.[13-15] Contagious bovine pleuropneumonia is caused by *M. mycoides mycoides* small-colony type (SC type). *M. mycoides mycoides* large-colony type is pathogenic for sheep and goats but not for cattle. *M. mycoides mycoides* (SC type) survives well only *in vivo* and is quickly inactivated when exposed to normal external environmental conditions. The pathogen does not survive in meat or meat products and does not survive outside the animal in nature for more than a few days. Many of the routinely used disinfectants will effectively inactivate the organism.

CBPP is predominantly a disease of the genus *Bos*; both bovine and zebu cattle are naturally infected. There are many reported breed differences with respect to susceptibility. In general, European breeds tend to be more susceptible than indigenous African breeds. In zoos, the infection has been recorded in bison and yak. Although it has been reported that the domestic buffalo (*Bubalus bubalis*) is susceptible, the disease is difficult to produce experimentally in this species.

CBPP is endemic in most of Africa. It is a problem in parts of Asia, especially India and China. Periodically, CBPP occurs in Europe, and outbreaks within the last decade have occurred in Spain, Portugal, and Italy. The disease was eradicated from the United States in the nineteenth century, and it is not present currently in the Western hemisphere.

CBPP is spread by inhalation of droplets from an infected, coughing animal. Consequently, relatively close contact is required for transmission to occur. Outbreaks usually begin as the result of movement of an infected animal into a naive herd. There are limited anecdotal reports of fomite transmission, but fomites are not generally thought to be a problem.

Laboratory Safety and Containment Recommendations

Humans are not susceptible to infection by CBPP. The greatest risk of working with these mycoplasma is the escape of the organism into susceptible domestic bovine populations, which would necessitate USDA emergency procedures to contain and eradicate the diseases.

CBPP is considered a foreign animal disease in the United States. Due to the highly contagious nature of the agent and the severe economic consequences of disease presence in the United States, this organism should only be handled *in vitro* in a BSL-3 laboratory with enhancements as required by the USDA and *in vivo* in a USDA-approved BSL-3-Ag facility for loosely housed animals.

Special Issues

The importation, possession, or use of this agent is prohibited or restricted by law or by USDA regulations or administrative policies. A USDA/APHIS import or interstate movement permit is required to obtain this agent or any livestock or poultry product, such as blood, serum, or other tissues containing the agent.

Contagious Caprine Pleuropneumonia Agent (CCPP)

CCPP is an acute highly contagious disease of goats caused by a mycoplasma and characterized by fever, coughing, severe respiratory distress, and high mortality.[16-18] The principal lesion at necropsy is fibrinous pleuropneumonia. The causative agent of CCPP is considered to be *M. mycoides capri* (type strain PG-3) or a new mycoplasma *M. capricolum* subsp. *capripneumoniae* (designated F-38).[19-21] Neither of these agents occurs in North America.

M. mycoides mycoides has also been isolated from goats with pneumonia. This agent (the so-called large colony or LC variant of *M. mycoides mycoides*) usually produces septicemia, polyarthritis, mastitis, encephalitis, conjunctivitis, hepatitis, or pneumonia in goats. Some strains of this agent (LC variant) will cause pneumonia closely resembling CCPP, but the agent is not highly contagious and is not considered to cause CCPP. It does occur in North America. *M. capricolum capricolum*, a goat pathogen commonly associated with mastitis and polyarthritis in goats, can also produce pneumonia resembling CCPP, but it usually causes severe septicemia and polyarthritis. This agent (which does occur in the United States) is closely related to mycoplasma F-38 but can be differentiated from it using monoclonal antibodies.

CCPP is a disease of goats, and where the classical disease has been described, only goats were involved in spite of the presence of sheep and cattle. Mycoplasma F-38, the probable cause of the classic disease, does not cause disease in sheep or cattle. *M. mycoides capri*, the other agent considered a cause of CCPP, will result in a fatal disease in experimentally inoculated sheep and can spread from goats to sheep. It is however, not recognized as a cause of natural disease in sheep.

CCPP has been described in many countries of Africa, the Middle East, Eastern Europe, the former Soviet Union, and the Far East. It is a major scourge in many of the most important goat-producing countries in the world and is considered by many to be the world's most devastating goat disease.

CCPP is transmitted by direct contact through inhalation of infective aerosols. Of the two known causative agents, F-38 is far more contagious. Outbreaks of the disease often occur after heavy rains (e.g., after the monsoons in India) and after cold spells. This is probably because recovered carrier animals start shedding the mycoplasmas after the stress of sudden climatic change. It is believed that a long-term carrier state may exist.

Laboratory Safety and Containment Recommendations

Humans are not susceptible to infection by the agent that causes CCPP. The greatest risk of working with this mycoplasma is the escape of the organism into susceptible domestic caprine populations, which would necessitate USDA emergency procedures to contain and eradicate the diseases.

CCPP is considered a foreign animal disease in the United States. Due to the highly contagious nature of the agent and the severe economic consequences of disease in the United States, this organism should only be handled *in vitro* in a BSL-3 laboratory with enhancements as required by the USDA and *in vivo* in a USDA-approved BSL-3-Ag facility for loosely housed animals.

Special Issues

The importation, possession, or use of this agent is prohibited or restricted by law or by USDA regulations or administrative policies. A USDA/APHIS import or interstate movement permit is required to obtain this agent or any livestock or poultry product, such as blood, serum, or other tissues containing the agent.

Foot and Mouth Disease Virus (FMD)

FMD is a severe, highly communicable viral disease of cloven-hoofed animals (cattle, swine, sheep, and goats), causing fever, malaise, vesicular lesions in affected livestock and in some cases death in young animals due to myocardial lesions.[22] It can also affect a variety of wild ruminants (e.g., deer, bison). FMD is one of the most devastating diseases of livestock, causing large economic losses when introduced to FMD-free countries. The etiologic agent, FMD virus (FMDV), is a member of the *aphtovirus* genus, family *picornaviridae* with seven serotypes

(A, O, C, Asia1, SAT1, SAT2 and SAT3).[23] Humans are considered accidental hosts for FMDV and rarely become infected or develop clinical disease. Historically, humans have been exposed to large quantities of FMDV both during natural outbreaks among large herds of animals and in laboratory settings. Despite this, there has been an extremely low incidence of human infections reported and many have been anecdotal. Reports of fever, headaches and vesicles in the skin (especially at an accidental inoculation site) and oral mucosa have been associated with documented FMDV infections. The symptoms can be easily mistaken with those of Hand, Foot and Mouth Disease caused by coxsackie A viruses. On the other hand, humans have been shown to carry virus in their throats for up to three days after exposure to aerosols from infected animals, potentially making them carriers of FMDV. Humans and their clothing and footwear have been implicated as fomites for transmission of FMDV during outbreaks. Therefore, most FMDV laboratories impose a five day period of contact avoidance with susceptible species for personnel working with the viruses.

Laboratory Safety and Containment Recommendations

Laboratory practices for FMDV are principally designed to prevent transmission to susceptible livestock, but also to protect workers. The greatest risk of working with FMD is the escape of the organism into susceptible animal populations, which would necessitate USDA emergency procedures to contain and eradicate the disease.

The virus is considered a cause of a foreign animal disease in the United States. Due to the highly contagious nature and the severe economic consequences of disease presence in the United States, this virus should only be handled *in vitro* in a BSL-3 laboratory with enhancements as required by the USDA (see Section IV of this Appendix) and *in vivo* in USDA-approved BSL-3-Ag animal facilities. Infected animals are handled with standard protection (gloves, protective clothing). Change of clothing, personal showers and clearing of the throat and nose are required upon exiting contaminated areas in order to minimize virus transmission to susceptible species. Laboratory workers should have no contact with susceptible hosts for five days after working with the agent. In the United States, the Plum Island Animal Disease Center in New York is the only laboratory authorized to possess and work with this agent.

Special Issues

FMDV is a select agent. Possession, transfer and use of this agent requires application of procedures as detailed in the Agricultural Bioterrorism Protection Act of 2002 and codified in 9 CFR Part 121. All rules concerning the possession, storage, use, and transfer of select agents apply. Please review Appendix F of this document for further instructions regarding select agents. Law prohibits research with FMD on the United States mainland.

Heartwater Disease Agent (HD)

HD is a non-contagious disease of domestic and wild ruminants caused by *Ehrlichia ruminantium*.[24] *E. ruminantium* (formerly *Cowdria ruminantium*) is a member of the family *Rickettsiaceae* characterized by organisms that are obligate intracellular parasites. These organisms often persist in the face of an immune response due to their protected intracellular status. Rickettsias in natural conditions are found in mammals and blood-sucking arthropods. Ticks of the genus *Amblyomma* transmit *E. ruminantium*. HD occurs primarily in Africa, but has been recognized in the West Indies since the 1980's. The pathogen is transmitted by ticks of the genus *Amblyomma*, most importantly *A. variegatum* (tropical bont tick). This tick has wide distribution in Africa and is present on several Caribbean islands. Three North American tick species, *A. maculatum* (Gulf Coast tick), *A. cajennese*, and *A. dissimile*, can transmit the organism, causing concern that competent vectors could transmit *E. ruminantium* in the United States.

Severe HD comprises fever, depression, rapid breathing, and convulsions in cattle, sheep, goats and water buffalo. Whitetail deer also are susceptible to *E. ruminantium* infection and develop severe clinical disease. HD has not been diagnosed in the United States but occurs in numerous Caribbean islands, as well as in most countries of Africa, south of the Sahara Desert.

Laboratory Safety and Containment Recommendations

E. ruminantium can be found in whole blood, brain and experimentally in liver and kidney. It is not a human pathogen. Humans are not susceptible to infection with the agent that causes HD. The greatest risk of working with this agent is the escape of this organism (or infected ticks) into a susceptible domestic bovine population, which would necessitate USDA emergency procedures to contain and eradicate the disease.

HD is considered a foreign animal disease in the United States. *E. ruminantium* should be handled *in vitro* in BSL-3 laboratory facilities. Animal work should be conducted in ABSL-3 animal facilities or in ABSL-2 animal facilities with special modifications such as tick dams (where applicable).

Special Issues

The importation, possession, or use of this agent is prohibited or restricted by law or by USDA regulations or administrative policies. A USDA/APHIS import or interstate movement permit is required to obtain this agent or any livestock or poultry product, such as blood, serum, or other tissues containing the agent.

Infectious Salmon Anemia (ISA) Virus

ISA is a disease of Atlantic salmon (*Salmo salar*) caused by an orthomyxovirus in the family *Orthomyxoviridae*, genus *Isavirus*. Both wild and cultured Atlantic salmon are susceptible to infection, as are brown trout (*Salmo trutta*), rainbow trout (*Oncorhynchus mykiss*) and herring. The first clinical cases of ISA in Atlantic salmon were reported from Norway in 1984. Since then, ISA has been observed in Canada (1996), Scotland (1998), Chile (1999), Faroe Islands (2000) and the U.S. (2001).[25,26] There is significant molecular difference between virus isolates (i.e., "Norwegian", "Scottish" and "North American").[27] Clinical signs of ISA include severe anemia, swelling and hemorrhaging in the kidney and other organs, pale gills, protruding eyes, darkening of the posterior gut, fluid in the body cavity and lethargy. The infection is systemic and most noted in blood and mucus, muscle, internal organs and feces. The principal target organ for ISA virus (ISAV) is the liver. Signs usually appear two to four weeks after the initial infection.

Reservoirs of ISAV infection are unknown, but the spread of infection may occur due to the purchase of subclinically infected smolts, from farm to farm, and from fish slaughterhouses or industries where organic material (especially blood and processing water) from ISAV-infected fish is discharged without necessary treatment.[28]

Laboratory Safety and Containment Recommendations

Humans are not susceptible to ISAV infection. The greatest risk of working with this virus is the escape of the organism into a susceptible fish population, which would necessitate USDA emergency procedures to contain and eradicate the disease.

ISA is considered a reportable disease in the United States. ISAV should be handled *in vitro* in BSL-2 laboratory facilities with enhancements as required by USDA. Animal inoculations should be handled in ABSL-3 animal facilities with special modifications as required. Recommended precautions include incineration of fish, tissues, blood and materials (gloves, laboratory coats, etc.) used in the collection and processing of fish samples. All surfaces exposed to potentially infected fish should be disinfected with 0.04 to 0.13% acetic acid, chlorine dioxide at 100 parts/million for five minutes or sodium hypochlorite 30 mg available chlorine/liter for two days or neutralized with sodium thiosulfate after three hours. General principles of laboratory safety should be practiced in handling and processing fish samples for diagnostic or investigative studies. Laboratory managers should evaluate the need to work with ISAV and the containment capability of the facility before undertaking work with the virus or suspected ISAV-infected fish.

Special Issues

The importation, possession, or use of this agent is prohibited or restricted by law or by USDA regulations or administrative policies. A USDA/APHIS import or interstate movement permit is required to obtain this agent or any livestock or poultry product, such as blood, serum, or other tissues containing the agent.

Lumpy Skin Disease (LSD)Virus

LSD is an acute to chronic viral disease of cattle characterized by skin nodules that may have inverted conical necrosis (sit fast) with lymphadenitis accompanied by a persistent fever.[29-31] The causative agent of LSD is a capripoxvirus in the family *Poxviridae*, genus *Capripoxvirus*. The prototype strain of LSD virus (LSDV) is the Neethling virus. LSDV is one of the largest viruses (170-260 nm by 300-450 nm) and there is only one serotype. The LSDV is very closely related serologically to the virus of sheep and goat pox (SGP) from which it cannot be distinguished by routine virus neutralization or other serological tests. The virus is very resistant to physical and chemical agents, persists in necrotic skin for at least 33 days and remains viable in lesions in air-dried hides for at least 18 days at ambient temperature.

LSD is a disorder of cattle. Other wild ungulates have not been infected during epizootics in Africa. Lumpy skin disease was first described in Northern Rhodesia in 1929. Since then, the disease has spread over most of Africa in a series of epizootics and most recently into the Middle East. Biting insects play the major (mechanical) role in the transmission of LSDV. Direct contact seems to play a minor role in the spread of LSD.

Laboratory Safety and Containment Recommendations

Human beings are not susceptible to infection by LSDV. The greatest risk of working with this virus is the escape of the organism into susceptible domestic animal populations, which would necessitate USDA emergency procedures to contain and eradicate the diseases.

Lumpy skin disease is considered a foreign animal disease in the United States. Due to the highly contagious nature of the agent and the severe economic consequences of disease in the United States, this organism should only be handled *in vitro* in a BSL-3 laboratory with enhancements as required by the USDA and *in vivo* in a USDA-approved BSL-3-Ag facility for loosely housed animals.

Special Issues

The importation, possession, or use of this agent is prohibited or restricted by law or by USDA regulations or administrative policies. A USDA/APHIS import or interstate movement permit is required to obtain this agent or any livestock or poultry product, such as blood, serum, or other tissues containing the agent.

Malignant Catarrhal Fever Virus (MCFV) (Exotic Strains)

Alcelaphine herpesvirus 1 (AlHV-1) is a herpesvirus of the *Rhadinovirus* genus in the *Gammaherpesvirinae* subfamily.[32] Common names for AlHV-1 include wildebeest-associated malignant catarrhal fever virus (MCFV), African form MCFV, and exotic MCFV. It also was previously called bovine herpesvirus[3]. As a typical herpesvirus, AlHV-1 is a linear double-stranded DNA, enveloped virus. The virus can be propagated in certain primary or secondary cell cultures such as bovine thyroid and testis cells. The isolation of AlHV-1 requires the use of viable lymphoid cells from the diseased animal or cell-free virus in ocular/nasal secretions from wildebeest calves during a viral shedding period. Like other herpesviruses, AlHV-1 is fragile and quickly inactivated in harsh environments (for example, desiccation, high temperatures, and UV/sunlight), and by common disinfectants.

Wildebeest-associated MCF caused by AlHV-1 is also known as the African form of MCF, malignant catarrh, or snotsiekte (snotting sickness). The disease primarily affects many poorly adapted species of *Artiodactyla* that suffer very high case mortality (>95%) but low case morbidity (<7%). Wildebeest are the reservoir for AlHV-1 and the virus does not cause any significant disease in its natural host. Wildebeest-associated MCF primarily occurs in domestic cattle in Africa and in a variety of clinically susceptible ruminant species in zoological collections where wildebeest are present. Virtually all free-living wildebeest are infected with the virus and calves less than four months of age serve as the source of virus for transmission. The disease can be experimentally transmitted between cattle only by injection with infected viable cells from lymphoid tissues of affected animals. The disease cannot be transmitted by natural means from one clinically susceptible host to another, because there is essentially no cell-free virus in tissues or secretions of diseased animals. MCF is not a contagious disease.

Laboratory Safety and Containment Recommendations

There is no evidence that AlHV-1 can infect humans. Virus can be grown in several bovine cell lines at relatively low titers (ranging from 103 to 105 $TCID_{50}$). Infectivity in blood and tissues of affected animals is generally associated with viable lymphoid cells. The virus can be easily inactivated by wiping down surfaces with common disinfectants (such as bleach and sodium hypochlorite) and by autoclaving virus-contaminated materials.

This organism should only be handled *in vitro* in a BSL-3 laboratory with enhancements as required by the USDA and *in vivo* in a USDA-approved ABSL-3 animal facility with enhancements.

Special Issues

The importation, possession, or use of this agent is prohibited or restricted by law or by USDA regulations or administrative policies. A USDA/APHIS import or interstate movement permit is required to obtain this agent or any livestock or poultry product, such as blood, serum, or other tissues containing the agent.

Menangle Virus (MenV)

MenV caused a single outbreak of reproductive disease in an Australian swine operation. Clinical signs included stillborn, deformed, mummified piglets and a drop in the farrowing rate. Transmission between pigs has been postulated to be of a fecal-oral nature. A serological survey of fruit bats living near the swine operation revealed the presence of antibodies to MenV. Fruit bats are considered to be the natural host of the virus and their proximity to the affected premises led to an incidental infection in the pig population.[33,34]

MenV is a member of the family *Paramyxoviridae*, subfamily *Paramyxovirinae*. Other members of this family include Hendra virus, Nipah virus and Tioman virus of which Hendra and Nipah have been found to be fruit bat-associated. This virus was isolated from stillborn piglets from a single outbreak of reproductive disease in a commercial swine operation in New South Wales, Australia in 1997.

Occupational Infections

There was serological evidence of MenV infection in two people that had close contact with infected pigs on the affected premises. They demonstrated clinical signs similar to those seen with influenza such as chills, fever, drenching sweats, headache and rash. Both workers recovered fully from their illness.

Laboratory Safety and Containment Recommendations

Laboratory practices for MenV are principally designed to prevent transmission to susceptible livestock, but also to protect workers. The greatest risk of working with MenV is the escape of the organism into susceptible animal populations, which would necessitate USDA emergency procedures to contain and eradicate the disease.

MenV is considered cause of a foreign animal disease in the United States and is a human pathogen. Due to the severe economic consequences of disease presence in the United States, this organism should only be handled *in vitro* in a BSL-3 laboratory with enhancements as required by the USDA and *in vivo* in a USDA-approved ABSL-3 animal facility with enhancements.

Special Issues

The importation, possession, or use of this agent is prohibited or restricted by law or by USDA regulations or administrative policies. A USDA/APHIS import or interstate movement permit is required to obtain this agent or any livestock or poultry product, such as blood, serum, or other tissues containing the agent.

Newcastle Disease (ND) Virus

ND is one of the most serious infectious diseases of poultry worldwide. It is primarily a respiratory disease, but nervous and enteric forms occur. All bird species are probably susceptible to infection with ND virus (NDV). The severity of the disease caused by any given NDV strain can vary from an unapparent infection to 100% mortality. The chicken is the most susceptible species. The bio-containment requirements for working with a particular strain are based on the virulence of the virus as determined by chicken inoculation and more recently by determination of amino acid sequence of the fusion protein cleavage site (as defined by the World Organization for Animal Health).[35] The virus is shed in respiratory secretions and in feces. Natural transmission among birds occurs by aerosol inhalation or by consumption of contaminated feed or water.[36,37]

NDV is classified in the *Avulavirus* genus within the family *Paramyxoviridae,* subfamily *Paramyxovirinae,* in the order *Mononegavirales*. All NDV isolates are of a single serotype avian paramyxovirus type 1 (APMV-1) that includes the antigenic variants isolated from pigeons called pigeon paramyxovirus[1]. All strains are readily propagated in embryonated chicken eggs and a variety of avian and mammalian cell cultures although special additives may be required to propagate the low virulence (lentogenic) viruses in some cell types.[35-37]

Occupational Infections

The most common infection is a self-limiting conjunctivitis with tearing and pain that develops within 24 hours of an eye exposure by aerosol, splash of infective fluids, or eye contact with contaminated hands. The occurrence of upper respiratory or generalized symptoms is rare.[38]

Laboratory Safety and Containment Recommendations

NDV isolates may be recovered from any infected bird, but on occasion may be recovered from humans infected by contact with infected poultry. Humans treated with live NDV in experimental cancer therapies, or those who are exposed by laboratory contamination also are sources of the virus.[38] The greatest risk is for susceptible birds that may be exposed to NDV. If isolates of moderate to high virulence for chickens are used for human cancer therapies, those isolates are probably of greater risk for inadvertent exposure of birds and poultry than they are to the humans handling or being treated with those viruses.

ND (produced by moderate or highly virulent forms of the virus) is considered a foreign animal disease in the United States. Due to the highly contagious nature of the agent and the severe economic consequences of disease presence in the United States, this organism should only be handled *in vitro* in a BSL-3 laboratory with enhancements as required by the USDA and *in vivo* in a USDA-approved BSL-3-Ag facility for loosely housed animals. Laboratory workers should have no contact with susceptible hosts for five days after working with the agent. Laboratory and animal studies with low virulence viruses or diagnostic accessions should be handled at BSL-2.

Special Issues

Velogenic strains of NDV are USDA select agents. Possession, transfer and use of this agent requires application of procedures as detailed in 9 CFR Part 121, Agricultural Bioterrorism Protection Act of 2002; Possession, Use and Transfer of Biological Agents and Toxins. All rules concerning the possession, storage, use, and transfer of select agents apply. Please review Appendix F of this document for further instructions regarding select agents. An importation or interstate movement permit for NDV must be obtained from USDA/APHIS/VS.

Peste Des Petits Ruminants Virus (PPRV)

PPRV causes disease variously termed stomatitis pneumoenteritis complex, kata, goat plague and pseudorinderpest. The virus affects sheep and especially goats, and is regarded as the most important disease of goats and possibly sheep in West Africa where they are a major source of animal protein. The disease is reported from sub-Saharan Africa north of the equator, the Arabian Peninsula, the Middle East, and the Indian Subcontinent. The virus has particular affinity for lymphoid tissues and epithelial tissue of the gastrointestinal and respiratory tracts, causing high fever, diphtheritic oral plaques, proliferative lip lesions, diarrhea, dehydration, pneumonia and death. In susceptible populations morbidity is commonly 90% and mortality 50-80%, but can reach 100%.[39]

PPRV is a member of the family *Paramyxoviridae*, subfamily *Paramyxovirinae*, genus *Morbillivirus*, and species *peste-des-petits-ruminants virus*. Other important morbilliviruses include measles virus, rinderpest virus and canine distemper virus. As in all morbilliviruses, it is pleomorphic, enveloped, about 150 nm in diameter and contains a single molecule of linear, non-infectious, negative sense ssRNA.[40]

The virus is environmentally fragile and requires close direct contact for transmission. Outbreaks typically occur after animal movement and commingling during seasonal migrations or religious festivals. Sources of virus include tears, nasal discharge, coughed secretions, and all secretions and excretions of incubating and sick animals. There is no carrier state, and animals recovering from natural infection have lifetime immunity.

Laboratory Safety and Containment Recommendations

PPRV is not known to infect humans in either laboratory or field settings. The greatest risk of working with PPRV is the escape of the organism into a susceptible sheep or goat population, which would necessitate USDA emergency procedures to contain and eradicate the disease.

The virus is considered cause of a foreign animal disease in the United States. Due to the highly contagious nature of the agent and the severe economic consequences of disease presence in the United States, this organism should only be handled *in vitro* in a BSL-3 laboratory with enhancements as required by the USDA and *in vivo* in a USDA-approved BSL-3-Ag facility for loosely housed animals. Laboratory workers should have no contact with susceptible hosts for five days after working with the agent.

Special Issues

The importation, possession, or use of this agent is prohibited or restricted by law or by USDA regulations or administrative policies. A USDA/APHIS import or interstate movement permit is required to obtain this agent or any livestock or poultry product, such as blood, serum, or other tissues containing the agent.

Rinderpest Virus (RPV)

Rinderpest (RP) is a highly contagious viral disease of domestic cattle, buffaloes, sheep, goats and some breeds of pigs and a large variety of wildlife species.[41] It is characterized by fever, oral erosions, diarrhea, lymphoid necrosis and high mortality. The disease is present in the Indian subcontinent, Near East and sub-Saharan Africa including Kenya and Somalia.

RPV is a single stranded RNA virus in the family *Paramyxoviridae*, genus *Morbillivirus*. It is immunologically related to canine distemper virus, human measles virus, peste des petits ruminants virus, and marine mammal morbilliviruses. There is only one serotype of RPV including several strains with a wide range of virulence.[42]

RPV is a relatively fragile virus. The virus is sensitive to sunlight, heat, and most disinfectants. It rapidly inactivates at pH 2 and 12. Optimal pH for survival is 6.5 – 7.0. Glycerol and lipid solvents inactivate this virus.

Spread of RPV is by direct and indirect contact with infected animals. Aerosol transmission is not a significant means of transmission. Incubation period varies with strain of virus, dosage, and route of exposure. Following natural exposure, the incubation period ranges from 3 to 15 days but is usually 4 to 5 days.

Laboratory Safety and Containment Recommendations

There are no reports of RPV being a health hazard to humans. The greatest risk of working with RPV is the escape of the organism into susceptible animal populations, which would necessitate USDA emergency procedures to contain and eradicate the disease.

The virus is considered cause of a foreign animal disease in the United States. Due to the highly contagious nature of the agent and the severe economic consequences of disease presence in the United States, this organism should only be handled *in vitro* in a BSL-3 laboratory with enhancements as required by the USDA and *in vivo* in a USDA-approved BSL-3-Ag facility for loosely housed animals. Laboratory workers should have no contact with susceptible hosts for five days after working with the agent.

Special Issues

The importation, possession, or use of this agent is prohibited or restricted by law or by USDA regulations or administrative policies. A USDA/APHIS import or interstate movement permit is required to obtain this agent or any livestock or poultry product, such as blood, serum, or other tissues containing the agent.

Sheep and Goat Pox Virus (SGPV)

Sheep and goat pox (SGP) is an acute to chronic disease of sheep and goats characterized by generalized pox lesions throughout the skin and mucous membranes, a persistent fever, lymphadenitis, and often a focal viral pneumonia with lesions distributed uniformly throughout the lungs. Subclinical cases may occur. The virus that causes SGP is a capripoxvirus (SGPV), one of the largest viruses (170 – 260 nm by 300 – 450 nm) in the *Poxviridae* family, genus *Capripoxvirus*. It is closely related to the virus that causes lumpy skin disease. The SGPV is very resistant to physical and chemical agents.[43-45]

SGPV causes clinical disease in sheep and goats. The virus replicates in cattle but does not cause clinical disease. The disease has not been detected in wild ungulate populations. It is endemic in Africa, the Middle East, the Indian subcontinent, and much of Asia.

Contact is the main means of transmission of SGPV. Inhalation of aerosols from acutely affected animals, aerosols generated from dust contaminated from pox scabs in barns and night holding areas, and contact through skin abrasions either by fomites or by direct contact are the natural means of transmitting SGPV. Insect transmission (mechanical) is possible. The virus can cause infection experimentally by intravenous, intradermal, intranasal, or subcutaneous inoculation.

Laboratory Safety and Containment Recommendations

Humans are not susceptible to infection by these poxviruses. The greatest risk of working with these agents is the escape of the organism into susceptible domestic animal populations, which would necessitate USDA emergency procedures to contain and eradicate the diseases.[46]

These viruses are considered cause of a foreign animal disease in the United States. Due to the highly contagious nature of the agent and the severe economic consequences of disease presence in the United States, this organism should only be handled *in vitro* in a BSL-3 laboratory with enhancements as required by the USDA and *in vivo* in a USDA-approved ABSL-3 animal facility with enhancements.

Special Issues

The importation, possession, or use of this agent is prohibited or restricted by law or by USDA regulations or administrative policies. A USDA/APHIS import or interstate movement permit is required to obtain this agent or any livestock or poultry product, such as blood, serum, or other tissues containing the agent.

Spring Viremia of Carp Virus (SVCV)

Spring Viremia of Carp virus (SVCV) is a rhabdovirus in the family *Rhabdoviridae*, genus *Vesiculovirus* that infects a broad range of fish species and causes high mortality in susceptible hosts in cold water. It is a World Organization for Animal Health Office International des Épizooties (OIE) reportable disease. Infections have occurred in common and koi carp (*Cyprinus carpio*), grass carp (*Crenopharyngodon idellus*), silver carp (*Hypophthalmichthys molitix*), bighead (*Aristichthys nobilis*), cruian carp (*Carassius carassius*), goldfish (*C. auratus*), roach (*Rutilus rutilus*), ide (*Leuciscus idus*), tench (*Tinca tinca*) and sheatfish (*Silurus glanis*). Long indigenous to Europe, the Middle East and Asia, the disease was reported recently in South and North America. In the spring of 2002, SVCV was isolated from koi carp farmed in North Carolina. That year the virus was detected in fish in several lakes and rivers in Wisconsin, including the Mississippi River. SVCV causes impairment in salt-water balance in fish resulting in edema and hemorrhages.

Reservoirs of SVCV are infected fish and carriers from either cultured, feral or wild fish populations.[47] Virulent virus is shed via feces, urine, and gill, skin and mucus exudates. Liver, kidney, spleen, gill and brain are the primary organs containing the virus during infection.[48] It is surmised that horizontal transmission occurs when waterborne virus enters through the gills. Vertical transmission may be possible, especially via ovarian fluids. This virus may remain infective for long periods of time in water or mud. Once the virus is established in a pond or farm, it may be difficult to eradicate without destruction of all fish at the farm.[25,28,49]

Laboratory Safety and Containment Recommendations

Human beings are not susceptible to SVCV infection. The greatest risk of working with SVCV is the escape of the organism into a susceptible fish population, which would necessitate USDA emergency procedures to contain and eradicate the disease.

SVC is considered a reportable disease in the United States. SVCV should be handled *in vitro* in BSL-2 laboratory facilities with enhancements as required by USDA. Animal inoculations should be handled in ABSL-3 animal facilities with special modifications as required. The OIE Diagnostic Manual for Aquatic Animal Disease has specifications for surveillance programs to achieve and maintain health status of aquaculture facilities.[48] Recommendations for preventing the disease and spread of disease include the use of a water source free of virus, disinfection of eggs and equipment, and proper disposal of dead fish.

Special Issues

The importation, possession, or use of this agent is prohibited or restricted by law or by USDA regulations or administrative policies. A USDA/APHIS import or interstate movement permit is required to obtain this agent or any livestock or poultry product, such as blood, serum, or other tissues containing the agent.

Swine Vesicular Disease Virus (SVDV)

Swine vesicular disease virus (SVDV) is classified in the genus *Enterovirus*, the family *Picornaviridae*, and is closely related to the human enterovirus coxsackievirus B5.[50] The virus is the causative agent of SVD, a contagious disease of pigs characterized by fever and vesicles with subsequent erosion in the mouth and on the snout, feet, and teats.[51,52] The major importance of SVD is that it clinically resembles FMD, and any outbreaks of vesicular disease in pigs must be assumed to be FMD until proven otherwise by laboratory tests.

Occupational Infections

SVDV can cause an "influenza-like" illness in man[1] and human infection has been reported in laboratory personnel working with the virus.[53,54] The virus may be present in blood, vesicular fluid, and tissues of infected pigs. Direct and indirect contacts of infected materials, contaminated laboratory surfaces, and accidental autoinoculation, are the primary hazards to laboratory personnel.

Laboratory Safety and Containment Recommendations

Laboratory practices for SVDV are principally designed to prevent transmission to susceptible livestock, but also to protect workers. Gloves are recommended for the necropsy and handling of infected animals and cell cultures. The greatest risk of working with SVD is the escape of the organism into susceptible animal populations, which would necessitate USDA emergency procedures to contain and eradicate the disease.[55]

SVD is considered a foreign animal disease in the United States. Due to the severe economic consequences of disease presence in the United States, SVDV should only be handled *in vitro* in a BSL-3 laboratory with enhancements as required by the USDA and *in vivo* in a USDA-approved ABSL-3 animal facility with enhancements.

Special Issues

The importation, possession, or use of this agent is prohibited or restricted by law or by USDA regulations or administrative policies. A USDA/APHIS import or interstate movement permit is required to obtain this agent or any livestock or poultry product, such as blood, serum, or other tissues containing the agent.

References

1. Hess WR. African swine fever virus. Virol Monogr. 1971;9:1-33.
2. Mebus CA. African swine fever. Adv Virus Res. 1988;35:251-69.
3. Mebus CA. African swine fever. In: Mebus CA, editor. Foreign animal diseases. Richmond (VA): U.S. Animal Health Association; 1998. p. 52-61.
4. M'Fadyean J. African horse sickness. J Comp Path Ther. 1900;13:1-20.
5. Erasmus BJ. African horse sickness. In: Mebus CA, editor. Foreign animal diseases. Richmond (VA): U.S. Animal Health Association; 1998. p. 41-51.
6. Della-Porta AJ, O'Halloran ML, Parsonson M, et al. Akabane disease: isolation of the virus from naturally infected ovine foetuses. Aust Vet J. 1977;53:51-2.
7. St. George TD. Akabane. In: Mebus CA, editor. Foreign animal diseases. Richmond (VA): U.S. Animal Health Association; 1998. p. 62-70.
8. Bowne JG. Bluetongue disease. Adv Vet Sci Comp Med. 1971;15:1-46.
9. Stott JL. Bluetongue and epizootic hemorrhagic disease. In: Mebus CA, editor. Foreign animal diseases. Richmond (VA): U.S. Animal Health Association; 1998. p. 106-17.
10. Van Oirschot JT, Terpstra C. Hog cholera virus. In: Pensaert MB, editor. Virus infections of porcines. New York: Elsever Science Publishers; 1989. p. 113-30.
11. Van Oirschot JT. Classical swine fever. In: Straw BE, D'Allaire S, Mengelng W, et al. Diseases of swine. 8th ed. Ames (IA): Iowa State University Press; 1999. p. 159-72.
12. Dulac DC. Hog cholera. In: Mebus CA, editor. Foreign animal diseases. Richmond (VA): U.S. Animal Health Association; 1998. p. 273-82.
13. Thiaucourt F. Contagious bovine pleuropneumonia. In: Manual of standards for diagnostic tests and vaccines for terrestrial animals: mammals, birds and bees. Paris: Office International des Épizooties; 2004. p. 163-74.
14. Contagious bovine pleuropneumonia. In: Aiello SE, Mays A, editors. The Merck veterinary manual. 8th ed. Whitehouse Station (NJ): Merck and Company; 1998. p. 1078-9.

15. Brown, C. Contagious bovine pleuropneumonia. In: Mebus CA, editor. Foreign animal diseases. Richmond (VA): U.S. Animal Health Association; 1998. p. 154-60.

16. Cottew GS. Overview of mycoplasmoses in sheep and goats. Isr J Med Sci. 1984;20:962-4.

17. DaMassa AJ, Wakenell PS, Brooks DL. Mycoplasmas of goats and sheep. J Vet Diagn Invest. 1992;4:101-13.

18. Mare JC. Contagious caprine pleuropneumonia. In: Mebus CA, editor. Foreign animal diseases. Richmond (VA): U.S. Animal Health Association; 1998. p. 161-9.

19. Cottew GS, Breard A, DaMassa AJ. Taxonomy of the *Mycoplasma mycoides* cluster. Isr. J Med Sci. 1987;23:632-5.

20. Christiansen C, Erno H. Classification of the F38 group of caprine Mycoplasma strains by DNA hybridization. J Gen Microbiol. 1982;128:2523-6.

21. DaMassa AJ, Holmberg CA, Brooks DL. Comparison of caprine mycoplasmosis caused by *Mycoplasma capricolum*, *M. mycoides* subsp. *mycoides*, and M. putrefasciens. Isr J Med Sci. 1987;23:636-40.

22. House J. Foot and mouth disease. In: Mebus CA, editor. Foreign animal diseases. Richmond (VA): U.S. Animal Health Association; 1998. p. 213-24.

23. Kitching RP, Barnett PV, Mackay DKJ, et al. Foot and Mouth Disease. In: Manual of standards for diagnostic tests and vaccines for terrestrial animals: mammals, birds and bees. Paris: Office International des Épizooties; 2004. p. 111-128.

24. Mare J. Heartwater. In: Mebus CA, editor. Foreign animal diseases. Richmond (VA): U.S. Animal Health Association; 1998. p. 253-64.

25. Lee CS, O'Bryen PJ, editors. Biosecurity in aquaculture production systems: exclusion of pathogens and other undesirables. Baton Rouge (LA): World Aquaculture Society; 2003. p. 293.

26. Miller O, Cipriano RC. International response to infectious salmon anemia: prevention, control and eradication. Proceedings of a Symposium; 2002 Sep 3-4; New Orleans, LA. Washington, DC: US Department of Agriculture, US Department of the Interior and the US Department of Commerce; 2003.

27. Office International des Épizooties. Infectious salmon anaemia. In: Manual of diagnostic tests for aquatic animals. 4th ed. Paris: Office International des Épizooties; 2003. p. 152-61.

28. Bruno DW, Alderman DJ, Schlotfeldt HJ. What should I do? A practical guide for the marine fish farmer. Dorset (England): European Association of Fish Pathologists; 1995.

29. Kitching RP. Lumpy skin disease. In: Manual of standards for diagnostic tests and vaccines for terrestrial animals: mammals, birds and bees. Paris: Office International des Épizooties; 2004. p. 175-84.

30. Lumpy skin disease. In: Aiello SE, Mays A, editors. The Merck veterinary manual. 8th ed. Whitehouse Station (NJ): Merck and Company; 1998; p. 621-2.

31. House JA. Lumpy skin disease. In: Mebus CA, editor. Foreign animal diseases. Richmond (VA): U.S. Animal Health Association; 1998. p. 303-10.

32. Heuschele WP. Malignant catarrhal fever. In: Mebus CA, editor. Foreign animal diseases. Richmond (VA): U.S. Animal Health Association; 1998. p.311-321.

33. Mackenzie JS, Chua KB, Daniels PW, et al. Emerging viral disease of Southeast Asia and the Western Pacific. Emerg Infect Dis. 2001;7:497-504.

34. Kirkland PD, Daniels PW, Nor MN, et al. Menangel and Nipah virus infections of pigs. Vet Clin North Am Food Anim Pract. 2002;18:557-71.

35. Alexander DJ. Newcastle disease. 2004. In: Manual of standards for diagnostic tests and vaccines for terrestrial animals: mammals, birds and bees. Paris: Office International des Épizooties; 2004. p. 270-82.

36. Alexander DJ. Newcastle disease. In: Swayne DE, Glisson JR, Jackwood MW, et al, editors. A laboratory manual for the isolation and identification of avian pathogens. 4th ed. Kennett Square (PA): American Association of Avian Pathologists; 1998. p. 156-63.

37. Alexander DJ. Newcastle disease, other paramyxoviruses and pneumovirus infections. In: Saif YM, Barnes YM, Glisson HJ, et al, editors. Diseases of poultry. 11th ed. Ames (IA): Iowa State Press: 2003. p. 63-87.

38. Swayne DE, King DJ. Avian Influenza and Newcastle disease. J Am Vet Med Assoc. 2003;222;1534-40.

39. Saliki JT. Peste des petits ruminants. In: Manual of standards for diagnostic tests and vaccines for terrestrial animals: mammals, birds and bees. Paris: Office International des Épizooties; 2004. p. 344-52.

40. Diallo A. Peste des petits ruminants. In: Manual of standards for diagnostic tests and vaccines for terrestrial animals: mammals, birds and bees. Paris: Office International des Épizooties; 2004. p. 153-62.

41. Mebus CA. Rinderpest. In: Mebus CA, editor. Foreign animal diseases. Richmond (VA): U.S. Animal Health Association; 1998. p. 362-71.

42. Taylor WP, Roeder P. Rinderpest. In: Manual of standards for diagnostic tests and vaccines for terrestrial animals: mammals, birds and bees. Paris: Office International des Épizooties; 2004. p. 142-52.

43. Davies FG. Characteristics of a virus causing a pox disease in sheep and goats in Kenya, with observations on the epidemiology and control. J Hyg (Lond). 1976;76:163-71.

44. Davies FG. Sheep and goat pox. In: Gibbs EPK, editor. Virus diseases of food animals. Vol 2. London: Academic Press; 1981. p. 733-48.

45. House JA. Sheep and goat pox. In: Mebus CA, editor. Foreign animal diseases. Richmond (VA): U.S. Animal Health Association; 1998. p. 384-91.

46. Kitching RP, Carn V. Sheep and goat pox. In: Manual of standards for diagnostic tests and vaccines for terrestrial animals: mammals, birds and bees. Paris: Office International des Épizooties; 2004. p. 211-20.

47. Goodwin AE, Peterson JE, Meyers TR, et al. Transmission of exotic fish viruses: the relative risks of wild and cultured bait. Fisheries. 2004;29:19-23

48. Office International des Épizooties. Spring viremia of carp. In: Manual of diagnostic tests for aquatic animals. 4th ed. Paris: Office International des Épizooties; 2003. p. 108-14.

49. Schlotfeldt HJ, Alderman DJ. What should I do? A practical guide for the freshwater fish farmer. Dorset (England): European Association of Fish Pathologists; 1995.

50. Fenner FJ, Gibbs EPJ, Murphy FA, et al. Picornaviridae. In: Fenner FJ, editor. Veterinary virology. 2nd ed. San Diego, CA: Academic Press; 1993. p. 403-23.

51. Mebus CA. Swine vesicular disease. In: Mebus CA, editor. Foreign animal diseases. Richmond (VA): U.S. Animal Health Association; 1998. p. 392-95.

52. Office International des Épizooties. Swine vesicular disease. In: Manual of diagnostic tests for aquatic animals. 4th ed. Paris: Office International des Épizooties; 2003.

53. Brown F, Goodridge D, Burrows R. Infection of man by swine vesicular disease virus. J Comp Path. 1976;86:409-14.

54. Graves JH. Serological relationship of swine vesicular disease virus and Coxsackie B5 virus. Nature. 1973;245:314-5.

55. Kitching RP, Mackay DKJ, Donaldson AI. Swine Vesicular Disease. In: Manual of standards for diagnostic tests and vaccines for terrestrial animals: mammals, birds and bees. Paris: Office International des Épizooties; 2004. p. 136-41.

VI. Additional Information:

U.S. Department of Agriculture
Animal and Plant Health Inspection Service
Veterinary Services, National Center for Import and Export
4700 River Road, Unit 133
Riverdale, Maryland 20737-1231
Telephone: (301) 734-5960
Fax: (301) 734-3256
Internet: *http://www.aphis.usda.gov/animal_health/permits*

Further information on Plant Select Agents, or permits for field release of genetically engineered organisms may be obtained from:

U.S. Department of Agriculture Animal
and Plant Health Inspection Service
Plant Protection and Quarantine, Permits, Agricultural Bioterrorism
4700 River Road, Unit 2
Riverdale, Maryland 20737-1231
Telephone: (301) 734-5960
Internet: *http://www.aphis.usda.gov/programs/ag_selectagent/index.shtml*

Appendix E—Arthropod Containment Guidelines (ACG)

An ad hoc committee of concerned vector biologists including members of the American Committee Medical Entomology (ACME), a subcommittee of the American Society of Tropical Medicine and Hygiene (ASTMH), and other interested persons drafted the "Arthropod Containment Guidelines." The ACG provide principles of risk assessment for arthropods of public health importance. The risk assessment and practices are designed to be consistent with the *NIH Guidelines* for recombinant DNA research and BMBL. Arthropods included are those that transmit pathogens; however, those arthropods that cause myiasis, infestation, biting, and stinging are not included. The ACG also specifically exclude most uses of *Drosophila* spp.

The ACG were published in Vector Borne and Zoonotic Diseases.1 They are freely downloadable from *www.liebertonline.com* and at the AMCE Web site: *www.astmh.org*.

The ACG recommend biosafety measures specific for arthropods of public health importance considering that:

- Arthropods present unique containment challenges not encountered with microbial pathogens.

- Arthropod containment has not been covered specifically in BMBL or the NIH Guidelines.

The ACG contain two sections of greatest interest to most researchers:

1. The Principles of Risk Assessment that discusses arthropods in the usual context (e.g., those known to contain a pathogenic agent, those with uncertain pathogens, and those with no agent).

2. They also consider the following:

   - Biological containment is a significant factor that reduces the hazards associated with accidental escape of arthropods.

   - Epidemiological context alters the risks of an escape and its impact on the location or site in which the work is performed.

   - The phenotype of the vector, such as insecticide resistance; and

   - genetically modified arthropods with an emphasis on phenotypic change.

Four Arthropod Containment Levels (ACL 1 – 4) add increasingly stringent measures and are similar to biosafety levels. The most flexible level is ACL-2 that covers most exotic and transgenic arthropods and those infected with pathogens requiring BSL-2 containment. Like BMBL, each level has the following form:

- standard practices;

- special practices;

- equipment (primary barriers);

- facilities (secondary barriers).

The ACG does not reflect a formal endorsement by ACME or ASTMH. The guidelines are subject to change based on further consideration of the requirements for containment of arthropods and vectors.

References

1.   American Committee of Medical Entomology; American Society of Tropical Medicine and Hygiene. Arthropod containment guidelines. A project of the American Committee of Medical Entomology and American Society of Tropical Medicine and Hygiene. Vector Borne Zoonotic Dis. 2003;3:61-98.

Appendix F—Select Agents and Toxins

The *Public Health Security and Bioterrorism Preparedness and Response Act of 2002, Subtitle A of Public Law 107-188 (42 U.S.C. 262a)*, requires DHHS to regulate the possession, use, and transfer of biological agents or toxins (i.e., select agents and toxins) that could pose a severe threat to public health and safety. The *Agricultural Bioterrorism Protection Act of 2002, Subtitle B of Public Law 107-188 (7 U.S.C. 8401)*, requires the USDA to regulate the possession, use, and transfer of biological agents or toxins (i.e., select agents and toxins) that could pose a severe threat to animal or plant health, or animal or plant products. These Acts require the establishment of a national database of registered entities, and set criminal penalties for failing to comply with the requirements of the Acts. In accordance with these Acts, DHHS and USDA promulgated regulations requiring entities to register with the CDC or the APHIS if they possess, use, or transfer a select agent or toxin (42 CFR Part 73, 7 CFR Part 331, and 9 CFR Part 121). CDC and APHIS coordinate regulatory activities for those agents that would be regulated by both agencies ("overlap" select agents).

The Attorney General has the authority and responsibility to conduct electronic database checks (i.e., the security risk assessments) on entities that apply to possess, use, or transfer select agents, as well as personnel that require access to select agents and toxins. The FBI, Criminal Justice Information Services Division (CJIS), has been delegated authority for conducting these security risk assessments.

The regulations provide that, unless exempted, entities must register with CDC or APHIS if they possess, use, or transfer select agents or toxins. The current list of select agents and toxins is available on the CDC and APHIS Web sites (see below). The regulations set out a procedure for excluding an attenuated strain of a select agent or toxin and exemptions for certain products and for select agents or toxins identified in specimens presented for diagnosis, verification, or proficiency testing.

The regulations also contain requirements to ensure that the select agents and toxins are handled safely and secured against unauthorized access, theft, loss, or release. For example, entities and their personnel must undergo a security risk assessment by CJIS as part of their registration; entities must limit access to select agents and toxins and develop and implement biosafety, security, and incident response plans. In addition, all select agents or toxins must be transferred in accordance with the regulations and any theft, loss, or release of a select agent or toxin must be reported to CDC or APHIS.

For additional information concerning the select agent regulations, contact CDC or APHIS. Information is also available at the following Web sites: *www.selectagents.gov*; *http://www.aphis.usda.gov/programs/ag_selectagent/index.shtml*.

Appendix G—Integrated Pest Management (IPM)

IPM is an important part of managing a research facility. Many pests, such as flies and cockroaches, can mechanically transmit disease pathogens and compromise the research environment. Even the presence of innocuous insects can contribute to the perception of unsanitary conditions.

The most common approach to pest control has been the application of pesticides, either as a preventive or remedial measure. Pesticides can be effective and may be necessary as a corrective measure, but they have limited long-term effect when used alone. Pesticides also can contaminate the research environment through pesticide drift and volatilization.

To control pests and minimize the use of pesticides, it is necessary to employ a comprehensive program approach that integrates housekeeping, maintenance, and pest control services. This method of pest control is often referred to as IPM. The primary goal of an IPM program is to prevent pest problems by managing the facility environment to make it less conducive to pest infestation. Along with limited applications of pesticides, pest control is achieved through proactive operational and administrative intervention strategies to correct conditions that foster pest problems.

Prior to developing any type of IPM program, it is important to define an operational framework for IPM services that helps promote collaboration between IPM specialists and facility personnel. This framework should incorporate facility restrictions and operational and procedural issues into the IPM program. An effective IPM program is an integral part of the facility's management. An IPM policy statement should be included in the facility's standard operating procedures to increase awareness of the program.

Training sources for the principles and practices of structural (indoor) IPM programs are available through university entomology departments, county extension offices, the Entomological Society of America, state departments of agriculture, state pest control associations, the National Pest Control Association, suppliers of pest control equipment, and IPM consultants and firms. Several universities offer correspondence courses, short courses, and training conferences on structural pest management.

IPM is a strategy-based service that considers not only the cost of the services, but also the effectiveness of the program's components. Each IPM program is site-specific, tailored to the environment where applied.

Laboratory IPM services will be different from those in an office building or an animal care facility. Interrelated components of "Environmental pest management" follow.

Facility Design

IPM issues and requirements should be addressed in a research facility's planning, design, and construction. This provides an opportunity to incorporate features that help exclude pests, minimize pest habitat, and promote proper sanitation in order to reduce future corrections that can disrupt research operations.

Monitoring

Monitoring is the central activity of an IPM program and is used to minimize pesticide use. Traps, visual inspections, and staff interviews identify areas and conditions that may foster pest activity.

Sanitation and Facility Maintenance

Many pest problems can be prevented or corrected by ensuring proper sanitation, reducing clutter and pest habitat, and by performing repairs that exclude pests. Records of structural deficiencies and housekeeping conditions should be maintained to track problems and determine if corrective actions were completed and in a timely manner.

Communication

A staff member should be designated to meet with IPM personnel to assist in resolving facility issues that impact on pest management. Reports communicated verbally and in writing concerning pest activity and improvement recommendations for personnel, practices and facility conditions should be provided to the designated personnel. Facility personnel should receive training on pest identification, biology, and sanitation, which can promote understanding and cooperation with the goals of the IPM program.

Recordkeeping

A logbook should be used to record pest activity and conditions pertinent to the IPM program. It may contain protocols and procedures for IPM services in that facility, Material Safety Data Sheets on pesticides, pesticide labels, treatment records, floor plans, survey reports, etc.

Non-pesticide Pest Control

Pest control methods such as trapping, exclusion, caulking, washing, and freezing can be applied safely and effectively when used in conjunction with proper sanitation and structural repair.

Pest Control with Pesticides

Preventive applications of pesticides should be discouraged, and treatments should be restricted to areas of known pest activity. When pesticides are applied, the least toxic product(s) available should be used and applied in the most effective and safe manner.

Program Evaluation and Quality Assurance

Quality assurance and program review should be performed to provide an objective, ongoing evaluation of IPM activities and effectiveness to ensure that the program does, in fact, control pests and meet the specific needs of the facility program(s) and its occupants. Based upon this review, current IPM protocols can be modified and new procedures implemented.

Technical Expertise

A qualified entomologist can provide helpful technical guidance to develop and implement an IPM program. Pest management personnel should be licensed and certified by the appropriate regulatory agency.

Safety

IPM minimizes the potential of pesticide exposure to the research environment and the staff by limiting the scope of pesticide treatments.

References

1. Robinson WH. Urban entomology: insect and mite pests in the human environment. New York: Chapman and Hall; 1996.
2. Bennett GW, Owens JM, editors. Advances in urban pest management. New York: Van Nostrand Reinhold Company; 1986.
3. Olkowski W, Daar S, Olkowski H. Common sense pest control: least-toxic solutions for your home, garden, pests and community. Newton (CT): The Taunton Press, Inc.; 1991.
4. National Pest Control *Association [http://www.pestworld.org]*. Fairfax (VA): The Association; [cited 2006 Sept 25]. Available from: *http://www.pestworld.org*.
5. *Biocontrol Network [http://www.biconet.com]*. Brentwood (TN): Biocontrol Network; c1995-2006 [updated 2006 Sept; cited 2006 Sept 25]. Available from: *http://www.biconet.com*.

Appendix H—Working with Human, NHP and Other Mammalian Cells and Tissues

Although risk of laboratory infection from working with cell cultures in general is low, risk increases when working with human and other primate cells, and primary cells from other mammalian species. There are reports of infection of laboratory workers handling primary rhesus monkey kidney cells,[1] and the bloodborne pathogen risks from working with primary human cells, tissues and body fluids are widely recognized.[2,3] OSHA has developed a bloodborne pathogens standard that should be applied to all work in the laboratory with human blood, tissues, body fluids and primary cell lines.[4] Procedures have also been published to reduce contamination of cell cultures with microorganisms.[5,6]

Potential Laboratory Hazards

Potential laboratory hazards associated with human cells and tissues include the bloodborne pathogens HBV, HIV, HCV, HTLV, EBV, HPV and CMV as well as agents such as *Mycobacterium tuberculosis* that may be present in human lung tissue. Other primate cells and tissues also present risks to laboratory workers.[7] Cells immortalized with viral agents such as SV-40, EBV adenovirus or HPV, as well as cells carrying viral genomic material also present potential hazards to laboratory workers. Tumorigenic human cells also are potential hazards as a result of self-inoculation.[8] There has been one reported case of development of a tumor from an accidental needle-stick.[9] Laboratory workers should never handle autologous cells or tissues.[1] NHP cells, blood, lymphoid and neural tissues should always be considered potentially hazardous.

Recommended Practices

Each institution should conduct a risk assessment based on the origin of the cells or tissues (species and tissue type), as well as the source (recently isolated or well-characterized). Human and other primate cells should be handled using BSL-2 practices and containment. All work should be performed in a BSC, and all material decontaminated by autoclaving or disinfection before discarding.[6,10,11,12] BSL-2 recommendations for personnel protective equipment such as laboratory coats, gloves and eye protection should be rigorously followed. All laboratory staff working with human cells and tissues should be enrolled in an occupational medicine program specific for bloodborne pathogens and should work under the policies and guidelines established by the institution's Exposure Control Plan.[4] Laboratory staff working with human cells and tissues should provide a baseline serum sample, be offered hepatitis B immunization, and be evaluated by a health care professional following an exposure incident. Similar programs should be considered for work with NHP blood, body fluids, and other tissues.

References

1. Doblhoff-Dier O, Stacey G. Cell lines: applications and biosafety. In: Fleming D, Hunt D, editors. Biological safety: principles and practices. Washington, DC: ASM Press; 2000. p. 221-39.
2. Centers for Disease Control and Prevention. Update: universal precautions for prevention of transmission of human immunodeficiency virus, hepatitis B virus and other bloodborne pathogens in healthcare settings. MMWR Morb Mortal Wkly Rep. 1988;37:377-82, 387-8.
3. Centers for Disease Control and Prevention. Guidelines for prevention of transmission of human immunodeficiency virus and hepatitis B virus to healthcare and public safety workers. MMWR Morb Mortal Wkly Rep. 1989;38;No.SU-06.
4. Occupational exposure to bloodborne pathogens. Final Rule. Standard interpretations: applicability of 1910.1030 to established human cell lines, 29 C.F.R. Sect. 1910.1030 (1991).
5. McGarrity GJ, Coriell LL. Procedures to reduce contamination of cell cultures. In Vitro. 1971;6:257-65.
6. McGarrity GJ. Spread and control of mycoplasmal infection of cell culture. In Vitro. 1976;12:643-8.
7. Caputo JL. Safety procedures. In: Freshney RI Freshney MG, editors. Culture of immortalized cells. New York: Wiley-Liss; 1996.
8. Weiss RA. Why cell biologists should be aware of genetically transmitted viruses. Natl Cancer Inst Monogr.1978;48:183-9.
9. Gugel EA, Sanders ME. Needle-stick transmission of human colonic adenocarcinoma (letter). N Engl J Med. 1986;315:1487.
10. Barkley WE. Safety considerations in the cell culture laboratory. Methods Enzymol. 1979;58:36-43.
11. Grizzle WE, Polt S. Guidelines to avoid personnel contamination by infective agents in research laboratories that use human tissues. J Tissue Cult Methods. 1988;11:191-9.
12. Caputo JL. Biosafety procedures in cell culture. J of Tissue Cult Methods. 1988;11:233-7.

Appendix I—Guidelines for Work with Toxins of Biological Origin

Biological toxins comprise a broad range of poisons, predominantly of natural origin but increasingly accessible by modern synthetic methods, which may cause death or severe incapacitation at relatively low exposure levels.[1,2] Laboratory safety principles are summarized herein for several toxins currently regulated as "Select Agent Toxins," including BoNT, SE, ricin and selected LMW toxins. Additional details are provided in the agent summary statements.

General Considerations for Toxin Use

Laboratory work with most toxins, in amounts routinely employed in the biomedical sciences, can be performed safely with minimal risk to the worker and negligible risk to the surrounding community. Toxins do not replicate, are not infectious, and are difficult to transmit mechanically or manually from person to person. Many commonly employed toxins have very low volatility and, especially in the case of protein toxins, are relatively unstable in the environment; these characteristics further limit the spread of toxins.

Toxins can be handled using established general guidelines for toxic or highly-toxic chemicals with the incorporation of additional safety and security measures based upon a risk assessment for each specific laboratory operation.[3,4] The main laboratory risks are accidental exposure by direct contamination of mouth, eyes or other mucous membranes; by inadvertent aerosol generation; and by needle-sticks or other accidents that may compromise the normal barrier of the skin.

Training and Laboratory Planning

Each laboratory worker must be trained in the theory and practice of the toxins to be used, with special emphasis on the nature of the practical hazards associated with laboratory operations. This includes how to handle transfers of liquids containing toxin, where to place waste solutions and contaminated materials or equipment, and how to decontaminate work areas after routine operations, as well as after accidental spills. The worker must be reliable and sufficiently adept at all required manipulations before being provided with toxin.

A risk assessment should be conducted to develop safe operating procedures before undertaking laboratory operations with toxins; suggested "pre-operational checklists" for working with toxins are available.[4] For complex operations, it is recommended that new workers undergo supervised practice runs in which the exact laboratory procedures to be undertaken are rehearsed without active toxin. If toxins and infectious agents are used together, then both must be considered when containment equipment is selected and safety procedures are developed. Likewise, animal safety practices must be considered for toxin work involving animals.

Each laboratory that uses toxins should develop a specific chemical hygiene plan. The National Research Council has provided a review of prudent laboratory practices when handling toxic and highly toxic chemicals, including the development of chemical hygiene plans and guidelines for compliance with regulations governing occupational safety and health, hazard communication, and environmental protection.[5]

An inventory control system should be in place to account for toxin use and disposition. If toxins are stored in the laboratory, containers should be sealed, labeled, and secured to ensure restricted access; refrigerators and other storage containers should be clearly labeled and provide contact information for trained, responsible laboratory staff.

Laboratory work with toxins should be done only in designated rooms with controlled access and at pre-determined bench areas. When toxins are in use, the room should be clearly posted: "Toxins in Use—Authorized Personnel Only." Unrelated and nonessential work should be restricted from areas where stock solutions of toxin or organisms producing toxin are used. Visitors or other untrained personnel granted laboratory access must be monitored and protected from inadvertently handling laboratory equipment used to manipulate the toxin or organism.

Safety Equipment and Containment

Routine operations with dilute toxin solutions are conducted under BSL-2 conditions with the aid of personal protective equipment and a well-maintained BSC or comparable engineering controls.[6] Engineering controls should be selected according to the risk assessment for each specific toxin operation. A certified BSC or chemical fume hood will suffice for routine operations with most protein toxins. Low molecular weight toxin solutions, or work involving volatile chemicals or radionucleotides combined with toxin solutions, may require the use of a charcoal-based hood filter in addition to HEPA filtration.

All work with toxins should be conducted within the operationally effective zone of the hood or BSC, and each user should verify the inward airflow before initiating work. When using an open-fronted fume hood or BSC, workers should wear suitable laboratory PPE to protect the hands and arms, such as laboratory coats, smocks, or coveralls and disposable gloves. When working with toxins that pose direct percutaneous hazards, special care must be taken to select gloves that are impervious to the toxin and the diluents or solvents employed. When conducting liquid transfers and other operations that pose a potential splash or droplet hazard in an open-fronted hood or BSC, workers should wear safety glasses and disposable facemask, or a face shield.

Toxin should be removed from the hood or BSC only after the exterior of the closed primary container has been decontaminated and placed in a clean secondary container. Toxin solutions, especially concentrated stock solutions, should be transported in leak/spill-proof secondary containers. The interior of the hood or BSC should be decontaminated periodically, for example, at the end of a series of related experiments. Until thoroughly decontaminated, the hood or BSC should be posted to indicate that toxins remain in use, and access should remain restricted.

Selected operations with toxins may require modified BSL-3 practices and procedures. The determination to use BSL-3 is made in consultation with available safety staff and is based upon a risk assessment that considers the variables of each specific laboratory operation, especially the toxin under study, the physical state of the toxin (solution or dry form), the total amount of toxin used relative to the estimated human lethal dose, the volume of the material manipulated, the methodology, and any human or equipment performance limitations.

Inadvertent Toxin Aerosols

Emphasis must be placed on evaluating and modifying experimental procedures to eliminate the possibility of inadvertent generation of toxin aerosols. Pressurized tubes or other containers holding toxins should be opened in a BSC, chemical fume hood, or other ventilated enclosure. Operations that expose toxin solutions to vacuum or pressure, for example sterilization of toxin solutions by membrane filtration, should always be handled in this manner, and the operator should also use appropriate respiratory protection. If vacuum lines are used with toxin, they should be protected with a HEPA filter to prevent entry of toxins into the line.

Centrifugation of cultures or materials potentially containing toxins should only be performed using sealed, thick-walled tubes in safety centrifuge cups or sealed rotors. The outside surfaces of containers and rotors should be routinely cleaned before each use to prevent contamination that may generate an aerosol. After centrifugation, the entire rotor assembly is taken from the centrifuge to a BSC to open it and remove its tubes.

Mechanical Injuries

Accidental needle-sticks or mechanical injury from "sharps" such as glass or metal implements pose a well-known risk to laboratory workers, and the consequences may be catastrophic for operations involving toxins in amounts that exceed a human lethal dose.

Only workers trained and experienced in handling animals should be permitted to conduct operations involving injection of toxin solutions using hollow-bore needles. Discarded needles/syringes and other sharps should be placed directly into properly labeled, puncture-resistant sharps containers, and decontaminated as soon as is practical.

Glassware should be replaced with plastic for handling toxin solutions wherever practical to minimize the risk of cuts or abrasions from contaminated surfaces. Thin-walled glass equipment should be completely avoided. Glass Pasteur pipettes are particularly dangerous for transferring toxin solutions and should be replaced with disposable plastic pipettes. Glass chromatography columns under pressure must be enclosed within a plastic water jacket or other secondary container.

Additional Precautions

Experiments should be planned to eliminate or minimize work with dry toxin (e.g., freeze-dried preparations). Unavoidable operations with dry toxin should only be undertaken with appropriate respiratory protection and engineering controls. Dry toxin can be manipulated using a Class III BSC, or with the use of secondary containment such as a disposable glove bag or glove box within a hood or Class II BSC. "Static-free" disposable gloves should be worn when working with dry forms of toxins that are subject to spread by electrostatic dispersal.

In specialized laboratories, the intentional, controlled generation of aerosols from toxin solutions may be undertaken to test antidotes or vaccines in experimental animals. These are extremely hazardous operations that should only be conducted after extensive validation of equipment and personnel, using non-toxic simulants. Aerosol exposure of animals should be done in a certified Class III BSC or hoodline. While removing exposed animals from the hoodline, and for required animal handling during the first 24 h after exposure, workers should take additional precautions, including wearing protective clothing (e.g., disposable Tyvek suit) and appropriate respiratory protection. To minimize the risk of dry toxin generating a secondary aerosol, areas of animal skin or fur exposed to aerosols should be gently wiped with a damp cloth containing water or buffered cleaning solution before the animals are returned to holding areas.

For high-risk operations involving dry forms of toxins, intentional aerosol formation, or the use of hollow-bore needles in conjunction with amounts of toxin estimated to be lethal for humans, consideration should be given to requiring the presence of at least two knowledgeable individuals at all times in the laboratory.[7]

Decontamination and Spills

Toxin stability varies considerably outside of physiological conditions depending upon the temperature, pH, ionic strength, availability of co-factors and other characteristics of the surrounding matrix. Literature values for dry heat inactivation of toxins can be misleading due to variations in experimental conditions, matrix composition, and experimental criteria for assessing toxin activity. Moreover, inactivation is not always a linear function of heating time; some protein toxins possess a capacity to re-fold and partially reverse inactivation caused by heating. In addition, the conditions for denaturizing toxins in aqueous solutions are not necessarily applicable for inactivating dry, powdered toxin preparations.

General guidelines for laboratory decontamination of selected toxins are summarized in Tables 1 and 2, but inactivation procedures should not be assumed to be 100% effective without validation using specific toxin bioassays. Many toxins are susceptible to inactivation with dilute sodium hydroxide (NaOH) at concentrations of 0.1-0.25N, and/or sodium hypochlorite (NaOCl) bleach solutions at concentrations of 0.1-0.5% (w/v). Use freshly prepared bleach solutions for decontamination; undiluted, commercially available bleach solutions typically contain 3-6% (w/v) NaOCl.

Depending upon the toxin, contaminated materials and toxin waste solutions can be inactivated by incineration or extensive autoclaving, or by soaking in suitable decontamination solutions (Table 2). All disposable material used for toxin work should be placed in secondary containers, autoclaved and disposed of as toxic waste. Contaminated or potentially contaminated protective clothing and equipment should be decontaminated using suitable chemical methods or autoclaving before removal from the laboratory for disposal, cleaning or repair. If decontamination is impracticable, materials should be disposed of as toxic waste.

In the event of a spill, avoid splashes or generating aerosols during cleanup by covering the spill with paper towels or other disposable, absorbent material. Apply an appropriate decontamination solution to the spill, beginning at the perimeter and working towards the center, and allow sufficient contact time to completely inactivate the toxin (Table 2).

Decontamination of buildings or offices containing sensitive equipment or documents poses special challenges. Large-scale decontamination is not covered explicitly here, but careful extrapolation from the basic principles may inform more extensive clean-up efforts.

Select Agent Toxins

Due diligence should be taken in shipment or storage of any amount of toxin. There are specific regulatory requirements for working with toxins designated as a "Select Agent" by the DHHS and/or the USDA. Select agents require registration with CDC and/or USDA for possession, use, storage and/or transfer. Importation of this agent may require CDC and/or USDA importation permits. Domestic transport of the agent may require a permit from USDA/ APHIS/VS. A DoC permit may be required for the export of the agent to another country. See Appendix C for additional information.

Table 1. Physical Inactivation of Selected Toxins

Toxin	Steam Autoclave	Dry Heat (10 min)	Freeze-thaw	Gamma Irradiation
Botulinum neurotoxin	Yes [a]	> 100° C [b]	No [c]	Incomplete [d]
Staphylococcal Enterotoxin	Yes [e]	> 100° C; refolds [f]	No [g]	Incomplete [h]
Ricin	Yes [i]	> 100° C [i]	No [j]	Incomplete [k]
Microcystin	No [l]	> 260° C [m]	No [n]	ND
Saxitoxin	No [l]	> 260° C [m]	No [n]	ND
Palytoxin	No [l]	> 260° C [m]	No [n]	ND
Tetrodotoxin	No [l]	> 260° C [m]	No [n]	ND
T-2 mycotoxin	No [l]	> 815° C [m]	No [n]	ND
Brevetoxin (PbTx-2)	No [l]	> 815° C [m]	No [n]	ND

Notes:

ND indicates "not determined" from available decontamination literature.

[a] Steam autoclaving should be at >121°C for 1 h. For volumes larger than 1 liter, especially those containing *Clostridium botulinum* spores, autoclave at >121°C for 2 h to ensure that sufficient heat has penetrated to kill all spores.[8,9]

[b] Exposure to 100°C for 10 min. inactivates BoNT. Heat denaturation of BoNT as a function of time is biphasic with most of the activity destroyed relatively rapidly, but with some residual toxin (e.g., 1-5%) inactivated much more slowly.[10]

[c] Measured using BoNT serotype A at -20°C in food matrices at pH 4.1 – 6.2 over a period of 180 days.[11]

[d] Measured using BoNT serotypes A and B with gamma irradiation from a ^{60}Co source.[12,13]

[e] Protracted steam autoclaving, similar to that described for BoNT, followed by incineration is recommended for disposal of SE-contaminated materials.

[f] Inactivation may not be complete depending upon the extent of toxin re-folding after denaturation. Biological activity of SE can be retained despite heat and pressure treatment routinely used in canned food product processing.[14]

[g] SE toxins are resistant to degradation from freezing, chilling or storage at ambient temperature.[15] Active SEB in the freeze-dried state can be stored for years.

[h] References [15,16]

[i] Dry heat of >100°C for 60 min in an ashing oven or steam autoclave treatment at >121°C for 1 h reduced the activity of pure ricin by >99%.[17] Heat inactivation of impure toxin preparations (e.g., crude ricin plant extracts) may vary. Heat-denatured ricin can undergo limited refolding (<1%) to yield active toxin.

[j] Ricin holotoxin is not inactivated significantly by freezing, chilling or storage at ambient temperature. In the liquid state with a preservative (sodium azide), ricin can be stored at 4°C for years with little loss in potency.

[k] Irradiation causes a dose-dependent loss of activity for aqueous solutions of ricin, but complete inactivation is difficult to achieve; 75 MRad reduced activity 90%, but complete inactivation was not achieved even at 100 MRad.[18] Gamma irradiation from a laboratory ^{60}Co source can be used to partially inactivate aqueous solutions of ricin, but dried ricin powders are significantly resistant to inactivation by this method.

[l] Autoclaving with 17 lb pressure (121-132° C) for 30 min failed to inactivate LMW toxins.[17,19] All burnable waste from LMW toxins should be incinerated at temperatures in excess of 815°C (1,500° F).

[m] Toxin solutions were dried at 150° C in a crucible, placed in an ashing oven at various temperatures for either 10 or 30 min, reconstituted and tested for concentration and/or activity; tabulated values are temperatures exceeding those required to achieve 99% toxin inactivation.[17]

[n] LMW toxins are generally very resistant to temperature fluctuations and can be stored in the freeze-dried state for years and retain toxicity.

Table 2. Chemical Inactivation of Selected Toxins

Toxin	NaOCl (30 min)	NaOH (30 min)	NaCOl + NaOH (30 min)	Ozone Treatment
Botulinum neurotoxin	> 0.1% [a]	> 0/25 N	ND	Yes [b]
Staphylococcal Enterotoxin	> 0.5% [c]	> 0.25 N	ND	ND
Ricin	> 1.0% [d]	ND	> 0.1% + 0.25N [e]	ND
Saxitoxin	≥ 0.1% [e]	ND	0.25% + 0.25N [e]	ND
Palytoxin	≥ 0.1% [e]	ND	0.25% + 0.25N [e]	ND
Microcystin	≥ 0.5% [e]	ND	0.25% + 0.25N [e]	ND
Tetrodotoxin	≥ 0.5% [e]	ND	0.25% + 0.25N [e]	ND
T-2 mycotoxin	≥ 2.5% [e,f]	ND	0.25% + 0.25N [e]	ND
Brevetoxin (PbTx-2)	≥ 2.5% [e,f]	ND	0.25% + 0.25N [e]	ND

Notes:

ND indicates "not determined" from available decontamination literature.

[a] Solutions of NaOCl (#0.1%) or NaOH (> 0.25 N) for 30 min inactivate BoNT and are recommended for decontaminating work surfaces and spills of *C. botulinum* or BoNT. Chlorine at a concentration of 0.3-0.5 mg/L as a solution of hypochlorite rapidly inactivates BoNT (serotypes B or E tested) in water.[20] Chlorine dioxide inactivates BoNT, but chloramine is less effective.[21]

[b] Ozone (> 2 mg/L) or powdered activated charcoal treatment also completely inactivate BoNT (serotypes A, B tested) in water under defined condition.[20,22]

[c] SEB is inactivated with 0.5% hypochlorite for 10-15 mi.[23]

[d] Ricin is inactivated by a 30 min exposure to concentrations of NaOCl ranging from 0.1-2.5%, or by a mixture of 0.25% NaOCl plus 0.25 N NaOH.[17] In general, solutions of 1.0% NaOCl are effective for decontamination of ricin from laboratory surfaces, equipment, animal cages, or small spills.

[e] The minimal effective concentration of NaOCl was dependent on toxin and contact time; all LMW toxins tested were inactivated at least 99% by treatment with 2.5% NaOCl, or with a combination of 0.25% NaOCl and 0.25N NaOH.[17]

[f] For T-2 mycotoxin and brevetoxin, liquid samples, accidental spills, and nonburnable waste should be soaked in 2.5% NaOCl with 0.25% N NaOH for 4 h. Cages and bedding from animals exposed to T-2 mycotoxin or brevetoxin should be treated with 0.25% NaOCl and 0.025 N NaOH for 4 h. Exposure for 30 min to 1.0% NaOCl is an effective procedure for the laboratory (working solutions, equipment, animal cages, working area and spills) for the inactivation of saxitoxin or tetrodotoxin. Decontamination of equipment and waste contaminated with select brevetoxins has been reviewed.[19]

Alternate methods of chemical decontamination: 1 N sulfuric or hydrochloric acid did not inactivate T-2 mycotoxin and only partially inactivated microcystin-LR, saxitoxin, and brevetoxin (PbTx-2). Tetrodotoxin and palytoxin were inactivated by hydrochloric acid, but only at relatively high molar concentrations. T2 was not inactivated by exposure to 18% formaldehyde plus methanol (16 h), 90% freon-113 + 10% acetic acid, calcium hypochlorite, sodium bisulfate, or mild oxidizing.17 Hydrogen peroxide was ineffective in inactivating T-2 mycotoxin. This agent did cause some inactivation of saxitoxin and tetrodotoxin, but required a 16 h contact time in the presence of ultraviolet light.

References

1. Franz DR. Defense against toxin weapons. In: Sidell FR, Takafuji ET, Franz DR, editors. Medical aspects of chemical and biological warfare. Vol 6. Textbook of military medicine, part 1: warfare, weaponry, and the casualty. Washington, DC: Office of the Surgeon General at TMM Publications, Borden Institute, Walter Reed Army Medical Center; 1997. p. 603-19.
2. Millard CB. Medical defense against protein toxin weapons: review and perspective. In: Lindler LE, Lebeda FJ, Korch GW, editors. Biological weapons defense: infectious diseases and counterbioterrorism. Totowa, NJ: Humana Press; 2005. p. 255-84.
3. Hamilton MH. The biological defense safety program--technical safety requirements. In: Series The Biological Defense Safety Program--Technical Safety Requirements. Department of Defense--Department of Army, 32CFR Part 627; 1993. p. 647-95.
4. Johnson B, Mastnjak R, Resnick IG. Safety and health considerations for conducting work with biological toxins. In: Richmond J, editor. Anthology of biosafety II: facility design considerations. Vol. 2. Mundelein, IL: American Biological Safety Association; 2000. p. 88-111.
5. Committee on Prudent Practices for Handling, Storage, and Disposal of Chemicals in Laboratories; Board on Chemical Sciences and Technology; Commission on Physical Sciences, Mathematics, and Applications; National Research Council. Prudent practices in the laboratory: handling and disposal of chemicals. Washington, DC: National Academy Press; 1995. p. xv:427.
6. Kruse RH, Puckett WH, Richardson JH. Biological safety cabinetry. Clin Microbiol Rev. 1991;4:207-41.
7. Morin R, Kozlovac J. Biological toxins. In: Fleming DO, Hunt DL, editors. Biological safety principles and practice. 3rd editon. Washington, DC: American Society for Microbiology; 2000. p. 261-72.
8. Balows A. Laboratory diagnosis of infectious diseases: principles and practice. New York: Springer-Verlag: 1988.
9. Hatheway C. Botulism. In: Balows A, Hausler W, Ohashi M, et al, editors. Laboratory diagnosis of infectious diseases: principles and practice. Vol 1. New York: Springer-Verlag; 1988. p. 111-33.
10. Siegel LS. Destruction of botulinum toxins in food and water. In: Hauschild AHW, Dodds KL, editors. Clostridium botulinum: ecology and control in foods. New York: Marcel Dekker, Inc.; 1993. p. 323-41.
11. Woolford A, Schantz EJ, Woodburn M. Heat inactivation of botulinum toxin type A in some convenience foods after frozen storage. J Food Sci. 1978;43:622-4.
12. Dack GM. Effect of irradiation on *Clostridium botulinum* toxin subjected to ultra centrifugation. Report No. 7. Natick, MA: Quartermaster Food and Container Institute for the Armed Forces; 1956.

13. Wagenaar R, Dack GM. Effect in surface ripened cheese of irradiation on spores and toxin of *Clostridium botulinum* types A and B. Food Res. 1956;21:226-34.
14. Bennett R, Berry M. Serological reactivity and in vivo toxicity of *Staphylococcus aureus* enterotoxin A and D in select canned foods. J Food Sci. 1987;52:416-8.
15. Concon J. Bacterial food contaminants: bacterial toxins. In: Food toxicology. Vol. B. Food science and technology. New York: Marcel Dekker, Inc.; 1988. p. 771-841.
16. Modi NK, Rose SA, Tranter HS. The effects of irradiation and temperature on the immunological activity of staphylococcal enterotoxin A. Int J Food Microbiol. 1990;11:85-92.
17. Wannemacher R, Bunner D, Dinterman R. Inactivation of low molecular weight agents of biological origin. In: Symposium on agents of biological origin. Aberdeen Proving Grounds, MD: US Army Chemical Research, Development and Engineering Center; 1989. p. 115-22.
18. Haigler HT, Woodbury DJ, Kempner ES. Radiation inactivation of ricin occurs with transfer of destructive energy across a disulfide bridge. Proc Natl Acad Sci USA. 1985;82:5357-9.
19. Poli MA. Laboratory procedures for detoxification of equipment and waste contaminated with brevetoxins PbTx-2 and PbTx-3. J Assoc Off Anal Chem. 1988;71:1000-2.
20. Notermans S, Havelaar A. Removal and inactivation of botulinum toxins during production of drinking water from surface water. Antonie Van Leeuwenhoek. 1980;46:511-14.
21. Brazis A, Bryant A, Leslie J, et al. Effectiveness of halogens or halogen compounds in detoxifying Clostridium botulinum toxins. J Am Waterworks Assoc. 1959;51:902-12.
22. Graikoski J, Blogoslawski W, Choromanski J. Ozone inactivation of botulinum type E toxin. Ozone: Sci Eng. 1985;6:229-34.
23. Robinson JP. Annex 2. Toxins. In: Public health response to biological and chemical weapons: WHO guidance. 2nd edition. Geneva: World Health Organization; 2004. p. 214-28.

Appendix J—NIH Oversight of Research Involving Recombinant Biosafety Issues

The NIH locus for oversight of recombinant DNA research is the Office of Biotechnology Activities (OBA), which is located within the Office of Science Policy, in the Office of the Director of the NIH. The OBA implements and manages the various oversight tools and information resources that NIH uses to promote the science, safety and ethics of recombinant DNA research. The key tools of biosafety oversight are the *NIH Guidelines*, IBCs, and the Recombinant DNA Advisory Committee (RAC). The NIH also undertakes special initiatives to promote the analysis and dissemination of information key to our understanding of recombinant DNA, including human gene transfer research. These initiatives include a query-capable database and conferences and symposia on timely scientific, safety, and policy issues. The NIH system of oversight is predicated on ethical and scientific responsibilities, with goals to promote the exchange of important scientific information, enable high-quality research, and help advance all fields of science employing recombinant DNA.

The *NIH Guidelines* promote safe conduct of research involving recombinant DNA by specifying appropriate biosafety practices and procedures for research involving the construction and handling of either recombinant DNA molecules or organisms and viruses that contain recombinant DNA. Recombinant DNA molecules are defined in the *NIH Guidelines* as those constructed outside of a living cell by joining natural or synthetic DNA segments to DNA molecules that can replicate in a living cell. The *NIH Guidelines* are applicable to all recombinant DNA work at an institution that receives any funding from the NIH for recombinant DNA research. Compliance with the *NIH Guidelines* is mandatory for investigators conducting recombinant DNA research funded by the NIH or performed at or sponsored by any public or private entity that receives any NIH funding for recombinant DNA research. This broad reach of the *NIH Guidelines* is intended to instill biosafety practices throughout the institution, which is necessary if the practices are to be effective.

The *NIH Guidelines* were first published in 1976 and are revised as technological, scientific, and policy developments warrant. They outline the roles and responsibilities of various entities associated with recombinant DNA research, including institutions, investigators, biological safety officers, and the NIH (see Section IV of the *NIH Guidelines*). They describe four levels of biosafety and containment practices that correspond to the potential risk of experimentation and require different levels of review for recombinant DNA research, based on the nature and risks of the activity. These include:

1. Review by the RAC, and approval by the NIH Director and the IBC.

2. Review by the NIH OBA and approval by the IBC.

3. Review by the RAC and approvals by the IBC and Institutional Review Board.

4. Approval by the IBC prior to initiation of the research.

5. Notification of the IBC simultaneous with initiation of the work.

See Section III of the *NIH Guidelines* for additional details. In all instances, it is important to note that review by an IBC is required.

The federally mandated responsibilities for an IBC are articulated solely in the *NIH Guidelines*. Their membership, procedures, and functions are outlined in Section IV-B-2. Institutions, ultimately responsible for the effectiveness of IBCs, may define additional roles and responsibilities for these committees in addition to those specified in the Guidelines. To access the NIH Guidelines see the following Web site: *http://oba.od.nih.gov/rdna/nih_guidelines_oba.html*.

The Recombinant DNA Advisory Committee is a panel of national experts in various fields of science, medicine, genetics, and ethics. It includes individuals who represent patient perspectives. The RAC considers the current state of knowledge and technology regarding recombinant DNA research and advises the NIH Director and OBA on basic and clinical research involving recombinant DNA and on the need for changes to the *NIH Guidelines*.

Additional information on OBA, the *NIH Guidelines*, and the NIH RAC can be found at: *http://oba.od.nih.gov*.

Appendix K—Resources

Resources for information, consultation, and advice on biohazard control, decontamination procedures, and other aspects of laboratory and animal safety management include:

AAALAC International
Association for Assessment and Accreditation
of Laboratory Animal Care International
5283 Corporate Drive
Suite 203
Fredrick, MD 21703-2879
Telephone: (301) 696-9626
Fax: (301) 696-9627
Web site: *http://www.aaalac.org*

American Biological Safety Association (ABSA)
American Biological Safety Association
1200 Allanson Road
Mundelein, IL 60060-3808
Telephone: (847) 949-1517
Fax: (847) 566-4580
Web site: *http://www.absa.org*

CDC Etiologic Agent Import Permit Program
Centers for Disease Control and Prevention
Etiologic Agent Import Permit Program
1600 Clifton Road, NE Mailstop: F-46
Atlanta, GA 30333
Telephone: (404) 718-2077
Fax: (404) 718-2093
Web site: *http://www.cdc.gov/od/eaipp*

CDC Office of Health and Safety
Centers for Disease Control and Prevention
Office of Health and Safety
Mailstop: F-05
1600 Clifton Road, NE
Atlanta, GA 30333
Telephone: (404) 639-7233
Fax: (404) 639-2294
Web site: *www.cdc.gov/biosafety*

CDC Select Agent Program
Centers for Disease Control and Prevention
Division of Select Agents and Toxins
Mailstop: A-46
1600 Clifton Road, NE
Atlanta, GA 30333
Telephone: (404) 718-2000
Fax: (404) 718-2096
Web site: *www.selectagents.gov*

Clinical and Laboratory Standards Institute
940 West Valley Road, Suite 1400
Wayne, PA 19087
Telephone: (610) 688-0100
Fax: (610) 688-0700
Web site: *http://www.clsi.org*

College of American Pathologists
1350 I St. N.W. Suite 590
Washington, DC 20005-3305
Telephone: (800) 392-9994
Telephone: (202) 354-7100
Fax: (202) 354-7155
Web site: *http://www.cap.org*

Department of the Army
Biological Defense Safety Program
Department of Defense
32 CFR Parts 626, 627
Web site: *www.gpo.gov*

National B-Virus Resource Laboratory
National B Virus Resource Laboratory
Attention: Dr. Julia Hillard
Viral Immunology Center
Georgia State University
50 Decatur Street
Atlanta, GA 30303
Telephone: (404) 651-0808
Fax: (404) 651-0814
Web site: *http://www2.gsu.edu/~wwwvir*

NIH Division of Occupational Health and Safety
National Institutes of Health
Division of Occupational Health and Safety
Building 13, Room 3K04
13 South Drive, MSC 5760
Bethesda, MD 20892
Telephone: (301) 496-2960
Fax: (301) 402-0313
Web site: *http://dohs.ors.od.nih.gov/index.htm*

NIH Office of Biotechnology Activities
National Institutes of Health
Office of Biotechnology Activities
6705 Rockledge Drive
Suite 750, MSC 7985
Bethesda, MD 20892
Telephone: (301) 496-9838
Fax: (301) 496-9839
Web site: *http://oba.od.nih.gov*

NIH Office of Laboratory Animal Welfare (OLAW)
National Institutes of Health
Office of Laboratory Animal Welfare (OLAW)
RKL 1, Suite 360, MSC 7982
6705 Rockledge Drive
Bethesda, MD 20892-7982
Telephone: (301) 496-7163
Web site: *http://grants.nih.gov/grants/olaw/olaw.htm*

Occupational Safety and Health Administration
U.S. Department of Labor
200 Constitution Avenue NW
Washington, DC 20210
Telephone: (800) 321-6742
Web site: *http://www.osha.gov/index.html*

USDA-APHIS National Center for Import / Export
USDA Animal and Plant Health Inspection Service Veterinary Services National
Center for Import and Export
4700 River Road, Unit 40
Riverdale, MD 20737
Web site: *http://www.aphis.usda.gov/import_export/index.shtml*

USDA Agriculture Select Agent Program
USDA Agriculture Select Agent Program
Animal and Plant Health Inspection Service
U.S. Department of Agriculture
4700 River Road, Unit 2, Mailstop 22
Riverdale, MD 20737
Web site: *http://www.aphis.usda.gov/programs/ag_selectagent/index.shtml*

USDA National Animal Disease Center
U.S. Department of Agriculture
National Animal Disease Center
P.O. Box 70
2300 Dayton Road
Ames, IA 50010
Telephone: (515) 663-7200
Fax: (515) 663-7458
Web site: *http://www.ars.usda.gov/main/site_main.htm?modecode=36-25-30-00*

USDA Plant Select Agents & Plant Protection and Quarantine
Animal and Plant Health Inspection Service
Plant Protection and Quarantine, Permits, Agricultural Bioterrorism
U.S. Department of Agriculture
4700 River Road, Unit 133
Riverdale, MD 20737
Telephone: (877) 770-5990
Fax: (301) 734-5786
Web site: *http://www.aphis.usda.gov/programs/ag_selectagent/index.shtml*;
and *http://www.aphis.usda.gov/permits/index.shtml*

US Department of Transportation
Hazardous Materials Center
Pipeline & Hazardous Materials Center
U.S. Department of Transportation
400 7th Street, S.W.
Washington, DC 20590
Telephone: (800) 467-4922
Web site: *http://www.phmsa.dot.gov/hazmat*

US Department of Commerce
Export Administration Program
Bureau of Industry and Security (BIS)
Export Administration Regulations (EAR)
U.S. Department of Commerce
14th Street and Constitution Avenue, N.W.
Washington, DC 20230
Telephone: (202) 482-4811
Web site: *http://www.access.gpo.gov/bis/index.html*

World Health Organization Biosafety Program
World Health Organization Biosafety Program
Avenue Appia 20
1211 Geneva 27
Switzerland
Telephone: (+ 41 22) 791 21 11
Fax: (+ 41 22) 791 3111
Web site: *http://www.who.int/ihr/biosafety/en*

Appendix L—Acronyms

A1HV-1	Alcelaphine Herpesvirus-1
ABSA	American Biological Safety Association
ABSL	Animal Biosafety Level
ACAV	American Committee on Arthropod-Borne Viruses
ACIP	Advisory Committee on Immunization Practices
ACG	Arthropod Containment Guidelines
ACL	Arthropod Containment Levels
ACME	American Committee of Medical Entomology
AHS	African Horse Sickness
AHSV	African Horse Sickness Virus
AKAV	Akabane Virus
APHIS	Animal and Plant Health Inspection Service
APMV-1	Avian Paramyxovirus Type 1
ASF	African Swine Fever
ASFV	African Swine Fever Virus
ASHRAE	American Society of Heating, Refrigerating, and Air-Conditioning Engineers
ASTMH	American Society of Tropical Medicine and Hygiene
BCG	Bacillus Calmette-Guérin
BDV	Border Disease Virus
BMBL	Biosafety in Microbiological and Biomedical Laboratories
BoNT	Botulinium neurotoxin
BSC	Biological Safety Cabinet
BSE	Bovine Spongiform Encephalopathy
BSL	Biosafety Level
BSL-3-Ag	BSL-3-Agriculture
BSO	Biological Safety Officer
BTV	Bluetongue Virus
BVDL	Bovine Viral Diarrhea Virus
CAV	Constant Air Volume
CBPP	Contagious Bovine Pleuropneumonia
CCPP	Contagious Caprine Pleuropneumonia
CETBE	Central European Tick-Borne Encephalitis
CDC	Centers for Disease Control and Prevention
CHV-1	Cercopithecine Herpesvirus-1
CJD	Creutzfeldt-Jakob Disease
CJIS	Criminal Justice Information Services Division
CNS	Central Nervous System
CSF	Cerebrospinal Fluid
CSFV	Classical Swine Fever Virus
DHHS	Department of Health and Human Services
DoC	Department of Commerce
DOD	Department of Defense
DOL	Department of Labor

DOT	Department of Transportation
EBV	Epstein-Barr Virus
EEE	Eastern Equine Encephalomyelitis
EPA	Environmental Protection Agency
EtOH	Ethanol
FDA	Food and Drug Administration
FFI	Fatal Familial Insomnia
FMD	Foot and Mouth Disease
FMDV	Foot and Mouth Disease Virus
GI	Gastrointestinal Tract
GSS	Gerstmann-Straussler-Scheinker Syndrome
HEPA	High Efficiency Particulate Air
HBV	Hepatitis B Virus
HCMV	Human Cytomegalovirus
HCV	Hepatitis C Virus
HD	Heartwater Disease
HDV	Hepatitis D Virus
HFRS	Hemorrhagic Fever with Renal Syndrome
HHV	Human Herpes Virus
HHV-6A	Human Herpes Virus -6A
HHV-6B	Human Herpes Virus -6B
HHV-7	Human Herpes Virus -7
HHV-8	Human Herpes Virus -8
HIV	Human Immunodeficiency Virus
HPAI	Highly Pathogenic Avian Influenza
HPAIV	Highly Pathogenic Avian Influenza Virus
HPS	Hantavirus Pulmonary Syndrome
HSV-1	Herpes Simplex Virus-1
HSV-2	Herpes Simplex Virus-2
HTLV	Human T-Lymphotropic Viruses
HVAC	Heating, Ventilation, and Air Conditioning
IACUC	Institutional Animal Care and Use Committee
IATA	International Air Transport Association
IBC	Institutional Biosafety Committee
ICAO	International Civil Aviation Organization
ID	Infectious Dose
ID_{50}	Number of organisms necessary to infect 50% of a group of animals
IgG	Immunoglobulin
ILAR	Institute for Laboratory Animal Research
IND	Investigational New Drug
IPM	Integrated Pest Management
IPV	Inactivated Poliovirus Vaccine
ISA	Infectious Salmon Anemia

ISAV	Infectious Salmon Anemia Virus
LAI	Laboratory-Associated Infections
LCM	Lymphocytic Choriomeningitis
LCMV	Lymphocytic Choriomeningitis Virus
LD	Lethal Dose
lfm	Linear Feet Per Minute
LGV	Lymphogranuloma Venereum
LMW	Low Molecular Weight
LSD	Lumpy Skin Disease
LSDV	Lumpy Skin Disease Virus
MCF	Malignant Catarrhal Fever
MenV	Menangle Virus
MMWR	Morbidity and Mortality Weekly Report
MPPS	Most Penetrating Particle Size
NaOCl	Sodium Hypochlorite
NaOH	Sodium Hydroxide
NBL	National Biocontainment Laboratory
NCI	National Cancer Institute
ND	Newcastle Disease
NDV	Newcastle Disease Virus
NHP	Nonhuman Primate
NIH	National Institutes of Health
NIOSH	National Institute for Occupational Safety and Health
OBA	NIH Office of Biotechnology Activities
OIE	World Organization for Animal Health
OPV	Oral Poliovirus Vaccine
OSHA	Occupational Safety and Health Administration
PAPR	Positive Air-Purifying Respirator
PBT	Pentavalent Botulinum Toxoid Vaccine
PPD	Purified Protein Derivative
PPM	Parts Per Million
PPRV	Pest des Petits Ruminants Virus
Prp	Prion Protein
RAC	Recombinant DNA Advisory Committee
RBL	Regional Biocontainment Laboratory
RP	Rinderpest
RPV	Rinderpest Virus
RVF	Rift Valley Fever
RVFV	Rift Valley Fever Virus
SALS	Subcommittee on Arbovirus Laboratory Safety
SARS	Severe Acute Respiratory Syndrome
SARS-CoV	SARS-Associated Coronavirus
SCID	Severe Combined Immune Deficient
SC type	Small-Colony type

SE	Staphylococcal Enterotoxins
SEA	SE Serotype A
SEB	SE Serotype B
SIV	Simian Immunodeficiency Virus
SGP	Sheep and Goat Pox
SGPV	Sheep and Goat Pox Virus
SOP	Standard Operating Procedure
SVCV	Spring Viremia of Carp Virus
SVD	Swine Vesicular Disease
SVDV	Swine Vesicular Disease Virus
TLV	Threshold Limit Values
TME	Transmissible Mink Encephalopathy
TSE	Transmissible Spongiform Encephalopathy
UV	Ultraviolet
USAMRIID	U.S. Army Medical Research Institute of Infectious Diseases
USDA	U.S. Department of Agriculture
USPS	United States Postal Service
UPS	Uninterrupted Power Supply
VAV	Variable Air Volume
VEE	Venezuelan Equine Encephalitis
VS	Veterinary Services
VZV	Varicella-Zoster Virus
WEE	Western Equine Encephalomyelitis
WHO	World Health Organization
WNV	West Nile Virus

Index

A

African horse sickness (AHSV) 246, 350, 353-55, 373, 401
African swine fever virus (ASFV) 246, 350, 352, 373, 401
Agriculture Agents 343, 352
Agriculture pathogen biosafety 343
Akabane virus 350, 354, 401
Allergic reactions 192
animal biosafety levels 61
Animal Biosafety Levels (ABSL) 60
    Animal Biosafety Level 1 61
    Animal Biosafety Level 2 67
    Animal Biosafety Level 3 61, 75
    Animal Biosafety Level 4 85
Animal facilities 27, 60, 91, 173, 243, 361-63, 372
Anthrax iii, 104, 123-24, 161, 273
Arboviruses 2, 5, 233, 236, 238-40, 246, 265-267
arenaviruses 234, 237
armadillos 144-45, 166
Arthropod containment 61, 160, 169, 377-78, 401
Ascaris 192-93

B

B virus (herpesvirus simiae) 1, 12, 19, 25, 29, 120-21, 204-10, 229, 232, 243,
   384, 397, 402
Bacillus anthracis 123, 350
Bacillus subtilis 25, 330
Bacterial agents 123
Besnoitia besnoiti 350
Biological safety cabinets (BSC) 15, 18, 22, 25-27, 36, 38, 41, 44-45, 49, 51-52,
   54, 57, 72, 75, 84, 86-87, 92-95, 97, 100, 103, 124-25, 128-30, 146, 150,
   155-56, 160, 173-74, 176-78, 185, 187, 194, 200, 210, 221, 225, 239, 274,
   278, 290-313, 315-19, 321-25, 330, 332, 335, 347, 383, 386-88, 401
    Class I 23, 27, 59, 103, 219, 290-93, 299, 311-13, 315, 321
    Class II 23, 27, 38, 41, 44, 51, 53-54, 57, 59, 75, 84, 86, 93-94, 96-97, 100,
       103, 173-74, 176-78, 219, 235, 274, 278, 290, 292-300, 302, 307-09,
       311-13, 315-18, 322-24, 347, 388
    Class III 23, 26, 41, 44-45, 48-50, 52, 54, 59, 84-86, 92-95, 97, 103, 290,
       297-98, 311-13, 319, 324, 347, 388
    Positive-pressure personnel suit 26
Biosafety levels (BSL) 4, 9-10, 24, 27, 30, 59, 60-61, 103-04, 234, 239-40, 290,
   294, 378, 401
    Biosafety Level 1 (BSL-1) 4-5, 23, 25, 30, 32-33, 59, 219

Biosafety Level 2 (BSL-2) 4, 7, 17, 23, 25, 27, 33, 35, 37, 59, 67, 104, 124-25, 127-29, 131, 133, 135, 137-41, 143-46, 148-50, 152-53, 155-57, 159-60, 170, 172-75, 177-78, 182, 185, 187, 189, 191, 194, 196, 198, 200-01, 203-04, 207, 210, 212, 216-17, 219, 221, 223, 225, 234-35, 237, 241, 243, 246, 269, 272, 274, 278, 284, 286, 305, 311, 363, 368, 372, 378, 383, 386

Biosafety Level 3 (BSL-3) 4, 7, 16, 24, 26-27, 38, 41-42, 45, 53, 57, 59, 104, 124, 126-29, 133, 135, 139, 141, 146, 149-50, 153, 160, 170, 172, 174, 196, 198, 201, 204, 207, 210, 212-13, 216, 219, 221, 223, 225-26, 234-38, 241, 243, 245, 269, 272, 284, 286, 300, 305, 311, 331-32, 343-44, 347-53, 355-57, 359-62, 364-67, 369-71, 373, 387

Biosafety Level 4 (BSL-4) 4, 16, 26, 45, 48-49, 51, 53-59, 87, 96, 100, 103, 120, 202, 207, 219, 234-38, 294, 305, 311, 331-32, 343-44, 350-51

Biosecurity iii, 6, 28, 86, 104-13, 374

Bioterrorism 1, 124, 134, 161, 279, 291, 339, 351, 361, 368, 376, 379, 399

Blastomyces dermatitidis 1, 170

Bloodborne pathogens 13, 20, 25, 27-29, 205, 223, 232, 383-84

Bluetongue virus 350

BMBL iii-iv, ix, 1, 3-7, 9, 11, 16, 19, 104, 290-91, 298, 377-78, 401

Bordetella pertussis 125, 162

Borna disease virus 351

Botulinum neurotoxin 134-35, 268, 390-91

Botulism 134-35, 164, 268-69, 278-79, 392

Bovine pleuropneumonia agent 358

Bovine spongiform encephalopathy 282, 284, 286, 351, 401

Brucella 1, 2, 7, 126-27, 162, 351
    abortus 126-27, 162, 351
    canis 126, 162
    maris 126
    melitensis 2, 7, 126-27, 162, 351
    suis 126-27, 351

Brucellosis 126, 162

BSL-3-Ag xix, 53, 61, 96, 100, 213, 246, 265, 343-47, 350-51, 353, 357, 359, 360-61, 364, 367, 369, 370, 401

Burkholderia mallei 127, 351

Burkholderia pseudomallei 129, 163, 351

C

C virus 2, 13, 204, 228, 402

Cabinet laboratory 45, 49, 51, 53-54, 58, 85-86, 92, 94, 96, 101

Camelpox virus 351

Campylobacter 130, 163, 165
    coli 130
    fetus 130
    jejuni 130, 163
    upsaliensis 130

Dermatophytes
    Epidermophyton 176
    Microsporum 176
    Trichophyton 176
Diagnostic specimens 28, 202, 224, 241, 341
diphtheria 137-38, 164
Diphtheria 164
Disinfection 38-39, 44, 71, 80, 157, 282, 326-29, 331, 333-35, 372, 383
DOT Packaging Requirements 340

E

Eastern equine encephalitis virus xviii, 242, 351
Ebola 251, 330
emergency response 49, 91, 104, 110, 112-13, 336, 340
Encephalitis
    Russian spring-summer 236
    Russian Spring-Summer 260
Epidermophyton 176
equine encephalomyelitis 19, 116, 235, 237, 351
Escherichia coli 154, 168

F

Facility design 17, 22-23, 25-26, 30, 44-45, 55, 58, 60-61, 85, 98, 101, 290, 347, 381, 392
Fasciola 188-89
Filoviruses 237
Foot and mouth disease 351, 360-61, 374, 402
Francisella tularensis 1, 12, 138, 164, 351
Fungal Agents 170

G

Ganjam virus 265, 351
Germicidal Chemicals 330
Giardia 186-87
Giardia spp. 187
Gloves 4, 15, 23, 25, 32, 37, 42, 47, 50-51, 59, 63, 65, 69, 72-73, 77, 81-82, 89, 92-94, 103, 128-29, 133, 143-44, 149-50, 152-53, 155-57, 159-60, 175, 185, 189, 191, 204, 207, 212, 221-22, 225, 269, 272, 274, 278, 286-87, 293, 297, 300-01, 304, 312-13, 319, 344, 348, 361, 363, 372, 383, 386, 388
gonorrhoeae 148
Guidelines 3-4, 6, 8, 10, 12-13, 17, 19, 24-25, 28-29, 61, 121, 124, 160, 169, 206-07, 214, 219, 221, 227, 229-31, 235, 238, 284-86, 289, 324, 334, 343, 377-78, 383-86, 389, 394-95, 401

H

Hantaviruses 200, 234
Heartwater Disease Agent 362
Helicobacter pylori 139, 165
Hemorrhagic fever 26-27, 200, 227, 244, 402
Hendra virus 201-02, 228, 351, 366
HEPA filter 44, 50, 53-54, 57, 75, 92, 96, 100, 274, 291-97, 306-09, 311-13,
    315-22, 345, 348-49, 387
Hepatitis
    A virus 202-03
    B virus (HBV) 1-2, 13, 25, 27, 29, 204-05, 229, 232, 330, 383-84, 402
    C virus (HCV) 2, 13, 204-05, 228, 330, 383, 402
    D virus (HDV) 204-05, 402
    E virus 202-03
Herpesviruses 205, 208-10, 365
    Herpesvirus simiae 205, 20-10, 229
    Human herpesviruses 208-10
Herpesvirus simiae 205, 209-10, 229
    CHV-1 12, 401
Histoplasma 173, 351
    capsulatum 173-74
    farciminosum 351
Hog cholera 357, 373
Human Herpes Virus 208, 402
Human immunodeficiency virus (HIV) 13, 25, 27-28, 146-47, 177, 209, 221-23,
    232, 266, 330, 383-84, 402
Hypr 236, 253

I

Immunoprophylaxis 114
Importation and interstate shipment 28
infectious canine hepatitis virus 25
Influenza 5, 20, 211-14, 230, 271, 350, 366, 372, 375, 402
International transfer 336
Interstate shipment 28
Investigational New Drug (IND) 116, 122, 135, 196, 208, 237, 269, 402

J

Junin virus 236, 265

K

Kuru 283
Kyasanur Forest disease 255

L

Laboratory-associated infections (LAI) 1-2, 6-7, 9, 18, 20, 126, 130, 132, 137, 140, 142, 150, 157-58, 161, 168, 171, 179-80, 192, 194-95, 197-99, 203-04, 210-11, 215, 220, 228, 230, 235-36, 241, 300, 326, 403

Laboratory coat(s) 32, 36, 59, 63, 65, 69, 72, 77, 81, 103, 150, 189, 272, 274, 278, 287, 300, 363, 383, 386

Laboratory hazard(s) 11, 17, 130, 132, 137, 140, 142, 145, 148-49, 152-53, 155-56, 159, 172, 184, 187, 189, 191, 204, 212, 235-36, 243, 290, 298, 302, 383

Laboratory practice(s) 15, 22, 24, 104, 191, 224, 238-39, 361, 366, 372, 386

Laboratory security and emergency response 113

Legionella pneumophila 140, 165

Leishmania spp. 20, 182-85

Leprosy 144, 166

Leptospira interrogans 166

Leptospirosis 142, 166

Listeria monocytogenes 143, 166

Loose-housed animals 27, 213, 246, 343, 351

Louping ill virus 256, 350

Low molecular weight (LMW) toxin(s) 274-75, 277-78, 385-86, 390-91, 403

Lumpy skin disease virus (LSDV) 350, 364, 403

Lymphocytic choriomeningitis 13, 20, 215, 230, 403

Lymphogranuloma venereum (LGV) 131-33, 403

M

Macaque 12, 205-07, 222

Machupo 256

malaria 182-84

Malignant catarrhal fever virus 350, 365

Mammalian Cells and Tissues 383

Marburg 26, 256

mask 15, 36, 42, 72, 225, 272

medical surveillance 35, 40, 47, 59, 62, 68, 70, 76, 79, 86, 90, 103, 214

Menangle virus (MenV) 350, 366, 403

metacercariae 189

mice 13, 20, 131, 133, 141, 144-46, 166, 169, 173, 176, 215-16, 245, 266, 273, 276, 280, 285, 353-55

Microsporum 176

Molds xvi, 171, 176-77, 287

monkey pox 218-19

Mycobacterium
    avium 147-48
    bovis 145-46, 327, 330, 350
    chelonei 147
    fortuitum 147

Pest management 31, 35, 40, 46, 64, 70, 78, 89, 380-82, 402
Physical Inactivation of Selected Toxins 390
Physical security 110, 344
Pike and Sulkin 1-2
Pipetting 14-15, 30, 34, 36, 39, 45, 63, 69, 78, 87, 127, 242, 301
Plague 159-60, 169, 351, 368
Plasmodium spp. 182-84
Plastic-backed absorbent toweling 301
Poliovirus 216-18, 231, 330, 402-03
Pontiac fever 141
Poxviruses 218-19, 231, 370
Primary barriers 22-23, 27, 32, 36, 41, 49, 59, 64, 72, 80, 92, 103, 290, 306, 347, 378
Primates 60, 102, 121, 123, 156, 166, 197, 203, 206-07, 216-17, 222-23, 229, 234
Prion diseases 282-85, 289
Production quantities 124, 129, 149-50, 153, 204, 216, 221
Protozoal parasites 182, 186
Pseudomonas
    pseudomallei 129, 163
psittacosis 131, 132

Q

Q fever 19, 195-96, 199, 351

R

Rabies virus 220
Radiological Hazards 299
Regulations iii, vi, 6, 28, 31, 34, 40, 44, 60, 85, 104, 114, 120, 243, 265, 286, 288, 299, 336-39, 350, 353-56, 358-60, 362-66, 369-73, 379, 386, 400
Retroviruses 119, 221-23, 240, 266
Ricin toxin 272, 274, 280
Rickettsia
    akari 196
    australis 196
    belli 198
    canada 198
    conorii 196
    Coxiella burnetii 195
    montana 198
    mooseri 196
    Orientia 196, 197
    prowazekii 196, 198
    rhipicephali 198
    rickettsii 196-99
    siberica 196

Staphylococcal Enterotoxins (SE) 270-72, 279, 385, 390-91, 393, 404
Sterilization iii, vi, 285, 288, 326-29, 331, 333-35, 346, 350, 387
Strongyloides spp. 192
Suit laboratory 45, 51, 55-57, 85, 93, 98, 100, 294, 331
Surveillance 2, 35-36, 40-41, 47, 49, 59, 62, 68, 70-71, 76, 79-80, 86, 88, 90,
    103, 147, 164, 198, 202, 214, 222, 232, 240, 372
Swine vesicular disease virus (SVDV) 351, 372-73, 376, 404
Syphilis 157, 169

T

Taenia solium 190
Teschen disease virus 351
Tetanus 136-37, 164, 279
Theileria
    annulata 351
    bovis 351
    hirci 351
    lawrencei 351
Tissue preparation 287
Toxin Agents 268
Toxins of biological origin 268, 271, 273, 277, 385
Toxoplasma 25, 182-85
Trachoma 132
Training v, vii, 2, 6-7, 9, 15, 18, 22, 25-26, 30, 32-33, 35, 38, 40, 45-46, 48-49,
    61-62, 67-68, 75-76, 85, 89-91, 104-05, 108-09, 112, 117, 179, 206, 238, 290,
    308, 331, 380-81, 385
Transmissible spongiform encephalopathies (TSE) 282, 285, 289, 334, 404
Trematode parasites
    Fasciola 188
    Schistosoma 188
Treponema pallidum 157, 168
Trichophyton 176, 180, 330
Trypanosoma
    cruzi 182-85
    evansi 351
    vivax 351
Tuberculosis 1-2, 20, 26-27, 145-48, 166, 327-30, 383
Tularemia 138-39, 164
Typhoid fever 153
Typhus 197-99

U

Ultraviolet (UV) lamps 306, 310
Universal precautions 28-29, 232, 384

V

Vaccines 22, 45, 59, 85, 103, 116, 121, 124, 126-27, 135-38, 147, 151, 153, 156, 158-59, 196, 203, 205, 208, 216, 236-37, 240, 243, 246, 267, 269, 272, 274, 278, 356, 373-76, 388

Vaccinia 218-19, 231

Variola 218-19

Venezuelan equine encephalomyelitis 19, 235, 351

Vesicular exanthema virus 351

Vesicular stomatitis virus 351

Vibrio
    cholerae 158-59, 169
    parahaemolyticus 158-59, 169

Viral agent(s) 200, 209-10, 226, 265, 383

W

warning signs 59, 103

Wesselsbron disease virus 351

Western equine encephalomyelitis (WEE) virus 116, 237, 242, 264, 266, 404

West Nile virus (WNV) 240-41, 264, 267, 404

Y

Yellow fever 235-37, 265, 267

Yersinia pestis 159-60, 169

Z

Zoonotic viruses 5, 233

www.ingramcontent.com/pod-product-compliance
Lightning Source LLC
Chambersburg PA
CBHW050520190326
41458CB00005B/1606